NEUROPSYCHOLOGY OF EPILEPSY AND EPILEPSY SURGERY

OXFORD WORKSHOP SERIES:
AMERICAN ACADEMY OF CLINICAL NEUROPSYCHOLOGY

Series Editors

Susan McPherson, *Editor-in-Chief*
Ida Sue Baron
Richard Kaplan
Sandra Koffler
Greg J. Lamberty
Jerry Sweet

Volumes in the Series

Neuropsychology of Epilepsy and Epilepsy Surgery
Gregory P. Lee

The Business of Neuropsychology
Mark T. Barisa

Adult Learning Disabilities and ADHD
Robert L. Mapou

Board Certification in Clinical Neuropsychology
Kira E. Armstrong, Dean W. Beebe, Robin C. Hilsabeck, Michael W. Kirkwood

Understanding Somatization in the Practice of Clinical Neuropsychology
Greg J. Lamberty

Mild Traumatic Brain Injury and Postconcussion Syndrome
Michael A. McCrea

Ethical Decision Making in Clinical Neuropsychology
Shane S. Bush

American Academy of
Clinical Neuropsychology

NEUROPSYCHOLOGY OF EPILEPSY AND EPILEPSY SURGERY

Gregory P. Lee
Department of Neurology
School of Medicine
Medical College of Georgia
Augusta, Georgia

OXFORD WORKSHOP SERIES

OXFORD
UNIVERSITY PRESS

2010

UNIVERSITY PRESS

Oxford University Press, Inc., publishes works that further
Oxford University's objective of excellence
in research, scholarship, and education.

Oxford New York
Auckland Cape Town Dar es Salaam Hong Kong Karachi
Kuala Lumpur Madrid Melbourne Mexico City Nairobi
New Delhi Shanghai Taipei Toronto

With offices in
Argentina Austria Brazil Chile Czech Republic France Greece
Guatemala Hungary Italy Japan Poland Portugal Singapore
South Korea Switzerland Thailand Turkey Ukraine Vietnam

Copyright © 2010 by Oxford University Press, Inc.

Published by Oxford University Press, Inc.
198 Madison Avenue, New York, New York 10016

www.oup.com

Oxford is a registered trademark of Oxford University Press, Inc.

All rights reserved. No part of this publication may be reproduced,
stored in a retrieval system, or transmitted, in any form or by any means,
electronic, mechanical, photocopying, recording, or otherwise,
without the prior permission of Oxford University Press

Library of Congress Cataloging-in-Publication Data
Lee, Gregory P.
Neuropsychology of epilepsy and epilepsy surgery / by Gregory P. Lee.
p. ; cm. — (Oxford workshop series)
Includes bibliographical references.
ISBN-13: 978-0-19-537250-2
ISBN-10: 0-19-537250-6
1. Epilepsy. 2. Epilepsy—Psychological aspects. 3. Epilepsy—Surgery.
I. American Academy of Clinical Neuropsychology. II. Title. III. Series: Oxford workshop series.
[DNLM: 1. Epilepsy. 2. Epilepsy—surgery. 3. Neurophysiology—methods. WL 385 L478n 2010]
RC372.L43 2010
616.8'53—dc22
2009040355

For Susie, Stuart, and Claudia

Let us be grateful to people who make us happy, they are the charming gardeners who make our souls blossom.
Marcel Proust

Medicine can only cure curable disease, but then not always.
Chinese proverb

Surgeons must be very careful when they take the knife!,
underneath their fine incisions, stirs the Culprit / Life!
Emily Dickinson

Medicine heals doubts as well as diseases.
Karl Marx

Preface

Psychologists in North America have been studying the cognitive and behavioral consequences of epilepsy surgery since the technique was introduced over three-quarters of a century ago. As surgical treatments for intractable seizures have gained widespread acceptance as a highly successful treatment option, neuropsychologists and clinical psychologists have been called upon in ever-increasing numbers to participate in the assessment and treatment of all patients with epilepsy. Epilepsy and seizures affect approximately three million Americans across all age groups, and there are over 200,000 new cases of seizures and epilepsy each year. Because neuropsychologists are frequently called upon to evaluate patients with brain tumors, stroke, traumatic brain injury, and toxic exposure (all of which may cause epilepsy), most neuropsychologists will encounter patients with seizures at some point in their practice career. Moreover, psychologists may be among the first to recognize an undiagnosed seizure disorder that may require referral for further medical work-up and possible treatment. Thus, neuropsychologists and clinical psychologists should be familiar with the important aspects of epilepsy (namely, its causes, presentations, cognitive and emotional implications, treatments, and prognosis), and this book was written to serve this purpose.

Although several excellent comprehensive reference books describe the multiplicity of seizure disorders and epilepsy syndromes, and a host of texts deal with neuropsychological assessment, there is a scarcity of books discussing the overlap between the two subjects. Similarly, there are many excellent chapters devoted to neuropsychological assessment and intracarotid amobarbital (Wada) testing in epilepsy surgery books, but for the most part they do not cover the more general (nonsurgical) epilepsy topics. The aim of this volume is, therefore, to provide a reference resource for neuropsychologists, clinical psychologists, allied health professionals, and graduate students that covers the classification of epileptic disorders, medical diagnostic and treatment methods used in epilepsy, the cognitive and psychiatric effects of epilepsy, neuropsychological assessment of patients with seizure disorders, epilepsy surgery procedures, and the neuropsychologist's role in these endeavors.

Much of the material in this book (such as, the classification of seizures, side effects of particular medications, or certain surgical procedures) is not intended to be read through all in one sitting, but rather to serve as a reference for looking up specific information as the need arises; for example, when information is needed about your patient's particular seizure type or the cognitive side effects of the antiepileptic medications they are taking. Brief descriptions of the less common epilepsies are located in the Appendix for quick reference. In contrast, other sections of the book may be valuable for neuropsychologists to read in their entirety such as those sections dealing with assessment of epilepsy patients, depression in epilepsy, psychogenic nonepileptic seizures, or how neuropsychologists estimate risk for postoperative cognitive decline in seizure surgery patients. The book ends with a glossary of commonly used epilepsy terms that may be helpful when a brief dictionary definition of an unfamiliar technical word is required.

Acknowledgments

As with all academic endeavors, whether research or clinical, this work rests on a broad underlying foundation consisting of the mentorship, collaboration, influence, and support of a generous group of coworkers and supporters. The day-to-day work within a comprehensive epilepsy surgery program depends upon a host of health care professionals including epileptologists, neurosurgeons, neuropsychologists, nurses, EEG technicians, and a variety of support staff, and I have been privileged to have worked among them for the past 24 years at the Medical College of Georgia.

Although they have gone on to broader horizons at different institutions, I owe an enormous debt of gratitude to my two long-term friends and collaborators, Drs. Kimford J. Meador and David W. Loring. These two were the instigators and motivators of many different research projects and clinical innovations. Their ideas never stopped flowing, and I am indebted to their unceasing pursuit of productivity and excellence that continue to this day.

The Epilepsy Surgery Program at the Medical College of Georgia was established in 1981 by Dr. Herman F. Flanigin, a neurosurgeon trained by Dr. Wilder Penfield at the Montreal Neurologic Institute, with whom he published the early landmark paper on temporal lobectomy (Penfield W & Flanigin H. Surgical therapy of temporal lobe seizures. *Archives of Neurology & Psychiatry* 1950;64:491–500). I am grateful to Dr. Flanigin, first for hiring me and then for guiding me through the various roles a neuropsychologist serves in an epilepsy surgery program; as well as to his postdoctoral fellow and successor, Dr. Joseph R. Smith, for his commitment to exacting detail and indulgence in dealing with my faux pas in the operating room; and finally to our current fearless leader, Dr. Cole A. Giller—an epilepsy neurosurgeon who obtained his Ph.D. in mathematics from UC-Berkeley but is nevertheless willing to listen to what the imperfect science of neuropsychology has to offer.

Additional appreciation goes to all the epileptologists and electrophysiology fellows, past and present, at the Medical College of Georgia (MCG), who have taught me so much about patients with epilepsy over the years, but particularly to Drs. Anthony Murro, Kim Meador, Ki Lee, Jeff Politsky, Don King, Brian Gallagher, and especially to Dr. Yong D. Park for his constant generation of research ideas, never-ending commitment to education, and steadfast support of neuropsychology in the epilepsy surgery process.

Vital portions of the book would not have been as informative or up to date if not for many of the staff on the MCG epilepsy monitoring unit from the caring and efficient nursing staff to the always patient (particularly during extraoperative mapping) and skilled EEG technicians. Special credit should go to Dr. Margo Henderson and Dr. Dan Drane for giving me all of their information and videos on psychogenic nonepileptic seizures. The medical illustrators in the MCG Departments of Neurology and Neurosurgery, Michael Jensen and John Foerster, are gratefully acknowledged for creating and helping to organize the illustrations in the book.

I am grateful to all of the neuropsychology staff at the Medical College of Georgia for their patience and assistance with this project, including Dr. Morris Cohen, Adrienne Wilson, Dr. Ben Johnson-Markve, and Teresa Chaney; to all those in the Department of Occupational Therapy, especially Dr. Kathy Bradley and Marlene Moore; and to the Chairman of Neurology, Dr. David Hess.

A special thanks to my postdoctoral fellows, past and present, for keeping me educated and on my toes: Dr. Daniel Drane, Dr. Benjamin Johnson-Markve, Dr. Christie Clason, and Dr. Kathryn Viner. For their years of never-ending support, constant development of knowledge, friendship, and perhaps most of all for their music, I am deeply appreciative of, and indebted to, my colleagues and friends, Dr. Anthony C. Kneebone, Dr. Patrick M. Plenger, and Dr. Mary Walker Sprunt.

Finally, thank you to Dr. Susan McPherson for inviting me to write this book on behalf of the American Academy of Clinical Neuropsychology, and to Joan Bossert and Shelley Reinhardt at Oxford University Press for seeing it through.

Contents

	Continuing Education Credits *xxiii*
PART ONE	**MEDICAL ASPECTS OF EPILEPSY FOR NEUROPSYCHOLOGISTS**
Chapter 1	**Introduction** *3*

Definitions *3*
Epidemiology *5*
Etiology *7*
Role of the Neuropsychologist in Epilepsy *7*
Role of the Neuropsychologist in Epilepsy Surgery *9*

Chapter 2 **Classification of Epilepsy Disorders** *13*

Traditional Classification of Epileptic Seizures *13*
   Partial (or Focal) Seizures *15*
      Simple Partial (Focal) Seizures *18*
      Complex Partial (Focal) Seizures *19*
   Generalized Seizures (Convulsive or Nonconvulsive) *24*
      Absence seizures *24*
      Tonic-Clonic Seizures *27*
Duration of Seizures Across Different Seizure Types *31*
Epidemiology of the Traditional Seizure Disorders *31*
Etiology of the Traditional Seizure Disorders *33*
   Cerebrovascular Disease *34*
   Developmental and Congenital Conditions *35*
   Head (Traumatic Brain) Injury *35*
      Subdural Hematomas *36*
   Brain Tumor *37*
   Brain Infections *37*

　　　　　　　　　　Neurodegenerative Central Nervous System
　　　　　　　　　　　　　　　　Diseases 38

Chapter 3　　　　**Epilepsy Syndromes** *39*
　　　　　　　　ILAE Classification of the Epileptic Syndromes *39*
　　　　　　　　　Localization-related Epilepsies and Syndromes 43
　　　　　　　　　　Idiopathic Localization-related Epilepsies 43
　　　　　　　　　Symptomatic Localization-related Epilepsy
　　　　　　　　　　　Syndromes 43
　　　　　　　　　　Reflex Epilepsies (Syndromes Characterized
　　　　　　　　　　　by Seizures with Specific Modes of
　　　　　　　　　　　Precipitation) 43
　　　　　　　　　　Temporal Lobe Epilepsies 44
　　　　　　　　　　Mesial Temporal Lobe Epilepsy 45
　　　　　　　　　　Lateral (Neocortical) Temporal Lobe Epilepsy 46
　　　　　　　　　　Frontal Lobe Epilepsy 47
　　　　　　　　　　Parietal Lobe Epilepsy 48
　　　　　　　　　　Occipital Lobe Epilepsy 49
　　　　　　　　　Cryptogenic Localization-related Epilepsies 50
　　　　　　　　Generalized Epilepsies and Syndromes 50
　　　　　　　　　Idiopathic, Generalized Epilepsy Syndromes 50
　　　　　　　　　　Childhood Absence Epilepsy (Pyknolepsy) 52
　　　　　　　　　　Juvenile Absence Epilepsy 52
　　　　　　　　　Cryptogenic or Symptomatic Generalized Epilepsy
　　　　　　　　　　　Syndromes 53
　　　　　　　　　　Lennox-Gastaut Syndrome 53
　　　　　　　　　Symptomatic Generalized Epilepsy Syndromes 53
　　　　　　　　　Epilepsies and Syndromes Undetermined Whether
　　　　　　　　　　　Focal or Generalized With Both Generalized and
　　　　　　　　　　　Focal Features 54
　　　　　　　　　　Acquired Epileptic Aphasia (Landau-Kleffner
　　　　　　　　　　　Syndrome) 54
　　　　　　　　　　Epilepsy with Continuous Spike-Waves During Slow Wave
　　　　　　　　　　　Sleep (ECSWS) 55
　　　　　　　　Special Syndromes *56*
　　　　　　　　　Situation-related Seizures (Gelegenheitsanfälle) 56
　　　　　　　　　　Febrile Convulsions 56

Simple Febrile Seizures 57
Complex Febrile Seizures 57
Isolated Status Epilepticus 58
Epidemiology of Epilepsy Syndromes 60
Etiology of Epileptic Syndromes 60
Nonepileptic Seizures 61
Sudden Unexplained Death in Epilepsy 62

Chapter 4 Diagnostic Tests in Epilepsy 65

Electroencephalography 65
Interictal Scalp Recordings 66
Specific Electroencephalographic Patterns in the Partial Epilepsies 67
Periodic Lateralized Epileptiform Discharges 68
Intermittent Rhythmic δ Activity 68
Specific Electroencephalographic Patterns in Generalized Epilepsies 69
3-Hz Spike-and-Wave Pattern 69
Multiple Spike-and-Wave Pattern 69
Slow Spike-and-Wave Pattern 69
Generalized Paroxysmal Fast Activity 70
Structural Neuroimaging in Epilepsy 71

Chapter 5 Medical Treatment of Epilepsy 73

Pharmacologic Therapies 74
Older Antiepileptic Drugs 74
Newer Antiepileptic Drugs 75
Antiepileptic Drug Adverse Effects 78
Switching Antiepileptic Drugs and Generic Equivalents 79
Antiepileptic Drugs in Women 79
Pregnancy and Teratogenic Effects 80
Antiepileptic Drug Treatment in Children 81
Adverse Antiepileptic Drug Effects in Children 82
Cognitive and Behavioral Effects of Antiepileptic Drugs 83
Gabapentin (Neurontin) 85
Lamotrigine (Lamictal) 85

Levetiracetam (Keppra)   85
Oxcarbazepine (Trileptal)   86
Tiagabine (Gabitril)   86
Topiramate (Topamax)   86
Zonisamide (Zonegran)   87
Pregabalin (Lyrica)   87
Lacosamide (Vimpat)   87
Use of Antiepileptic Drugs for the Treatment of Bipolar Disorder   88
Suicidality and Antiepileptic Drugs   88
Compliance with Antiepileptic Drug Therapy   90
Ketogenic Diet   91
Mechanism of Action   91
The Diet Itself   92
Adverse Effects of the Ketogenic Diet   92
Efficacy of the Ketogenic Diet   92

Chapter 6   **Neuropsychological Assessment in Epilepsy**   95
Selection of Neuropsychological Tests   95
Factors Contributing to Cognitive Decline in Epilepsy   98
Etiology   99
Location and Extent of Lesion   99
Seizure Frequency and Severity   100
Age of Onset and Duration of Seizure Disorder   100
Seizure Type   101
Additional Considerations in Children   101
Confounding Factors in Test Interpretation   103
Medical Adverse Effects of Antiepileptic Drugs   103
Transient Cognitive Impairment   104
Postictal Neuropsychological Assessment   104
Cognitive Deficits in Epilepsy   105
Cognitive Effects of Recurrent Seizures   107
Attention   108
Attention-deficit Hyperactivity Disorder   109
Memory   110
Intelligence   112

    Language 113
    Visual-Perceptual and Spatial Functions 116
    Executive Functions 117
    Childhood Learning Disabilities 119
  Quality-of-Life Assessment in Epilepsy 119
    Adult Quality of Life in Epilepsy Measures 120
    Pediatric Quality of Life in Epilepsy Measures 122
    Adolescent Quality of Life in Epilepsy Measures 124
  Driving Issues in Epilepsy 125
    Seizure-free Period Requirements 125
    Exceptions to Seizure-free Period Requirements 126
  Case Example 127
    Preoperative Neuropsychological Test Results in Left Mesial Temporal Lobe Seizure Onset 127

Chapter 7 Psychological and Psychiatric Disorders in Epilepsy 133

  Risk Factors 134
  Mood Disorders 135
    Depression 135
      Prevalence 135
      Symptoms 135
      Pharmacologic Treatment 136
    Bipolar Disorder 137
      Symptoms and Prevalence 137
      Pharmacologic Treatment 137
  Anxiety Disorders 138
    Prevalence 138
    Symptoms by Subtype 139
      Panic attacks 139
      Obsessive-Compulsive Disorder 140
      Generalized Anxiety Disorder 140
    Pharmacologic Treatment 141
  Psychotic Disorders 141
    Prevalence 142
    Symptoms 142
    Pharmacologic Treatment 143

Personality Disorders   143
  Prevalence   143
  Symptoms   144
Interictal Behavior Syndrome of Temporal Lobe Epilepsy   145
Epilepsy-specific Psychological Disorders   145
  Psychoses of Epilepsy   146
    Interictal Psychosis of Epilepsy   146
    Alternative Psychosis   146
  Affective-Somatoform (Dysphoric) Disorders of Epilepsy   147
    Interictal Dysphoric Disorder   147
    Prodromal Dysphoric Disorder   147
    Postictal Dysphoric Disorder   147
    Alternative Affective-Somatoform Syndromes   147
  Personality Disorders   147
  Anxiety/Phobias   148
  Anticonvulsant-induced Psychiatric Disorders   148

Chapter 8   **Psychogenic Nonepileptic Seizures**   151
Diagnosis   151
Prevalence   153
Etiology   154
  Psychological Etiology   155
Symptoms   155
Psychological and Neuropsychological Assessment   158
  Personality Testing in PNES   159
  Cognitive Testing in PNES   159
Treatment   160

PART TWO   **SURGICAL TREATMENT OF EPILEPSY**

Chapter 9   **Neuropsychological Assessment in Epilepsy Surgery**   165
Preoperative Neuropsychological Assessment   167
  Purposes of Preoperative Neuropsychological Assessment   167

    Lateralization and Localization 167
    Risk for Postoperative Cognitive Impairment 169
    Establish a Baseline 171
    Prediction of Seizure Control 172
   Postoperative Neuropsychological Assessment 172
  Case Example 173
   Neuropsychological Results Pre- and Post-Right Anterior Temporal Lobectomy 173

Chapter 10 Other Neuropsychological Procedures in Epilepsy Surgery 183
  Intracarotid Amobarbital (Wada) Procedure 183
   Description of the Procedure 184
   Wada Language Assessment 185
    Recovery of Language After Amobarbital Injection 186
    Mixed or Atypical Language Representation 186
    Clinical Implications of Wada Language Representation 188
    Limitations of Clinical Interpretation with Atypical Language 189
    Excluding Patients from Preoperative Wada Language Evaluation 190
   Wada Memory Assessment 190
    Purpose of Wada Memory Assessment 190
   Which Drug to Select for Wada: Amobarbital (Amytal) or Methohexital (Brevital)? 193
   Wada Testing in Children 194
  Functional Magnetic Resonance Imaging 196
   Sensorimotor Mapping Using Functional Magnetic Resonance Imaging 197
   Language Mapping Using Functional Magnetic Resonance Imaging 197
   Functional Magnetic Resonance Imaging and Memory 198
  Electrocortical Stimulation Mapping 199
   Extraoperative Stimulation Mapping 200
    Stimulation Parameters and Methodology 200

          Language and Related Tests Used During
             Mapping 201
          Interpretation of Results 202
       Intraoperative Stimulation Mapping 203
          Stimulation Parameters and Methodology 204
Case Example 205
    Wada Testing Predicts Memory Decline in a Case with
        Right Hemisphere Language Dominance 205

Chapter 11   Medical Aspects of Epilepsy Surgery 211

Criteria for Surgical Evaluation 211
    Number of Drugs Failed 211
    Duration of Antiepileptic Drug Therapy 212
    Seizure Frequency 213
Diagnostic Evaluation for Epilepsy Surgery 213
    Noninvasive Video-Electroencephalography
        Monitoring 214
        Cognitive Assessment During Video-
            Electroencephalography-monitored Seizures 214
        Electroencephalography Analysis During Inpatient
            Monitoring 215
    Structural Neuroimaging in the Presurgical
        Evaluation 216
        Magnetic Resonance Imaging 216
        Magnetic Resonance Spectroscopy 218
    Diffusion Tensor Imaging 218
    Functional Neuroimaging in the Presurgical
        Evaluation 218
        Positron Emission Tomography 218
        Single-photon Emission Computed Tomography 219
        Subtraction Ictal SPECT Co-registered with MRI 219
        Magnetoencephalography 221
    Intracranial Electrodes: Invasive Video-EEG (Phase II)
        Monitoring with Grid, Strip, or Depth
        Electrodes 221
        Subdural Strip and Grid Electrodes 222
        Depth Electrodes 223
Epilepsy Surgery Procedures 224

Vagus Nerve Stimulation   225
   Vagus Nerve Stimulation Parameters   226
   Vagus Nerve Stimulation Mechanism of Action   226
   Vagus Nerve Stimulation Efficacy   227
   Adverse Effects of Vagus Nerve Stimulation   227
   Current Status of Vagus Nerve Stimulation   227
Anterior Temporal Lobectomy   227
   Outcome of Anterior Temporal Lobectomy   228
   Common Complications of Anterior Temporal Lobectomy   230
Frontal Lobectomy   231
   Outcome of Frontal Lobectomy   231
   Complications of Frontal Lobectomy   231
Parietal Lobectomy   232
   Outcome of Parietal Lobectomy   232
   Complications of Parietal Lobectomy   233
Occipital Lobectomy   233
   Outcome of Occipital Lobectomy   234
   Complications of Occipital Lobectomy   234
Lesionectomy   234
Hemispherectomy   235
   Outcome of Hemispherectomy   236
   Complications of Hemispherectomy   236
Corpus Callosotomy   237
   Outcome of Corpus Callosotomy   237
   Complications of Corpus Callosotomy   238
Multiple Subpial Transection   238
   Outcome of Multiple Subpial Transection   239
   Complications of Multiple Subpial Transection   239
Selective Amygdalohippocampectomy   240
   Outcome of Selective Amygdalohippocampectomy   240
   Implanted Electrical Brain Stimulators   241
Case Example   242
   Independent Bilateral Seizure Onset: Candidate for Responsive Neurostimulator Implant   242

Appendix I	**Traditional Classification of Epileptic Seizures: Description of Seizure Types Not Covered in Body of Text** 247

Simple Partial (or Focal) Seizures 247
   Simple Partial Seizures with Motor Signs 247
   Simple Partial Seizures with Somatosensory or Special Sensory Symptoms 248
   Simple Partial Seizures with Autonomic Symptoms or Signs 249
   Simple Partial Seizures with Psychic Signs 250
   Simple Partial Seizures with Cognitive Signs 250
   Simple Partial Seizures with Affective Signs 251
   Simple Partial Seizures with Illusions 252
   Simple Partial Seizures with Hallucinations 252
Generalized Seizures (Convulsive or Nonconvulsive) 252
   Atypical Absence Seizures 252
   Myoclonic Seizures 253
   Clonic Seizures 254
   Tonic Seizures 254
   Atonic (Astatic) Seizures 255

Appendix II	**Classification of Epilepsy Syndromes: Description of Seizure Syndromes Not Covered in Body of Text** 257

Idiopathic Localization-related Epilepsies 257
   Benign Childhood Epilepsy with Centrotemporal Spikes 257
   Childhood Epilepsy with Occipital Paroxysms 258
   Reading Epilepsy 258
   Hot Water Epilepsy 259
   Autosomal Dominant Nocturnal Frontal Lobe Epilepsy 259
Symptomatic Localization-related Epilepsy Syndromes 260
   Rasmussen Syndrome (Kojewnikow Syndrome): Chronic Progressive Epilepsia Partialis Continua of Childhood 260

　　　　　　　　　　Reflex Epilepsies　261
　　　　　　　　　　　　Photosensitive Seizures　261
　　　　　　　　　　　　Musicogenic Epilepsy　263
　　　　　　　　　　　　Eating Epilepsy　263
　　　　　　　　　　　　Startle Epilepsy　263
　　　　　　　　　　Frontal Lobe Epilepsies　264
　　　　　　　　　　　　Precentral Frontal Lobe Seizures　264
　　　　　　　　　　　　Premotor Frontal Lobe Seizures　264
　　　　　　　　　　　　Supplementary Motor Area Seizures　265
　　　　　　　　　　　　Dorsolateral Prefrontal Lobe Seizures　265
　　　　　　　　　　　　Orbitofrontal Seizures　265
　　　　　　　　　　　　Medial Frontal Lobe Seizures　266
　　　　　　　　　　　　Frontal Opercular Seizures　266
　　　　　　　　　　Parietal Lobe Epilepsies　266
　　　　　　　　　　　　Postcentral Gyrus Seizures　266
　　　　　　　　　　　　Superior Parietal Lobule Seizures　267
　　　　　　　　　　　　Inferior Parietal Lobule Seizures　267
　　　　　　　　　　　　Paracentral Parietal Lobe Seizures　268
　　　　　　　　Idiopathic Generalized Epilepsy Syndromes　268
　　　　　　　　　　Benign Neonatal Familial Convulsions　268
　　　　　　　　　　Benign Neonatal Convulsions (Nonfamilial)　268
　　　　　　　　　　Benign Myoclonic Epilepsy of Childhood　268
　　　　　　　　　　Juvenile Myoclonic Epilepsy　269
　　　　　　　　　　Epilepsy with Tonic-Clonic (Generalized Tonic-Clonic
　　　　　　　　　　　　Seizures) Seizures upon Awakening　270
　　　　　　　　Cryptogenic or Symptomatic Generalized Epilepsy
　　　　　　　　　　Syndromes　270
　　　　　　　　　　West Syndrome (Infantile Spasms)　270
　　　　　　　　　　Epilepsy with Myoclonic-Astatic Seizures　271
　　　　　　　　　　Epilepsy with Myoclonic Absences　271

Appendix III　　**Wada Assessment Procedures and Rating Criteria at the Medical College of Georgia**　273
　　　　　　　　Procedure for Measuring Wada Language Functions
　　　　　　　　　　at the Medical College of Georgia　274
　　　　　　　　　　Fluency of Speech　274
　　　　　　　　　　Aural Comprehension　274
　　　　　　　　　　Visual Naming　275

   Repetition 275
   Reading 276
   Paraphasias 276
  Procedure for Measuring Wada Memory at the
   Medical College of Georgia 276
   Wada Memory Stimuli 277
   Scoring of Wada Memory Items 277
  Summary of Behavioral Assessment During Wada
   Testing at the Medical College of Georgia 278

Glossary *291*

References *303*

Index *339*

Continuing Education Credits

Continuing Education
The American Academy of Clinical Neuropsychology (AACN) is offering continuing education (CE) through its book series with Oxford University Press, *Oxford Workshop Series: American Academy of Clinical Neuropsychology* (*Workshop Series*).

Target Audience
Clinical Neuropsychologists and Clinical Psychologists.

AACN Online System
Any licensed psychologist who reads a volume in the *Workshop Series* can earn up to three CE credits by completing an online quiz about the volume's content. A fee of $20 per credit ($15 for AACN members), payable by credit card online, will be charged for participation in this activity.

To Receive Online CE Credit
Read any volume in the *Workshop Series* in its entirety. Access the CE quiz online at the AACN website (www.theaacn.org), register for the specific *Workshop* book for which you wish to receive CE credit, and answer all questions on the quiz. The estimated time to read a book in the series is three hours. Credits will be awarded to individuals scoring 75% or better on the quiz. Three CE credits will be awarded for each volume. Participants will receive an immediate confirmation of credits earned by email.

CE Accreditation Statement
The AACN is approved by the American Psychological Association to sponsor continuing education for psychologists. The AACN maintains responsibility for this program and its content.

Questions
If you have questions regarding the CE program, please contact: webmaster@theaacn.org.

Author's Workshop Materials

To download materials from the author's workshop presentation, such as PowerPoints, visit www.AACNWorkshopSeries/Lee.

PART ONE

MEDICAL ASPECTS OF EPILEPSY FOR NEUROPSYCHOLOGISTS

I

Introduction

Epilepsy and epilepsy surgery have played an important role in the development of neuropsychology since the beginnings of clinical neuropsychology as a field of practice in the 1950s and 1960s. Molly Harrower and Brenda Milner, in collaboration with the distinguished neurosurgeon Wilder Penfield at the Montreal Neurological Institute, were among the first to use psychological tests to assess the effects of localized brain damage in patients with seizures (Barr, 2007). Since that time, the cognitive and psychosocial consequences of epilepsy and its treatment have become well-known, and neuropsychologists have come to be recognized as providing a valuable and distinctive role in evaluating patients with epilepsy.

Considering the prevalence of epilepsy alone, one might predict most neuropsychologists would encounter such patients fairly frequently and even more so when diseases causing symptomatic epilepsies are considered. Neuropsychologists commonly evaluate patients with brain tumors, strokes, traumatic brain injuries, brain infections, and poisonings (e.g., lead, substance abuse), and all of these conditions may cause epilepsy. Thus, most neuropsychologists will encounter patients who have epileptic seizures at some point in the course of their practice, and they may benefit from a resource, such as this book, that presents the complicated issues of seizure classification, diagnostic evaluation, medical treatment options, and neurobehavioral consequences of epilepsy.

Definitions

Epilepsy has been defined as a medical condition that involves recurrent seizures due to excessive disorderly discharges of cerebral neurons (Gastaut, 1973). A *seizure* is a single event that occurs when a strong surge of electrical activity causes the abnormal and excessive discharge of a set of neurons in the brain, resulting in a variety of clinical signs that are accompanied by electroencephalographic (EEG) changes. A seizure may be envisioned as a brief electrical storm within the brain. Seizures typically last a few seconds to a few minutes. Epilepsy is often called a "seizure disorder."

An individual is considered to have epilepsy after having two or more unprovoked seizures. *Unprovoked seizures* are seizures occurring with no clear antecedent cause and are contrasted with *provoked seizures,* in which the seizure occurs during or soon after some acute medical condition affecting the brain. Provoked seizures are also sometimes called "acute symptomatic" seizures. Common causes of provoked seizures include acute systemic, metabolic, or toxic insults, or these seizures occur in association with an acute central nervous system (CNS) insult such as infection, stroke, brain trauma, intracerebral hemorrhage, or acute alcohol intoxication withdrawal (ILAE, 1997).

The uncontrolled neuronal discharges causing seizures are thought to occur when the membrane stabilizing mechanisms within neurons are disrupted due to abnormal membrane structure or an imbalance between excitatory and inhibitory neurotransmitter influences (Browne and Holmes, 2004). Some have hypothesized that, in seizure disorders, the membrane resting potential of a population of neurons has been reset to from the normal −70mV firing threshold to some lower threshold, thus causing the cell to fire with much less stimulation than usual. If the uncontrolled neuronal discharges are confined to a circumscribed area of nerve cells, a focal seizure will occur. This has traditionally been called a *partial seizure*. In contrast to focal partial seizures, *generalized seizures* are characterized by the simultaneous onset of EEG changes and associated clinical phenomena in both cerebral hemispheres (ILAE, 1997).

A seizure may begin as a partial seizure and then spread to both cerebral hemispheres, thus evolving into a generalized seizure. This is referred to as a *partial complex seizure with secondary generalization*. On the other hand, when both cerebral hemispheres show simultaneous onset of epileptiform EEG changes and associated clinical phenomenon, this is referred to as *primary generalized epilepsy*. These epilepsy terms are taken from the traditional seizure classification system, which is based on the clinical symptoms (called *seizure semiology*) and associated EEG pattern during seizures to classify seizure types. These seizure disorders will be described in detail in Chapter 2.

Epidemiology

Recall that *incidence* is a measure of the number of new cases of a medical condition occurring in a defined population during a specified period of time, usually one year. The cumulative incidence of epilepsy (risk of developing epilepsy) by age 20 years is estimated to be about 1%, whereas by 75 years of age, 3%–4% of the U.S. population can be expected to have been diagnosed with epilepsy (Annegers, 1996). The incidence is highest under the age of 2 and over the age of 65, with approximately 200,000 new cases of epilepsy diagnosed in the United States each year. The Epilepsy Foundation of America (2008) has

Table 1.1 Estimates of incidence of seizures and epilepsy in the United States

Incidence of seizures in the U.S.:

- 300,000 people have a first convulsion each year
- 120,000 of them are under the age of 18
- Between 75,000 and 100,000 of them are children under the age of 5 who have experienced a febrile (fever-caused) seizure

Incidence of epilepsy in the U.S.:

- 200,000 new cases of epilepsy are diagnosed each year
- Incidence is highest under the age of 2 and over 65
- 45,000 children under the age of 15 develop epilepsy each year
- Males are slightly more likely to develop epilepsy than females
- Incidence is greater in African Americans and socially disadvantaged populations
- Trend shows decreased incidence in children and increased incidence in the elderly
- In 70% of new cases, no cause is apparent
- 50% of people with new cases of epilepsy will have generalized-onset seizures
- Generalized seizures are more common in children under the age of 10; afterwards more than half of all new cases of epilepsy will have partial seizures

Cumulative incidence (risk of developing epilepsy) in the U.S.:

- By 20 years of age, 1% of the population can be expected to have developed epilepsy
- By 75 years of age, 3% of the population can be expected to have been diagnosed with epilepsy, and 10% will have experienced some type of seizure

Adapted from Epilepsy Foundation of America website statistics (2008).

collated data on the incidence of seizures and epilepsy (seizure disorders) in the United States, and these facts of interest are presented in Table 1.1.

Incidence is contrasted with *prevalence,* which is the total number of existing cases of a disease in a specific population at one point in time. More than 3 million people in the United States have epilepsy, and 25 to 30 million Americans (10%) will have a seizure at some point in their lives. Thirty percent of patients with epilepsy are children under the age of 18, although there has been a recent decrease in the incidence of childhood cases and a simultaneous increase in the elderly (Hauser et.al, 1993; ILAE, 1997). The Epilepsy Foundation of America (2008) has collected information on the prevalence of epilepsy (seizure disorders) in the United States and estimates that more than 3 million people have active epilepsy. Moreover, prevalence tends to increase with age and is higher among racial minorities.

Epilepsy is more common among racial minorities, people from economically disadvantaged backgrounds, and those in developing countries. The World Health Organization estimates that about 50 million people worldwide are living with epilepsy (ILAE, 1997; Hauser et al., 1993). Epilepsy is the most common neurologic disorder in children and the third most common in adults after Alzheimer disease and stroke. Populations who are at higher risk for developing epilepsy, as reported by the Epilepsy Foundation of America (2008), are presented in Table 1.2.

Table 1.2 Estimated incidence of epilepsy among "at risk" populations

POPULATION	ESTIMATED % WITH EPILEPSY
Children with mental retardation	10%
Children with cerebral palsy	10%
Children with mental retardation and cerebral palsy	50%
Alzheimer disease	10%
Stroke	22%
Children of mothers with epilepsy	8.7%
Children of fathers with epilepsy	2.4%
Individuals with single, unprovoked seizure	33%

Adapted from Epilepsy Foundation of America website statistics (2008).

Etiology

Epileptic seizures are due to some perturbation of brain from either pathology within the brain itself or from systemic medical conditions that affect the brain secondarily. Seizures may also result from nonepileptic causes, as in cardiogenic seizures or psychogenic nonepileptic seizures. The cause of epileptic seizures is unknown in just under 70% of cases, whereas some neurologic etiology is identified in approximately 30% of patients (Hauser, 1990). However, the number of cases with an unknown cause is shrinking as modern neuroimaging technology evolves. *Symptomatic seizures* is the term applied to the 30% or so of patients whose seizures are linked to identifiable diseases or brain abnormalities. *Idiopathic* (no known cause, presumed genetic) or *cryptogenic* (undetected but presumed to be caused by a developmental lesion) seizures are diagnosed in the majority of cases when no cause for the seizures can be found.

The etiology of epileptic seizures differs across the lifespan and depends upon the age of seizure onset. For example in early childhood, birth injury, genetic metabolic disorders, and fever are most common, whereas in late life, vascular disease and neurodegenerative disorders are most often encountered. The most common causes of epilepsy for each age group have been reported by the Epilepsy Foundation of America (2008) and are listed in Table 1.3.

Role of the Neuropsychologist in Epilepsy

The role of the neuropsychologist in evaluating patients with epilepsy is essentially the same as it is with any other neurologic condition. The purpose of testing is guided by the referral question. The reason for referral often depends upon the age of the patient, with academic or vocational planning concerns being prominent in children and adolescents with epilepsy, and cognitive deterioration or job performance concerns being most common in adults. Similar to any assessment, medical records are reviewed, collateral information from teachers or family members is obtained, a clinical interview with the patient is conducted, and psychometric tests and questionnaires are administered. The goal of the evaluation is to establish a profile of the patient's cognitive strengths and weaknesses across multiple domains, in order to arrive at a neurobehavioral diagnosis and likely explanation for the chief complaints, as well as to assist in developing a comprehensive, individualized treatment plan. The clinical purposes of the neuropsychological assessment of epilepsy patients whose seizures are well-controlled on antiepileptic medications is

Table 1.3 Common potential causes of epilepsy by age of seizure onset

AGE GROUP	POTENTIAL CAUSE
Newborns:	Brain malformations Lack of oxygen during birth Low levels of blood sugar, blood calcium, blood magnesium, or other electrolyte disturbances Inborn errors of metabolism Intracranial hemorrhage Maternal drug use Infection
Infants and Children:	Fever (febrile seizures) Infections Brain tumor (rarely)
Children and Adults:	Congenital conditions (Down syndrome, Angelman syndrome, tuberous sclerosis and neurofibromatosis Genetic factors Head trauma Progressive brain diseases (rare)
Elderly:	Stroke Alzheimer disease Trauma

Adapted from Epilepsy Foundation of America website (2008).

really no different than the clinical purpose of testing in any patient with known or suspected neurological dysfunction. The major clinical purposes of neuropsychological assessment are to:

- Aid in the detection of neurological disorders
- Determine if a developmental learning disorder or other neurodevelopmental disorder exists and specify subtype, prognosis, and treatment strategies
- Diagnose neurobehavioral disorders and provide information about the course and prognosis of deficits

- Diagnose psychological/psychiatric disorders and evaluate impact on cognition and adaptive behavior
- Assess adaptive functioning in response to neuropathology or psychopathology and use this information to assist in educational and vocational planning
- Evaluate the cognitive and behavioral side effects of antiepileptic drugs (AEDs)
- Monitor changes in cognition and behavior over the course of the disease in chronic epilepsy

Within epilepsy populations, the pattern of neuropsychological test results is frequently determined by multiple influences, which include the underlying cause of the epilepsy in cases where this is known (e.g., tumor, stroke, prior head injury), the epileptic process itself, adverse side effects of AED treatment, and psychological-emotional adjustment effects. The neuropsychologist's task is to interpret the pattern of deficits across the neuropsychological tests, to understand the relative contributions of each of the potential etiological factors and assist with treatment planning. Deficit patterns occurring across neuropsychological tests can be suggestive of various sites of cerebral dysfunction and the various neurologic or psychologic processes that underlie the deficit pattern. An effort is made by the neuropsychologist to integrate neuropsychological test data, history, clinical interview, behavioral observations, and available laboratory and radiological evidence into one cohesive summary report that, (a) arrives at a neurobehavioral diagnosis or description, (b) discusses the neurological implications (e.g., course, prognosis, localization), and (c) informs other professionals about follow-up management and treatment issues.

Role of the Neuropsychologist in Epilepsy Surgery

The role of the neuropsychologist is more specialized when working in an epilepsy surgery context with patients who have seizures that are refractory to traditional medical therapies. The epilepsy surgery context is one of the few settings left in which neuropsychologists are asked to localize cerebral functions. This is accomplished in several ways. In the preoperative cognitive assessment, neuropsychologists attempt to lateralize, and if possible localize, the seizure focus by linking specific cognitive deficits to a particular brain region. Although simultaneous video-electroencephalography (EEG) recording, magnetic resonance imaging (MRI), magnetoencephalography (MEG), positron emission tomography (PET), and ictal single-photon emission computed tomography (SPECT) may all arguably be methods superior to neuropsychological testing for lesion localization, prognosis for postoperative seizure relief is improved when the functional localization of neuropsychological testing is consistent with the radiological and physiological measures. This prognostic significance is one justification for including neuropsychological testing in the preoperative workup for epilepsy surgery.

Another important role of preoperative testing is to predict the cognitive risks of surgery and, in conjunction with the epilepsy surgery team, the patient, and the family, an attempt is made to weigh these risks against the likelihood of curing the seizures. However, the role of neuropsychological assessment within an epilepsy surgery context goes beyond localization of the seizure focus and prediction of postoperative loss of normal cognitive functions.

Assessment within an epilepsy surgery program serves many of the same functions as with any other chronic neurological condition. The data derived from the assessment may be used to help formulate a diagnosis, assist in determining prognosis, monitor the evolution of epilepsy and its related conditions, evaluate the efficacy of various therapeutic interventions, help guide the neurosurgeon in deciding which brain areas to resect and which to spare, and assist health care providers and educators to adjust their treatment programs to the changing needs of the epilepsy patient (Lassonde, Sauerwein, Gallagher, et al., 2006). More specifically, neuropsychological evaluations within the epilepsy surgery context may be used to:

- Help lateralize and localize the seizure focus
- Predict risk for postoperative cognitive impairment
- Establish a baseline against which to measure change
- Help predict seizure relief outcome
- Diagnose psychiatric disorders and consider potential impact on ability to cooperate with epilepsy surgery process

Many neuropsychologists also play an important role in epilepsy surgery programs by assessing preoperative language lateralization and memory asymmetry in patients by means of the *intracarotid amobarbital (Wada) procedure* (Loring, Meader, Lee, et al., 1992). Language lateralization and hippocampal memory functions are assessed before surgery in an attempt to reduce the risk for postoperative language or memory impairments. In Wada testing, one hemisphere is temporarily anesthetized by injecting amobarbital (or another barbiturate drug) into the internal carotid artery. The patient is then presented with a series of language and memory tasks to determine whether the to-be-resected hemisphere mediates language and to gauge the risk for language and memory disorders after surgery.

Patients whose Wada assessment suggests language functions overlap with to-be-resected brain regions may then need to undergo *electrocortical stimulation* language mapping of these brain regions. Electrocortical stimulation mapping may be conducted either in the operating room at the time of surgery (but prior to the actual cortical resection) or by means of chronically implanted grid or strip electrodes at bedside. During stimulation mapping, electrodes on the cortical surface are systematically stimulated one at a time while the patient carries out various language tasks. In this way, active language sites are identified, and eloquent language cortex is charted. The role of the neuropsychologist is to administer, record, and interpret the results of electrical stimulation language testing and advise the epileptologist and neurosurgeon of the results and their implications.

Because epilepsy is a chronic disease that begins by and large in childhood, many epilepsy neuropsychologists specialize in both pediatric and adult neuropsychology. Even epilepsy neuropsychologists who specialize in adult neuropsychology must have a strong background in neurodevelopmental issues in order to appreciate the neurological, developmental, academic, and psychosocial impact of childhood epilepsy on their adult patients.

There is worldwide consensus that neuropsychologists should play an integral role at epilepsy surgery centers by evaluating and treating patients within a multidisciplinary team (Reynders and Baker, 2002). The Committee to Revise the Guidelines for Services, Personnel and Facilities at Specialized Epilepsy Centers of the National Association of Epilepsy Centers considers neuropsychologists to be key personnel, and has included neuropsychological services in defining the standard of care for patients at specialty epilepsy centers (Walczak et al., 2001).

The remainder of this book covers the basics of epilepsy classification, current state of diagnosis and treatment, neuropsychological assessment in epilepsy and epilepsy surgery settings, cognitive and psychiatric disorders commonly seen in patients with epilepsy, and the epilepsy surgery process with particular attention to the neuropsychologist's role. Since many of the seizure types and epilepsy syndromes are relatively rare, and thus less likely to be seen by neuropsychologists, descriptions of many of the less commonly encountered seizure disorders have been placed in the appendices. The appendices may thereby serve as a quick reference guide for clinicians who have patients with one of these more infrequent forms of epilepsy.

2

Classification of Epilepsy Disorders

Accurate diagnosis is essential in epilepsy because knowing the type of epilepsy will determine its proper treatment as well as guide recommendations based upon the prognosis. It is also important that any underlying disease causing the epilepsy be identified and appropriately treated in cases of symptomatic epilepsy. Recall that epilepsy is not a single disease or syndrome and may be caused by a variety of different medical conditions, each with its particular symptoms, severity level, and course (Engel et al., 2007). Epilepsy has been broadly defined as a condition characterized by recurrent (two or more) epileptic seizures that are unprovoked by an identified cause immediately preceding the event (Commission of Epidemiology and Prognosis of the ILAE, 1993).

Traditional Classification of Epileptic Seizures

There are currently two standardized systems for the classification of epilepsy seizures and syndromes in use today, and both have been proposed as helpful frameworks for organizing seizure classification by the International League Against Epilepsy (ILAE). The older classification schema of epileptic seizures was codified by the ILAE in 1981 and continues to be the most commonly used classification model worldwide (Commission on Classification and Terminology of the ILAE, 1981). The newer classification system is based upon epilepsy syndromes. These will be briefly reviewed in Chapter 3, and the syndrome subtypes will be discussed in more detail in Appendix II.

The traditional epileptic seizure classification model is based on the clinical symptoms observed during the seizure (seizure semiology) and the electroencephalographic (EEG) (ictal and interictal) manifestations. The ILAE's traditional classification of epileptic seizures divides them into the following types:

- *Partial (focal or localized) seizures*; abnormal neuronal epileptiform discharges localized in the cerebral cortex (Table 2.1)
- *Generalized seizures* (tonic, clonic, or tonic-clonic, myoclonic, or absence attacks); abnormal paroxysmal discharge of cerebral neurons that involves large portions of the cortex bilaterally from seizure onset to seizure cessation (Table 2.1)
- *Unclassified epileptic seizures*; these seizures cannot be classified as either partial or generalized due to either inadequate or incomplete data. Others do not clearly fit into any other seizure classification category, such as some neonatal seizures that are characterized by rhythmic eye movements, chewing, and swimming movements (Commission on Classification and Terminology of the ILAE, 1981).

Table 2.1 Simplified traditional classification of epileptic seizures and their associated signs and symptoms

Partial Seizures:

Seizure Type	Seizure Semiology	Typical Duration
Simple	Focal jerking Sensory phenomena	~3–120 seconds
Complex	Automatisms (e.g., lip smacking, picking at clothes, fumbling hands), Loss of awareness	~2–4 minutes

Generalized Seizures:

Seizure Type	Seizure Semiology	Typical Duration
Absence	Behavioral arrest Blank stare Repetitive blinking	~5–45 seconds
Tonic-clonic	Tonic extension of limbs Generalized convulsions Falls	~1–2 minutes

- *Prolonged or repetitive seizures* (status epilepticus); a single prolonged continuous seizure or an ongoing series of brief seizures with only short-lived intervals between them.

At the most basic level, most neuropsychologists have been taught to know the essential traditional seizure classification presented in Table 2.1. This simplified classification schema only includes two overarching categories of epilepsy (partial and generalized) and provides two exemplary seizure types under each category. Although simplified, these four seizure types can provide the framework upon which to build one's knowledge of the less commonly encountered forms of epilepsy, such as those contained in the appendices of this book. Moreover, every self-respecting neuropsychologist should have at least the basic level of epilepsy understanding provided in Table 2.1.

Partial (or Focal) Seizures
The partial epilepsies are characterized by seizures that begin in a restricted cortical area within one cerebral hemisphere as evidenced by clinical symptoms or localized EEG activity (Table 2.2). Partial seizures can be classified into one of the following three broad categories:

- *Simple partial seizures* with motor, somatosensory, autonomic, or psychic signs
- *Complex partial seizures* either with (a) impairment of consciousness at onset or (b) simple partial onset followed by impairment of consciousness
- *Partial seizures evolving into generalized tonic-clonic convulsions* with either (a) simple partial onset evolving to generalized tonic-clonic or (b) complex partial onset evolving to generalized tonic-clonic seizures.

Each of these three subtypes of partial seizures is then further divided by the nature of the symptoms during the seizure proper or by how the seizure symptoms progress from seizure onset to termination (Table 2.2).

Partial (focal) seizures are classified primarily by whether or not consciousness is impaired during the seizure and whether the partial seizure evolves into a generalized convulsion. *Consciousness* is defined here as the degree of awareness or responsiveness of the patient during a seizure. Responsiveness and awareness are evaluated through the patient's ability to carry out simple commands and by

Table 2.2 International League Against Epilepsy (ILEA) classification of partial (focal) seizures

CLINICAL SEIZURE TYPE	EEG SEIZURE TYPE	EEG INTERICTAL EXPRESSION

A. Simple partial seizures (consciousness not impaired)
1. With motor signs
  a) Focal motor without march
  b) Focal motor with march (jacksonian)
  c) Versive
  d) Postural
  e) Phonatory (vocalization or arrest of speech)
2. With somatosensory or special-sensory symptoms (simple hallucinations, e.g., tingling, light flashes, buzzing)
  a) Somatosensory
  b) Visual
  c) Auditory
  d) Olfactory
  e) Gustatory
  f) Vertiginous
3. With autonomic symptoms or signs (including epigastric sensation, pallor, sweating, flushing, piloerection, and pupillary dilatation)
	Local contralateral discharge starting over the corresponding area of cortical representation (not always recorded on the scalp)	Local contralateral discharge

(continued)

Table 2.2 (Continued)

4. With psychic symptoms (disturbance of higher cerebral function); these symptoms rarely occur without impairment of consciousness and are much more commonly experienced as complex partial seizures
  a) Dysphasic
  b) Dysmnesic (e.g., déjà-vu)
  c) Cognitive (e.g., dreamy states, distortions of time sense)
  d) Affective (fear, anger, etc.)
  e) Illusions (e.g., macropsia)
  f) Structured hallucinations (e.g., music, scenes)		

B. Complex partial seizures (with impairment of consciousness; may sometimes begin with simple symptomatology)
1. Simple partial onset followed by impairment of Consciousness
  a) With simple partial features (A1 to A4) followed
  b) With automatisms
	Unilateral or, frequently, bilateral discharge, diffuse or focal in temporal or fronto-temporal regions	Unilateral or bilateral generally asynchronous focus; usually in the temporal or frontal regions

(continued)

Table 2.2 (Continued)

CLINICAL SEIZURE TYPE	EEG SEIZURE TYPE	EEG INTERICTAL EXPRESSION

2. With impairment of consciousness at onset
  a) With impairment of consciously only
  b) With automatisms		

C. **Partial seizures evolving to secondarily generalized seizures** (may be generalized tonic-clonic, tonic, or clonic)	Discharges become secondarily and rapidly generalized	
1. Simple partial seizures (A) evolving to generalized seizures 2. Complex partial seizures (B) evolving to generalized seizures 3. Simple partial seizures evolving to complex partial seizures evolving to generalized seizures		

From Commission on Classification and Terminology, International League Against Epilepsy. Proposal for revised clinical and electrographic classification of epileptic seizures. *Epilepsia* 1981;22:489–501.

assessing the patient's degree of contact with, and memory for, events occurring during a seizure. When consciousness is not impaired during the seizure, it is classified as a *simple partial seizure*. When consciousness is impaired during a partial seizure, this would be considered a *complex partial seizure*.

Simple Partial (Focal) Seizures

Simple partial seizures arise when localized EEG discharges disrupt the neuronal activity of focal cortical areas for several seconds to a few minutes. Because awareness is preserved during simple partial seizures, patients are alert and maintain ongoing memory for events occurring during the seizure. These seizures may, however, spread to other areas that may subsequently cause alterations of consciousness and generalized convulsions. After the seizure spreads, it would no longer be classified as a simple partial event. Symptoms of simple partial seizures often have localizing significance, especially early in the seizure. The specific symptom depends upon which area of cortex is activated. Simple partial seizures are classified by whether motor, somatosensory, autonomic, psychic, cognitive, or affective signs develop or whether they are characterized by illusions or hallucinations. These specific simple partial seizure subtypes are detailed in Appendix I.

Complex Partial (Focal) Seizures
A complex partial seizure is a focal seizure with an impairment of consciousness. This so-called 'impairment of consciousness' is the primary feature distinguishing a complex from a simple partial seizure. During a seizure, patients with complex partial epilepsy have a decreased awareness of their surroundings and are unresponsiveness to external stimulation. After the seizure is over, there is a lack of recall for events that occurred during the seizure (Commission on Classification and Terminology of the ILAE, 1981). Reduced awareness is usually judged after a seizure has ended by the patient's inability to recall events that occurred during the seizure. Unresponsiveness is typically inferred from a patient's inability to follow commands or perform volitional activities during the seizure proper.

Auras. Complex partial seizures frequently begin as a simple partial seizure, which may be experienced as an aura. An *aura* is a simple partial (focal) seizure occurring in the context of complex partial seizures that serves as a signal or warning to the patient that a larger seizure is about to occur. An aura usually lasts from a few seconds to a few minutes and immediately precedes a more extensive epileptic seizure. After an aura, the seizure focus spreads to involve larger portions of the brain, and patients become unresponsive and unaware of their surroundings as the complex seizure develops. Since consciousness is not affected in a simple focal seizure, patients are aware of and retain memory for the aura itself.

The incidence of auras in epileptic outpatients and inpatients varies by sample and aura definition, but is generally reported to be between 56% and 70% of all patients with epilepsy (Lennox and Cobb, 1933; Sperling and O'Connor, 1990; Sirven et al., Sperling, et al., 1996). Among patients with complex partial seizures, auras are most commonly seen in seizures originating in the temporal lobes. Although auras have localizing significance, such clinical inferences are complicated by auras that change over time and by the presence of multiple auras occurring simultaneously in the same individual.

An aura may represent activation of a circumscribed cortical area by a seizure discharge, or it may spread across different functional regions of brain. Nonetheless, some cortical regions are associated with specific aura types. For example, patients with temporal lobe epilepsy most often have epigastric, psychic, or affective auras. Patients whose seizures originate in the frontal lobes often do not experience an aura. In patients with frontal lobe seizures who do experience an aura, cephalic auras and general bodily sensations are the most common. Somatosensory auras are most often reported in patients who have perirolandic epilepsy with centroparietal epileptic foci. As would be expected, patients with occipital lobe onset of complex partial seizure have the highest incidence of visual auras (Rasmussen, 1982; Palmini and Gloor, 1992). The commonly accepted aura types and their most frequent locations (adapted from Palmini and Gloor, 1992) are provided in Table 2.3. As may be seen in the table, many auras have localizing significance for the neuropsychologists, as well as for the epileptologist.

Table 2.3 Frequency (Ns and %s) of auras in patients with focal epilepsies

AURA TYPE	N	TEMPORAL	FRONTAL	PARIETO-OCCIPITAL
Somatosensory	32	1 (3%)	9 (28%)	22 (69%)
Epigastric	47	40 (85%)	3 (6%)	4 (9%)
Cephalic	22	8 (36%)	13 (59%)	1 (5%)
Diffuse warm sensation	10	1 (10%)	9 (90%)	0 (0%)
Psychic	51	46 (90%)	2 (4%)	3 (6%)
Elementary visual	13	1 (8%)	0 (0%)	12 (92%)
Elementary auditory	3	3 (100%)	0 (0%)	0 (0%)
Vertiginous	7	1 (14%)	2 (29%)	4 (57%)
Conscious confusion	11	6 (55%)	4 (36%)	1 (9%)
Total	196	104	42	47

Adapted from Palmini AL, Gloor P. The localizing value of auras in partial seizures: a prospective and retrospective study. *Neurology* 1992;42: 801–808.

Automatisms. Automatisms are repetitive involuntary motor actions that occur during a state of reduced awareness usually followed by amnesia for the event (Commission on Classification and Terminology of the ILAE, 1981). The term *automatism* comes from the notion that these automatic behaviors occurring during a seizure are beyond the patient's control. Automatisms may occur at any time during a seizure (beginning, middle, or end), typically last a few minutes, and involve such actions as lip-smacking, chewing or swallowing movements, emotional facial expressions, repetitive hand motions, or repetitive speech.

Patients usually have the same automatism(s) whenever they have one of their typical seizures. Thus, automatisms are generally consistent from seizure to seizure in the same individual. The classic view of automatisms is that they are release phenomena, in which higher cortical circuits are disrupted, thereby 'releasing' lower brain centers from the inhibitory control imposed by higher centers. However, some automatisms (e.g., hand automatisms or mouthing movements) may represent ictal activation of specific brain regions. Automatisms most commonly occur in complex partial seizures of temporal or frontal lobe onset, but also may occur in parietal or occipital lobe seizures that spread to the temporal lobes, as well as in other seizure types (e.g., absence seizures). Oxbury and Duchowny (2000) have divided automatisms into the following five subtypes:

1. *Oropharyngeal*. These are also referred to as *oroalimentary* automatisms and involve repetitive stereotyped movements of the mouth, jaw, lips or tongue. Often these consist of chewing, lip-smacking, repetitive swallowing or gulping, excessive salivation with drooling, or spitting. Oroalimentary automatisms tend to occur early in seizures of medial temporal lobe origin but may occur without loss of consciousness in seizures when the ictal discharge is restricted to the amygdala or anterior hippocampus (Kotagal and Loddenkemper, 2006).

2. *Expression of emotion*. These are also called *mimetic* automatisms and commonly include changes in facial expression with smiling, laughing, grimacing, pouting, or angry facial expressions and presumably internally experienced emotion as well. Sexual automatisms are relatively uncommon but may involve pelvic thrusting, masturbation, or playing with the genitals during complex partial seizures.

3. *Gestural.* Hand automatisms typically involve rapid repetitive movements of the hands such as picking at clothing, pulling at sheets, or fumbling with or grasping real objects or pantomimed actions. Hand automatisms suggest medial temporal lobe seizure onset similar to oropharyngeal automatisms. Repetitive eye blinking or fluttering may also be seen. Other motor phenomena, such as tonic or dystonic posturing of the limbs or face or versive head movement, which are commonly observed during complex partial seizures, should not be considered automatisms.
4. *Ambulatory.* These automatisms may include whole body movements in which the patient attempts to sit up, turn over, or get out of bed, and if standing, to walk or run. Truncal or whole body movement automatisms usually occur later in a seizure. Bicycling or pedaling movements of the legs may be observed in complex partial seizures arising from the frontal lobe, most often in the mesial frontal regions (Kotagal and Loddenkemper, 2006).
5. *Verbal.* Ictal speech, such as repetition of a single word or brief phrase, is usually seen in complex partial seizures of temporal lobe origin and has been more commonly observed in seizures beginning in the nondominant (right) temporal lobe. Speech and language disorders during the seizure (e.g., dysphasic disturbances, as well as simple speech arrest, mutism, or dysarthria) suggest a dominant hemisphere seizure focus. Paraphasic errors, effortful output, and naming difficulties may be seen during and after a complex seizure in the language-dominant hemisphere.

Since most patients with complex partial seizures have no memory of their automatisms, they must be described by a witness to the seizure or recorded on videotape (either by the family or during inpatient seizure monitoring).

Other Symptoms of Complex Partial Seizures. Complex partial seizures may begin with an aura and then proceed to a blank stare with an alteration of awareness, followed by automatisms if they occur. Following these early indicators of seizure onset, motor symptoms (e.g., tonic extension of the upper extremity, versive head turning) or autonomic signs (e.g., increased blood pressure, respiratory inhibition) of a seizure frequently develop. Tonic extension or flexion of the extremity (usually the arm) often occurs contralateral to the seizure focus and typically involves the proximal musculature. Tonic posturing may be seen with either temporal or extratemporal ictal discharges. In contrast, dystonic posturing (rotation of limb with a fixed, unnatural posture) more often involves the distal limbs (e.g., hands). Dystonic posturing is considered an excellent lateralizing sign as it almost always occurs on the opposite side of the body from the ictal discharge and often reflects involvement of the contralateral frontal lobe (Bleasel et al., 1997).

Versive movement of the head and eyes, although usually an early sign, may also be observed later in a complex partial seizure by a forced involuntary tonic or clonic deviation of the head and eyes to one side. Versive head and eye turning of the forced type was contralateral to the seizure focus in 90% of patients with temporal or frontal lobe seizures in one sample (McLachlan, 1987). In temporal lobe seizures, versive head and eye deviation often occur later in a seizure after the paroxysmal discharges have spread to the suprasylvian motor regions and often just before the seizure secondarily generalizes.

Autonomic symptoms may arise during a partial seizure and include increased heart rate and blood pressure, flushing, piloerection, decreased gastric motility, or increased peristalsis of the esophagus. Such autonomic ictal manifestations have been associated with seizures arising in the orbitofrontal and insular regions. The positive predictive values of various lateralizing signs, including automatisms and other motor phenomena, were reported in a group of patients with partial seizures from the Cleveland Clinic (Chee et al., 1993).

Complex partial seizures often begin with simple partial seizure symptoms (an aura) and then proceed to a blank stare with loss of awareness and responsiveness. Other complex partial seizures may begin without an aura. In those without an aura, the first sign a seizure is occurring may be development of an alteration of awareness, with or without automatisms. After these early seizure signs, dystonic posturing of the contralateral arm or leg, or deviation of the head and eyes away from the ictal discharges, may become evident. Ultimately, the complex partial seizures may spread to the other hemisphere and cause secondary generalization of the partial seizure into a full-blown tonic-clonic convulsive episode. The complex partial seizure itself, called the *ictus*, generally only lasts for several minutes. Most of the EEG disorganization is seen during the ictal portion of the seizure. A *postictal* period follows the seizure proper, which involves a state of inactivity, headache, disorientation, and inattention that gradually resolves over the course of 2–10 minutes (Lechtenberg, 1985).

Generalized Seizures (Convulsive or Nonconvulsive)
The initial clinical symptoms of a generalized seizure reflect the ictal manifestations of the seizure, which begins simultaneously in both cerebral hemispheres (Table 2.4). Motor symptoms are bilateral, and consciousness may be impaired early on in the seizure. Electroencephalographic changes are widespread in both hemispheres and are mirrored by the clinical manifestations of the seizure. Generalized seizures may either primarily or secondarily generalize. In *primary* generalized seizures, the EEG and clinical manifestations are generalized from the onset. In *secondarily* generalized seizures, the seizure begins as a focal-onset, partial seizure, but then spreads to involve both hemispheres and induces a generalized epileptic spell. Generalized seizure disorders may be convulsive or nonconvulsive and vary greatly with regard to the severity of the epilepsy. Typical absence and generalized tonic-clonic seizures are discussed in this chapter, while the other less commonly encountered generalized seizure types are briefly detailed in Appendix I. The ILAE classification of the generalized epilepsies is presented in Table 2.4 and uses the following categories:

    A1. Absence seizures (typical)
    A2. Atypical absence seizures (see Appendix I)
    B.   Myoclonic seizures (see Appendix I)
    C.   Clonic seizures (see Appendix I)
    D.   Tonic seizures (see Appendix I)
    E.   Tonic-clonic seizures
    F.   Atonic seizures (see Appendix I)

Absence seizures
Typical absence seizures consist of an abrupt cessation of all ongoing activity with impaired responsiveness that is associated with a characteristic ictal EEG pattern of generalized 3 Hz spike-and-wave complexes. Consciousness is impaired during the episode. The patient is motionless with a fixed blank stare and loss of contact with the environment. The eyes may roll upward and eyelids may flutter briefly. Automatisms, such as elevation of the eyelids, lip

Table 2.4 International League Against Epilepsy (ILAE) classification of generalized seizures (convulsive and nonconvulsive)

CLINICAL SEIZURE TYPE	EEG SEIZURE TYPE	EEG INTERICTAL EXPRESSION
A1. Absence seizures a) Impairment of consciousness only b) With mild clonic components c) With atonic components d) With tonic components e) With automatisms f) With autonomic components *(b through f may be used alone or in combination)*	Usually regular and symmetrical but may be 2-4 Hz spike-and-slow wave complexes and may have multiple spike-and-slow wave complexes. Abnormalities are bilateral	Background activity usually normal although paroxysmal activity (such as spikes or spike-and-slow wave complexes) may occur. This activity is usually regular and symmetrical
A2. Atypical absence May have: a) Changes in tone which are more pronounced than in A1 b) Onset and/or cessation which is not abrupt	EEG more heterogeneous, may include irregular spike-and-wave complexes, fast activity or other paroxysmal actions. Abnormalities are bilateral but often irregular and asymmetrical	Background usually abnormal paroxysmal activity (such as spikes or spike-and-slow wave complexes) frequently irregular and asymmetrical
B. Myoclonic seizures Myoclonic jerks (single or multiple)	Polyspike and wave or sometimes spike and wave or sharp and slow waves	Same as ictal
C. Clonic seizures	Fast activity (10 cycles/s or more and slow waves or occasional spike wave patterns	Spike and wave or polyspike and wave discharges

(continued)

Table 2.4 (Continued)

CLINICAL SEIZURE TYPE	EEG SEIZURE TYPE	EEG INTERICTAL EXPRESSION
D. Tonic seizures	Low voltage fast activity or a fast rhythm 9 10 cycles/s or more decreasing in frequency and increasing in amplitude	More or less rhythmic discharges of sharp and slow waves, sometimes asymmetrical, background often abnormal for age
E. Tonic-clonic seizures	Rhythm at 10 or more cycles/s decreasing in frequency and increasing in amplitude during tonic phase. Interrupted by slow waves during clonic phase	Polyspike and waves or spike and wave or, sometimes, sharp and slow-wave discharges
F. Atonic seizures (astatic) *Combinations of the above may Occur, e.g., B and F, B and D*	Polyspikes and wave or flattening or low-voltage fast activity	Polyspikes and slow wave

From Commission on Classification and Terminology, International League Against Epilepsy. Proposal for revised clinical and electrographic classification of epileptic seizures. *Epilepsia* 1981;22:489–501.

smacking, or swallowing movements, are common in absence attacks. Many absence attacks include clonic or myoclonic jerks or changes in postural tone. Typically, no aura signals the onset of the seizure and no postictal confusion or fatigue. Absence seizures usually last 5–10 seconds or so but can range between 1 and 45 seconds (Penry et al., 1975). After the brief attack is over, the child typically picks up and resumes the activity he was involved in just prior to the seizure. Most typical absence attacks can be precipitated in untreated patients with hyperventilation. Typical absence spells were called *petit mal* seizures in the past. Absence seizures most often begin in childhood, between the ages of 5 and

12 years, and often spontaneously stop during adolescence. Etiologically, typical absence seizures are primarily due to genetic factors.

Tonic-Clonic Seizures
Generalized tonic-clonic seizures (called *grand mal* seizures in the past) are the best known and most severe type of epileptic attack. Fisch and Pedley (1987) have indicated that all generalized tonic-clonic seizures have the following features:

1. loss of consciousness (onset may vary in relation to seizure progression),
2. a sequence of motor events that includes widespread tonic muscle contraction evolving to clonic jerking,
3. approximately symmetrical clinical and electroencephalographic, manifestations (as implied by the term generalized), and
4. postictal cerebral metabolic and behavioral suppression.

Generalized tonic-clonic seizures may begin with widespread behavioral and EEG changes typical of these seizures from the initial onset, may secondarily generalize from one cerebral hemisphere to involve both later in the course of a seizure, or alternatively, may evolve directly from another form of generalized seizures, such as an absence attack or myoclonic seizure.

Generalized tonic-clonic seizures (GTCSs) may be divided into several phases: a premonitory period, a tonic phase, a tonic-clonic phase, and a postictal period.

Premonitory symptoms. Although most patients with GTCS do not experience typical or consistent premonitory symptoms, some GTCS patients experience predictable premonitory changes, which may precede the seizure by hours or days, warning them of an impeding epileptic attack. Premonitory symptoms usually include such things as headache, changes in mood, or difficulties concentrating. Fisch and Pedley (1987) have provided a more complete listing of premonitory signs and symptoms commonly seen in GTCSs.

Precipitating Factors. Some generalized tonic-clonic seizures (GTCSs) may be anticipated by exposure to a variety of factors that have been shown to precipitate GTCSs, such as sleep deprivation or photic stimulation.

Idiopathic GTCSs that occur during sleep usually take place near the beginning or towards the end of sleep and only rarely during rapid eye movement (REM) sleep (Fisch and Olejniczak, 2006). Skipping or missing a dose of a patient's antiepileptic drug (AED) or abrupt withdrawal of AEDs for medical reasons can trigger a GTCS. Light-sensitive seizures can be facilitated by sleep deprivation (Fisch and Olejniczak, 2006).

Tonic Phase. GTCSs typically begin with a tonic phase characterized by an increase in tone across multiple muscle groups and upward deviation of the eyes. At first, there may be a very brief tonic flexor spasm that causes a forced expiration of air resulting in an 'epileptic cry." After the rapid spasm, a persistent tonic contraction is seen consisting of an arched back and extension of the arms and legs with labored breathing. Apnea begins early in the tonic phase and persists throughout the seizure. Blood pressure, heart rate, salivation, and bladder pressure all increase during the tonic phase and consciousness is lost at this time. The tonic phase usually lasts between 10 and 30 seconds.

Tonic-Clonic Phase. There is a gradual transition from the tonic phase to the clonic phase that consists of repeated, violent, bilaterally symmetric jerking of the arms and legs. These rhythmic flexor spasms (i.e., clonus) are characterized by whole body muscle contractions that alternate with abrupt muscle relaxation. The tonic-clonic muscle activity gradually decreases in frequency and amplitude over time until there is one final weak clonic jerk. Incontinence occurs as the clonic phase ends and the postictal phase begins. The tonic-clonic phase typically lasts from 30 to 90 seconds. The entire ictus or seizure proper (tonic plus tonic-clonic phase) will thus usually last only a minute or two.

Postictal Phase. After the final clonic jerk, respiration rate returns to normal after a few seconds, and there is sustained pupillary dilatation (Fisch and Olejniczak, 2006). In many patients, tonic muscle contractions may return during the immediate postictal period and last for several seconds to several minutes. Positron emission tomography (PET) scans show diffuse cortical hypometabolism during the postictal phase, in contrast to hypermetabolic activity during the ictal phase of a GTCS (Engel et al., 1982). During the postictal phase, the patient will gradually awaken by passing through stages of coma to stupor and obtundation to confusion to drowsiness. In other cases, the patient may fall asleep immediately following the seizure for varying periods of time and often awaken feeling tired with muscle soreness and headache. The postictal phase usually lasts between 5 and 15 minutes. Helmsteadter, Elger, and Lendt (1994) formally measured the time for complete reorientation to person, place, and time to take place. In secondarily generalized tonic-clonic seizures, the mean time for recovery of orientation was 18 minutes with a range from 4 to 45 minutes.

Ictal Electroencephalography in Generalized Tonic-Clonic Seizures. At the onset of a GTCS proper, a generalized attenuation of all electrographic activity occurs, leaving only very low-voltage fast (20–40 Hz range) electrical activity, referred to as *EEG desynchronization*. Electroencephalography recordings made during GTCSs in epileptic patients are always obscured by muscle and movement artifact resulting in an inability to see the activity of cortical neurons. Because of these confounding factors, knowledge of the EEG changes during the tonic-clonic phase of GTCSs has been obtained in studies of patients who are pharmacologically paralyzed at the time of the seizure, most often in patients undergoing electroconvulsive shock therapy (ECT). During the tonic-clonic period, bursts of high-amplitude surface negative waves occur at about 10 Hz (the so-called "recruiting rhythm") that alternate with slow waveforms. The EEG recruiting rhythm bursts are associated with the generalized muscle jerks or spasms. The slow EEG waveforms reflect the muscle relaxation portion of the spasms.

As the tonic-clonic seizure winds down, the recruiting rhythm bursts become less frequent and the slow waves become slower. After the final clonic muscle jerk, the EEG becomes flattened and characterized by very low voltage and very slow activity in the δ range (typical frequency between 0.5 and 2 Hz). The EEG at this point reflects electrographic cortical exhaustion and is said to be "isoelectric." This period of EEG suppression and cortical exhaustion usually lasts from several seconds to several minutes. After a minute or two, the EEG gradually increases in voltage and frequency as the patient recovers (Browne and Holmes, 2004). The return of the EEG to baseline levels will normally take place within 30 minutes following a seizure, although mild degrees of EEG slowing may be seen for up to 24 hours after a single, uncomplicated GTCS.

Immediate Complications of Generalized Tonic-Clonic Seizures. Generalized tonic-clonic seizures are violent events that increase patients' risk for injury or even death. The most common injuries that occur during

GTCSs are oral trauma, head trauma, and orthopedic injuries (Fisch and Olejniczak, 2006). Oral trauma usually involves laceration of the tongue, lips, or the inside of the mouth. Significant head injuries occur during seizures secondary to falls where the patient's head strikes hard surfaces, such as the floor, cement walkways, or furniture. Skull fractures, cerebral contusions, or subdural hematomas may result from head injuries suffered during GTCSs. Common orthopedic injuries include vertebral compression fractures (typically involving the thoracic vertebrae) and serious shoulder and knee joint injuries. Aspiration pneumonia, caused by aspiration of saliva or regurgitation of stomach contents, is a potentially life-threatening event in GTCSs. Patients are particularly vulnerable to aspirate during the postictal phase when the normal protective reflexes of the airway are inhibited (Fisch and Olejniczak, 2006). Sudden death may also occur during or shortly following GTCSs from a variety of causes including accidental death from injuries to cardiac arrest.

An abridged overview of the diagnostic decision tree for the traditional seizure disorders classification is presented in Figure 2.1 (with thanks to Dr. Michael Westerveld for first showing this to decision tree to me). The first decision is whether the EEG onset of the seizure is focal or generalized.

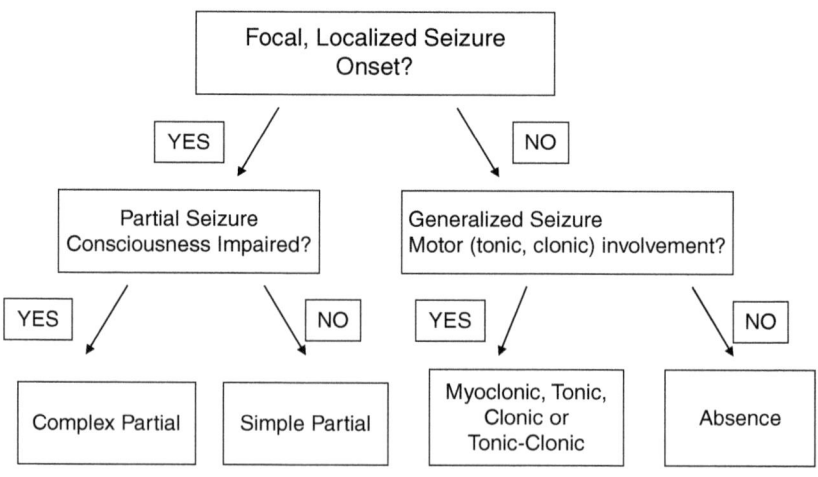

FIGURE 2.1. An abridged diagnostic decision tree for the traditional seizure classification schema.

Duration of Seizures Across Different Seizure Types

The length of a single seizure differs depending upon seizure type. In focal-onset seizures, the duration of the event increases as the severity and extent of the seizure increases. In general, tonic-clonic seizures with secondarily generalization are longer than complex partial seizures, which in turn typically last longer than simple partial seizures. Jenssen, Gracely, and Sperling (2006) measured the duration of 579 seizures recorded with continuous video and scalp EEG and found secondary generalized tonic-clonic seizures lasted a median 130 seconds (range 37–139 seconds), complex partial seizures lasted 78 seconds (range 8–298 seconds), primary generalized tonic-clonic seizures lasted 66 seconds (range 59–75 seconds), simple partial seizures lasted 28 seconds (range 3–180 seconds), and tonic seizures lasted a median 18.5 seconds (range 8–410 seconds). For complex partial seizures, there were no differences in seizure duration between those occurring while awake and those occurring during sleep, nor with those occurring between seizures that began in the left or right hemisphere.

Epidemiology of the Traditional Seizure Disorders

Epilepsy is one of the most common of all neurologic disorders worldwide. Epidemiological studies traditionally estimate the frequency of epilepsy in various populations using prevalence and incidence measures. Prevalence is the percentage of a population affected with epilepsy at a given point in time and is expressed as the number of cases per 1,000 people in the population. Prevalence estimates across many different populations range between 5 and 10 cases per 1,000 people (Sander, 2003; Annegers, 1996). Based on these prevalence estimates, there are approximately 2 million people in the United States with some form of epilepsy. There are higher prevalence rates among people from developing countries, rural regions, and lower socioeconomic groups in industrialized countries. In the United States, there is also a higher prevalence of epilepsy among African Americans as compared to whites (Haerer et al., 1986). Lifetime prevalence rates are higher than prevalence of active epilepsy. Most investigators estimate that around 4%–5% of the population will experience a nonfebrile seizure at some time in their lives (Sander, 2003; Bell and Sander, 2001).

With regard to the prevalence of seizure types, it has been estimated that approximately 60% of patients with epilepsy have some form of partial seizure disorder, and slightly less than 40% have generalized seizures (the remainder are unclassified). This figure includes both children and adults. Generalized seizure disorders are more common in children and partial seizures are more common in adults (see Table 2.1). Among patients with partial seizure disorders, the epileptogenic region has been localized to the temporal lobes in 50%–80% of cases, frontal lobes in 20%–30%, parietal lobes in 1%– 6%, and occipital lobes in 6%–8% (Manford et al., 1992; Sveinbjörnsdottir and Duncan, 1993).

Incidence is measured by taking the number of people who develop epilepsy during a 1-year period divided by the total person-time risk typically expressed as the number of cases per 100,000 people in the population. In industrialized countries, the incidence is generally considered to be around 50 per 100,000/year (Sander, 2003). In the developing world, these incidence estimates range between 100 and 190 per 100,000/year (Rwiza et al., 1992; Placencia et al., 1992).

Incidence rates have been estimated for each of the major seizure types using the traditional seizure classification system. Browne and Holmes (2004) have summarized the average annual incidence rates for the main traditional seizure types calculated from the Olmstead County, Minnesota data between 1945 and 1964 (Annegers, 1996). The incidence rates by traditional seizure type are presented in Table 2.5. Examination of Table 2.5 reveals simple partial epilepsy and generalized tonic-clonic seizures are the most common traditional seizure types.

Table 2.5 Incidence of epilepsy by traditional seizure type

SEIZURE TYPE	INCIDENCE RATE
Simple partial seizures	12.8
Complex partial seizures	10.4
Multiple or unclassified partial seizures	7.2
Generalized tonic-clonic seizures only	12.5
Incompletely generalized, with or without associated tonic-clonic activity	6.1
Absence seizures, with or without associated tonic-clonic	3.4
Other	4.5

Note: Mean annual rate per 100,000 population calculated for Olmstead County, MN, 1945–1965.
Adapted from Annegers JF. The epidemiology of epilepsy. In: E Wyllie, ed. *The treatment of epilepsy: Principles and practice* (2nd ed.). Philadelphia: Lippincott Williams & Wilkins, 1996:165–172.

The incidence rates of epilepsy are about 15% higher for males as compared to females for most of the traditional seizure types across the lifespan. An important exception to this occurs in absence seizures, which are twice as common in females as in males.

Age is an important factor to consider when looking at incidence rates of epilepsy because the distribution across the lifespan is currently bimodal. Incidence is highest among children less than 10 years old and in older (>60 years old) adults. Although this continues to be the case at present, incidence rates are changing as the population in the industrialized world grows older. Epilepsy used be considered primarily a childhood disorder, but is now becoming a disorder of the elderly in the developed countries due to increased life expectancy and overall aging of the population. In contrast, in the developing world, where health care services are less available and life expectancy is lower, incidence is still the highest among children.

Myoclonic seizures are the most common traditional seizure type during the first year of life and the incidence then drops off rapidly across the rest of the lifespan. The incidence of partial seizures (with or without secondary generalization) stays fairly constant from infancy to age 60 years at approximately 20 per 100,000, but after age 60 increases rapidly until there are about 82 per 100,000 cases by age 80 years. Absence seizures are relatively common during the first 10 years of life (at about 11 per 100,000), but new onset of absence seizures is uncommon after age 14 (Annegers, 1996). The incidence of generalized tonic-clonic seizures is 15 per 100,000 during the first year of life, and then gradually diminishes to 10 per 100,000 for children aged 10–14 years. The rate of onset of new generalized tonic-clonic seizures remains steady throughout adulthood until the rates begin to increase after age 60–65 years to approximately 35 per 100,000 by age 80.

Etiology of the Traditional Seizure Disorders

The cause of epilepsy is unknown (i.e., idiopathic) in roughly 40%–50% of cases. Most idiopathic cases of epilepsy are presumed to be due to genetic factors. The etiology and risk factors underlying the traditional seizure disorders differ with age, and in some cases differ depending upon geographical location. In childhood and adolescence, congenital, developmental, and genetic causes predominate. During late adolescence and early adulthood, head trauma is a frequent cause of epilepsy, while central nervous system infections and brain tumors are common etiologies during the middle adult years. Cerebrovascular disease and neurodegenerative conditions are the most common risk factors for epilepsy in people over the age of 60 (Sander, 2003). With regard to geographical location, endemic brain infections (e.g., malaria, neurocysticercosis, and toxocariasis) are more commonly encountered in the underdeveloped world and tropical environments.

After examination of all medical contacts for seizures in Olmstead County between 1935 and 1984, Hauser, Annegers, and Kurkland (1994) found that approximately 69% of the cases had an unknown cause of epilepsy. The presumed cause of epilepsy from this sample is shown in Table 2.6. The high percentage of idiopathic cases in the Rochester, Minnesota data has declined in more recent years as medical diagnostic technology has improved. Modern neuroimaging has led to an enhanced ability to detect brain lesions that were undetectable in earlier times, and this has caused a decline in the number of idiopathic cases of epilepsy.

Cerebrovascular Disease

As may be seen in Table 2.6, cerebrovascular disease is probably the most common cause of symptomatic epilepsy in the United States In the Rochester, Minnesota sample, cerebrovascular disease accounted for 13.2% of all epilepsies and 37% of acquired epilepsy with a known (or presumed) cause (Annegers, 1996). Although cerebrovascular disease is a condition of the elderly, it accounts for such a large percentage of the known causes of epilepsy due to increased life expectancy and an aging population. In the Rochester,

Table 2.6 Etiology of traditional seizure disorders: Incidence in Rochester, Minnesota, 1935–1984

PRESUMED ETIOLOGY	PERCENTAGE OF CASES
Idiopathic	68.7%
Cerebrovascular disease	13.2%
Developmental / congenital	5.5%
Head trauma	4.1%
Brain tumor	3.6%
Brain infection	2.6%
Degenerative conditions	1.8%
Other causes	0.5%

Adapted from Annegers JF. The epidemiology of epilepsy. In: E. Wyllie, ed. *The treatment of epilepsy: Principles and practice* (2nd ed.). Philadelphia: Lippincott, Williams, & Wilkins, 1996:165–172.

Minnesota sample, 55% of all newly diagnosed seizures in patients over age 65 were due to the acute or late effects of cerebrovascular disease (Annegers, 1996). The development of late (1–5 years) seizures after a stroke ranges between 3% after 1 year to 9% after 5 years.

Developmental and Congenital Conditions

Developmental and congenital conditions account for approximately 5.5% of all new cases of epilepsy in the Rochester, Minnesota, sample and for about 13% of all acquired epilepsy with a known (or presumed) cause (Annegers, 1996). These conditions, which include cerebral palsy and diffuse congenital disorders resulting in moderate or severe mental retardation, are among the most common causes of symptomatic epilepsy in early childhood; one-third of children with cerebral palsy, mental retardation, or both will also have epilepsy. Moreover, their risk of developing epilepsy is increased across the lifespan. The incidence of epilepsy in Down syndrome, for example, progressively increases as these patients age; 50% of patients with Down syndrome develop epilepsy by age 50–60 years (McVicker et al., 1994) presumably from neurodegenerative neurological processes.

Head (Traumatic Brain) Injury

As shown in Table 2.6, head injury only accounts for about 4% of all causes of epilepsy. This is because most traumatic brain injuries (TBIs) are mild and relatively few patients with severe TBIs survive. Traumatic brain injury accounts for about 13% of all acquired epilepsy with a known (or presumed) cause (Annegers, 1996).

Among civilian populations, the overall risk of developing epilepsy after a head injury with loss of consciousness, posttraumatic amnesia, or skull fracture is only about 2% at 5 years post head injury. Important risk factors that increase the likelihood of seizures include severity of injury, presence of intracranial mass lesions, and whether the dura mater surrounding brain was penetrated (Christensen et al., 2009). The presence of early seizures (i.e., in the first week or two after injury) is also a negative prognostic factor.

Annegers and colleagues (1980) examined the head injury risk factors leading to the development of late seizures and came to the following conclusions. Most civilian head injuries carry little, if any, increased risk for epilepsy. There was no detectable increased risk for individuals who suffered mild TBI, which was defined as a period of posttraumatic amnesia or loss of consciousness lasting less than 0.5 hour. This has the effect of lowering the overall incidence of post-TBI epilepsy since approximately 80% of all civilian head injuries are classified as mild. In moderate TBI (defined as loss of consciousness from 30 minutes to 24 hours or a nondepressed skull fracture), the risk is about 2%, lasting for only a few years following the injury. In civilian survivors of severe TBIs (defined as the presence of unconsciousness for more than 24 hours, intracranial mass lesions, or both), approximately 12% go on to develop seizures. The highest risk for posttraumatic seizures is 36%, which occurs in patients who had severe TBIs and experienced early seizures during the first week after injury. The period of increased risk in moderate and severe civilian TBIs continues for about 20 years after the head injury (see Figure 2.2 for seizure risk following TBI).

In contrast to civilian head injuries, there is a consistent increased risk for the development of seizures across time among military personnel with penetrating head injuries. These patients also have a high likelihood for developing epilepsy. For example, 50% of patients who survived penetrating head injuries in Vietnam went on to develop posttraumatic epilepsy (Salazar et al., 1985).

Subdural Hematomas

The risk of developing epilepsy with subdural hematomas depends upon the chronicity of the lesion and type of treatment. An acute subdural hematoma that requires surgical evacuation, either through a burr hole or open craniotomy, carries a very high risk of subsequent development of a seizure disorder; recent estimates suggest that 40%–61% of these patients will develop epilepsy within 6 months to 1 year following surgery. In contrast,

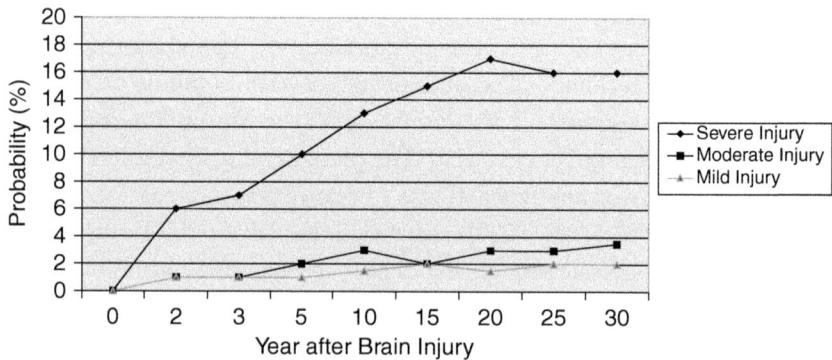

FIGURE 2.2. Cumulative probability of developing seizures in 4,541 patients with traumatic brain injuries by severity of injury. Adpated from: Annegers, Grabow, Groover, et al. (1980).

later-discovered chronic subdural hematomas cause secondary epilepsy in only 5%–10% of cases. These difference in epilepsy risk are most likely due to different mechanisms in the two conditions. Acute subdurals are more likely to be seen in younger patients with an arterial tear resulting in fast bleeding, while chronic subdurals are more often found in the elderly who have slow bleeding from the bridging veins following a fall. The acute condition causes greater neuronal damage, and thus, carries a higher epilepsy risk.

Brain Tumor
Brain tumors are associated with approximately 3.6% of all new cases of epilepsy in the Rochester, Minnesota, sample and in 6% of new cases in a large United Kingdom study (Sander et al., 1990). Brain tumors account for about 12% of all acquired epilepsy with a known (or presumed) cause (Annegers, 1996). Although epilepsy caused by brain tumor may occur at any age, it is most frequently seen in middle-age (25–64 years old). Brain tumors rarely cause seizures in children; the incidence was 1% in those less than 15 years old in the Rochester sample. This is due in part to posterior fossa tumors being more common in children since they are less likely to cause seizures than supratentorial tumors. New-onset seizures in a previously healthy adult with no family history of epilepsy should arouse suspicion of a brain tumor. In the UK study, 30% of adults with newly diagnosed seizures had a brain tumor.

With regard to specific tumor types, seizures are the most common initial presentation of low-grade gliomas. In a series of patients with gliomas from Montreal, seizures occurred in 92% of patients with oligodendrogliomas, 70% of patients with astrocytomas, and 37% of patients with glioblastoma multiforme tumors (Penfield et al.1940). A substantial number of patients with meningiomas also have seizures; approximately 30% before surgery and a slightly larger number following tumor resection (Chow et al., 1995).

Brain Infections
Brain infections may be the cause of seizures in 2%–7% of all cases of epilepsy. Viral encephalitis (e.g., mumps, herpes simplex, rabies, and rubella) carries the greatest risk of developing epilepsy after the infection, with a 10-fold increased risk that lasts for at least 15 years post-infection (Annegers, 1996). The risk of developing epilepsy within 20 years of suffering an episode of viral encephalitis is approximately 20% among patients who experienced early

seizures (early in the course of the infection) and 10% among patients who did not (Annegers et al., 1988).

Bacterial meningitis carries less risk for the development of seizures than encephalitis. The risk of developing epilepsy within 20 years of bacterial meningitis was about 13% in patients with early seizures and in about 2% of cases without early seizures in the Annegers and colleagues (1988) sample. Bacterial meningitis is most often encountered in children. The most common bacterial infection causing meningitis is *Haemophilus influenza*, followed by *Neisseria meningitides* and then *Streptococcus pneumonia*. Acute, or early seizures, are seen in approximately 31% of patients during the initial presentation of the meningitis, but seizure disorders are less often encountered later on; estimates range between 2% and 14% of cases (Sell, 1983). Many patients who have an intracranial abscess will present with seizures (perhaps around 33%), and the majority of surviving patients will suffer from postinfectious epilepsy usually within the first year after infection.

Neurodegenerative Central Nervous System Diseases
The final putative etiology listed in Table 2.6 continues to have the lowest incidence rates at present but will likely continue its upward trend as the population of the United States ages. Degenerative central nervous system (CNS) diseases accounted for just under 2% of all new cases of epilepsy, and 6% of acquired epilepsy with a known (or presumed) cause, in the Rochester, Minnesota, sample between 1934 and 1984 (Annegers, 1996). The degenerative disorders cause epilepsy later in life because most of these conditions are diseases of the elderly. For example, in the most common of the neurodegenerative disorders, Alzheimer disease, the risk of epilepsy is 10 times higher than that seen in healthy older individuals. It has been estimated that 10% of patients with Alzheimer disease will develop epilepsy as the disease progresses (Romanelli et al., 1990). The most common neurodegenerative disorder causing epilepsy among middle-aged (25–64 years old) individuals is multiple sclerosis (MS). It has been estimated that between 2% and 5% of patients with MS have epilepsy (Kinnunen and Wikstrom, 1986).

3

Epilepsy Syndromes

Although the traditional seizure classification discussed in Chapter 2 classifies seizures primarily into either focal ("partial") or generalized onset based on seizure semiology (observed clinical symptoms) and electroencephalography (EEG) pattern, the International League Against Epilepsy (ILAE) proposed to improve diagnosis and treatment by implementing a classification schema of epilepsy syndromes (Commission on Classification and Terminology of the ILAE, 1989). The need for an alternative diagnostic classification system arose because simply identifying the traditional seizure type communicated no information about accumulated knowledge regarding many different seizure variables. For example, diagnosing complex partial seizures communicates no information about the etiology, age of onset, severity of disease, course, prognosis, or genetics of the condition. As knowledge about genetics, etiology, and usual age of onset of particular seizure types has accumulated, the need for a more comprehensive diagnostic system emerged.

ILAE Classification of the Epileptic Syndromes

The ILAE (1989) defines an epilepsy syndrome as a cluster of signs and symptoms customarily occurring together; these include such items as type of seizure, etiology, anatomy, precipitating factors, age of onset, severity, chronicity, diurnal or circadian cycling, and sometimes prognosis.

The two overriding considerations used to classify the epileptic syndromes are:

- Whether the epilepsy is characterized by generalized seizures (*generalized epilepsies and syndromes*) or partial seizures (*localization-related, local, partial*), or
- Whether the epilepsy has a known etiology (*symptomatic* or "*secondary epilepsy*"), an unknown etiology (*idiopathic*), or a hidden, presumed symptomatic, etiology (*cryptogenic*).

Symptomatic epilepsies are identified when a lesion is discovered that is thought to be etiologically related to the seizures. Essentially, some identifiable lesion (e.g., trauma, infection, tumor, stroke) secondarily causes the recurrent epileptic attacks. *Idiopathic* epilepsies are those for which no cause can be identified after a thorough laboratory and radiological evaluation has been conducted, and thus, are presumed to be genetic in origin. *Cryptogenic* seizures are presumed to be caused by some type of brain pathology, but the pathology cannot be established using current diagnostic technology. In practice, cryptogenic epilepsy is almost always caused by some form of developmental cortical abnormality that formed during fetal development (e.g., cortical dysplasia, polymicrogyria, neuronal migrational disorder). Thus, symptomatic (or secondary) epilepsy is caused by an acquired postnatal lesion; idiopathic epilepsy is genetic or familial; and cryptogenic epilepsy is due to prenatal or intrauterine-acquired (developmental) conditions of the brain.

There are four major categories in the ILAE 1989 syndrome classification schema including:

1. Localization-related (focal, local, and partial) epilepsies and syndromes,
2. Generalized epilepsies and syndromes,
3. Epilepsies and syndromes undetermined as to whether they are focal or generalized, and
4. Special syndromes.

The specific syndromes across these four broad categories are summarized in Table 3.1.

Table 3.1 Classification of epileptic syndromes. Commission on Classification and Terminology of the International League Against Epilepsy (1989)

1. **Localization-related (focal, local, partial) epilepsies and syndromes**

    1.1. Idiopathic (with age-related onset)
    At present, the following syndromes are established, but more may be identified in the future:

    - Benign childhood epilepsy with centrotemporal spike
    - Childhood epilepsy with occipital paroxysms
    - Primary reading epilepsy

    1.2. Symptomatic

    - Chronic progressive epilepsia partialis continua of childhood (Kojewnikow or Rasmussen syndrome)
    - Syndromes characterized by seizures with specific modes of precipitation (reflex epilepsies)
    - Temporal lobe epilepsies
    - Frontal lobe epilepsies
    - Parietal lobe epilepsies
    - Occipital lobe epilepsies

    1.3. Cryptogenic
    Cryptogenic epilepsies are presumed to be symptomatic and the etiology is unknown. This category thus differs from the previous one by the lack of etiologic evidence (see definitions).

2. **Generalized epilepsies and syndromes**

    2.1. Idiopathic (with age-related onset—listed in order of age)

    - Benign neonatal familial convulsions
    - Benign neonatal convulsions
    - Benign myoclonic epilepsy of childhood
    - Childhood absence epilepsy (pyknolepsy)
    - Juvenile absence epilepsy
    - Juvenile myoclonic epilepsy (impulsive petit mal)
    - Epilepsy with tonic-clonic (GTCS) seizures on awakening
    - Other generalized idiopathic epilepsies not defined above
    - Epilepsies with seizures precipitated by specific modes of activation

    2.2. Cryptogenic or symptomatic (in order of age)

    - West syndrome (infantile spasms, Blitz-Nick-Salaam Krämpfe)
    - Lennox-Gastaut syndrome

(continued)

Table 3.1 (Continued)

- Epilepsy with myoclonic-astatic seizures
- Epilepsy with myoclonic absences

2.3. Symptomatic

2.3.1 Nonspecific etiology

- Early myoclonic encephalopathy
- Early infantile epileptic encephalopathy with suppression burst
- Other symptomatic generalized epilepsies not defined above

2.3.2 Specific syndromes

- Epileptic seizures may complicate many disease states. Under this heading are included diseases in which seizures are a presenting or predominant feature

3. **Epilepsies and syndromes undetermined whether focal or generalized**

3.1. With both generalized and focal features

- Neonatal seizures
- Severe myoclonic epilepsy in infancy
- Epilepsy with continuous spike-waves during slow-wave sleep
- Acquired epileptic aphasia (Landau-Kleffner syndrome)
- Other undetermined epilepsies not defined above

3.2. Without unequivocal generalized or focal features.
All cases with generalized tonic-clonic seizures in which clinical and EEG findings do not permit classification as clearly generalized or localization related; e.g., many cases of sleep-grand mal (GTCS) are considered not to have unequivocal generalized or focal features

4. **Special syndromes**

4.1. Situation-related seizures (Gelegenheitsanfälle)

- Febrile convulsions
- Isolated seizures or isolated status epilepticus
- Seizures occurring only when there is an acute metabolic or toxic event due to factors such as alcohol, drugs, eclampsia, nonketotic hyperglycemia

Adapted from Commission on Classification and Terminology of the International League Against Epilepsy. Proposal for revised classification of epilepsies and epileptic syndromes. *Epilepsia* 1989;30:389–399.

Localization-related Epilepsies and Syndromes

Localization-related epilepsies and syndromes are epileptic disorders in which seizure semiology (clinical symptoms) or EEG indicates a localized origin of the seizures. In most localization-related epilepsies, the epileptogenic lesions may be traced to a circumscribed portion of one cerebral hemisphere. However, in idiopathic age-related epilepsies with focal seizures, corresponding regions of both cerebral hemispheres may be involved (Commission on Classification and Terminology of the ILAE, 1989).

Idiopathic Localization-related Epilepsies

Idiopathic focal epilepsies are genetically determined, partial seizure disorders in which no clear structural brain pathology can be discovered. Their onset is age-related; most begin in childhood. These seizure syndromes tend to spontaneously remit over time. Clinically, no neurological or neuropsychological deficits or any history of an illness is etiologically related to the seizures. A family history of seizures is common. The characteristic seizures are usually the same from episode to episode within a single individual (Commission on Classification and Terminology of the ILAE, 1989). A description of the localization-related syndromes with a known genetic cause follows.

Symptomatic Localization-related Epilepsy Syndromes

These are focal-onset seizures that are symptomatic due to some underlying medically-documented epileptogenic neuropathology. The first two syndromes listed under this classification rubric, (1) Kojewnikow syndrome and (2) syndromes with specific modes of precipitation, are rare conditions that would most often be encountered in tertiary medical settings if at all. The remaining syndromes under this classification are categorized by location of the epileptogenic lesion, namely, whether the seizures originate in the temporal, frontal, parietal, or occipital lobe. Correct classification of the syndrome can only take place after a careful diagnostic work-up that includes detailed analysis of the seizure semiology, EEG characteristics, and evidence of brain pathology in conjunction with other considerations, such as age of onset.

Reflex Epilepsies (Syndromes Characterized by Seizures with Specific Modes of Precipitation)

These seizure disorders are currently referred to as the "reflex epilepsies." In the stimulus-sensitive reflex epilepsies, seizures are triggered by some specific environmental (e.g., flickering lights) or internal (e.g., mental calculation) stimuli. The mechanism of onset is thought to be due to hyperexcitable neurons in either primary sensory cortices or secondary association areas caused by structural or biochemical abnormalities. Reflex seizures are diagnosed by demonstrating that a particular afferent stimulus consistently provokes the same clinical seizure phenomena and EEG responses repeatedly. These specific seizure triggers need to be differentiated from more general, nonspecific precipitating factors (e.g., sleeplessness, alcohol withdrawal, hyperventilation, electrolyte imbalances) that may predispose or cause seizures, but should not be considered specific modes of seizure precipitation.

Reflex seizures have a prevalence of between 4% and 7% among patients with epilepsy (Zifkin et al., 1998). The seizure types subsumed under the reflex epilepsies are usually generalized seizures although complex partial seizures may also occur. Some of the reflex epilepsies may be idiopathic (for example, reading epilepsy and hot water epilepsy discussed earlier under the *idiopathic localization-related epilepsies*), some are symptomatic, and to a lesser extent, some are cryptogenic (e.g., eating epilepsy).

The stimulus that precipitates a reflex seizure is always the same for a given patient. Environmental stimuli may be either simple stimuli, such as flickering light (as in certain television or video game induced seizures) or a certain type of tactile stimulation, or the stimulus may be complex, as in reading or eating epilepsy. Simple stimuli evoke the EEG and clinical seizure phenomena quickly (within 1–3 seconds), whereas complex stimuli may take many minutes to precipitate a seizure (Panayiotopoulos, 2005). In a similar fashion, intrinsic stimuli may either be elementary (e.g., proprioceptive) or more elaborate (e.g., calculating, music, thinking). The most common types of stimuli that evoke reflex seizures include flickering lights, specific thoughts, music, eating, reading, hot water, and startle (Engel, 2001). Several commonly encountered reflex epilepsy subtypes, such as photosensitive seizures, musicogenic epilepsy, eating epilepsy, and startle epilepsy, are detailed in Appendix II.

Temporal Lobe Epilepsies

After the clinical seizure symptoms, EEG characteristics, age of onset, and evidence of brain pathology have been documented, many focal seizures may be classified into one of the epilepsy syndromes based upon where in the brain they originate. The common syndromes include temporal lobe epilepsy (TLE) (mesial temporal lobe epilepsy and lateral temporal lobe epilepsy), frontal lobe epilepsy, parietal lobe epilepsy, and occipital lobe epilepsy.

Mesial Temporal Lobe Epilepsy

The mesial temporal lobe structures include the amygdala, hippocampus, and entorhinal cortex. These structures are highly sensitive to any noxious stimuli that may cause mesial temporal lobe sclerosis, which is the most common pathology underlying mesial temporal lobe epilepsy. Mesial TLE is the most common form of temporal lobe seizure disorder. Moreover, the temporal lobes are the most frequent site of localization-related seizures.

Mesial temporal lobe seizures typically begin with an aura (i.e., a simple partial seizure), which is a seizure restricted to a circumscribed brain region producing a subjective sensation. The most common aura in mesial TLE is epigastric distress, often described as a rising uncomfortable sensation similar to nausea. Many auras are portrayed as difficult to describe bodily sensation or a "strange" or "weird" feeling. Auras of fear, gustatory and olfactory auras, and auras with experiential phenomenon (e.g., *déjà-vu, jamais vu*) are less common.

After the subjective warning of an aura, the first observable signs of the seizure proper commonly involves motionless staring, motor restlessness, oral-alimentary automatisms (e.g., lip smacking, facial grimacing, unformed vocalizations), and a non-forced head deviation. The single most common objective phenomenon seen in mesial TLE is oral-alimentary automatisms, which occur in 70% of patients (Gil-Nagel and Risinger, 1997).

In one sample of mesial TLE patients, these early signs were most often followed by either behavioral arrest or motor restlessness and head deviation or staring if these hadn't already occurred (Elger, 2000). In many cases, dystonic posturing of one or both of the arms begins around this time. Hypersalivation may also occur.

Loss of awareness occurs, but is difficult to accurately assess during the seizure. Loss of consciousness or awareness is evaluated by determining the patient's ability to carry out simple commands during the seizure, and by assessing the patient's memory for events that occurred during the seizure after its over. Loss of awareness or consciousness is more common in left mesial TLE than right mesial TLE.

The postictal state is characterized by gradual reorientation to one's surroundings. Some language dysfunction may occur, and less often, coughing. The typical progression of mesial TLE is reviewed in Table 3.2.

A specific seizure subtype of mesial temporal lobe seizures is the *syndrome of mesial TLE* with hippocampal sclerosis. Mesial temporal sclerosis (MTS) is characterized by hippocampal neuronal loss, relative hippocampal formation atrophy, mesial temporal lobe gliosis, and variable loss of neurons

Table 3.2 Typical seizures semiology of mesial temporal lobe epilepsy

STAGE	SYMPTOM
Initial sign	Aura (e.g., epigastric distress)
Early signs	Staring
	Behavioral arrest
	Automatisms (oroalimentary, manual)
	Motor restlessness
	Nonforced head deviation
Late signs	Arrest reaction (if not already seen)
	Restlessness (if not already seen)
	Staring (if not already seen)
	Dystonic posturing of arm(s)
Postictal signs	Reorientation / confusion
	Dysphasia (if dominant hemisphere)
	Coughing (uncommon)

Adapted from Elger CE. Semiology of temporal lobe seizures. In Oxbury JM, Polkey CE, Duchowny M. (eds.), *Intractable focal epilepsy.* London: W.B. Saunders, 2000:63–68.

in the amygdala, parahippocampal gyrus, and entorhinal cortex. Mesial temporal sclerosis is the most common pathology identified in tissue excised from adults with mesial TLE. There is often a strong family history of epilepsy in patients with the MTS syndrome. These patients will often have experienced a few seizures during infancy or early childhood, and then the seizures will remit for a number of years. Seizures then recur in late childhood or early adolescence. These later unprovoked partial seizures are refractory to antiepileptic drug treatment in up to one-third of patients, but fortunately, this seizure syndrome is highly amenable to successful seizure surgery (see Chapter 11).

Lateral (Neocortical) Temporal Lobe Epilepsy
Neocortical TLE is less common than mesial or medial TLE. Lateral TLE usually begins with an aura (i.e., simple partial seizure). The type of aura experienced at the beginning of a lateral TLE seizure depends upon the location of seizure onset in the cortex. Examples of common lateral TLE auras include auditory illusions or hallucinations, experiential auras (e.g., *déjà-vu*), dreamy states, visual misperceptions, gustatory auras, or dysphasia when the seizure focus is in the language

dominant hemisphere. Auras with experiential content, such as *déjà-vu* and *jamais-vu*, may occur exclusively in seizures with lateral neocortical temporal lobe onset (Gil-Nagel and Risinger, 1997). These simple partial seizure symptoms often evolve into complex partial seizures if the focus spreads to mesial temporal lobe or extra-temporal lobe structures. Patients with lateral neocortical TLE may show clonic movements of the facial muscles, grimacing, restlessness, or forced vocalizations. In one sample of such patients, rotation of the whole body (aversion) occurred, which was a symptom that was never seen in patients with mesial TLE. Furthermore, impairment of consciousness is apparently less pronounced in patients with lateral, than in mesial, TLE. A summary of the signs and symptoms differentiating mesial TLE from lateral TLE is provided in Table 3.3.

Frontal Lobe Epilepsy
The second most common site of localization-related seizure disorder onset is in the frontal lobes. This may be somewhat surprising since the frontal lobe is by far the largest in the brain. Williamson and Engel (1997) found the

Table 3.3 Signs and symptoms differentiating mesial from lateral temporal lobe epilepsy (TLE)

SIGNS AND SYMPTOMS	MESIAL TLE	LATERAL TLE
Epigastric auras, fear, and early oroalimentary automatisms	Predominate	Rare
Nonspecific auras, early focal motor, somatosensory, visual or auditory symptoms	Rare	Predominate
Contralateral hand dystonia	Common	Rarer
Early clonic activity following automatisms	Rare	Common
Generalized tonic-clonic seizures	Infrequent	Frequent
History of febrile seizures	Predominates	Rare
Interictal electroencephalogram	Ipsilateral anterior temporal spikes	Middle & posterior temporal spikes
Magnetic resonance imaging	Hippocampal sclerosis	Neocortical lesions e.g., malformations of cortical development

Adapted from Panayiotopoulos CF. *The epilepsies: seizures, syndromes and management.* Chipping Norton, Oxfordshire, UK: Bladon Medical Publishing, 2005.

following clinical characteristics tend to distinguish frontal lobe seizures from seizures arising in other cortical regions:

- Rapid secondary generalization
- Focal clonic motor activity
- Prominent asymmetric tonic posturing
- Explosive onset
- Sudden ending
- Minimal postictal confusion
- Frequent, brief seizures that often occur in clusters

Frontal lobe seizures frequently occur during sleep; most commonly in stage 2 sleep. All of the focal onset seizure types can be seen in frontal lobe epilepsy including simple partial, complex partial, secondarily generalized seizures, or any combination of these. Frontal lobe seizures have be classified on the basis of the anatomic location of the seizure focus within the frontal lobes. Frontal lobe seizures arising from the motor strip (precentral area) and premotor (including the supplementary motor area) regions have been well-defined, whereas seizures originating in the prefrontal (dorsolateral, mesial, and orbitofrontal) areas have been more difficult to characterize due to the strong interconnections between these and other brain regions and the associated rapid seizure spread to other areas. The major focal onset frontal lobe seizure syndromes, such as precentral frontal lobe seizures, premotor frontal lobe seizures, supplementary motor area (SMA) seizures, dorsolateral prefrontal lobe seizures, orbitofrontal lobe seizures, medial frontal lobe seizures, and frontal opercular seizures, are detailed in Appendix II.

Parietal Lobe Epilepsy

Seizures originating in the parietal lobes are quite rare; accounting for only 5% of all partial seizures in one epilepsy surgery series (Rasmussen, 1975). Seizure semiology depends upon where the epileptic focus is located within the parietal lobe and the pattern of seizure spread across the parietal and extra-parietal regions, if spread occurs. Many patients with parietal lobe seizures have no symptoms whatsoever that would suggest a parietal lobe seizure origin. Most parietal lobe seizures are of the simple partial type. Many of these patients have somatosensory symptomatology, which indicates neuronal excitation within the postcentral gyrus. In a comprehensive series of surgical parietal lobe epileptics from 1929 to 1988, Salanova and colleagues (1995) reported that 94% of patients experienced some form of aura, and almost all auras were contralateral to the side of seizure onset. Somatosensory auras (e.g., tingling, numbness) were most frequent; present in 64% of cases. Other auras included painful sensations, disturbance of body image, visual illusions, vertiginous sensations, and aphasia. Parietal lobe epilepsies may be classified by seizure foci emerging from the postcentral gyrus region, superior parietal lobule, inferior parietal lobule, and the paracentral region on the mesial surface of the parietal lobe. The major focal onset parietal lobe seizure syndromes, such as postcentral gyrus seizures, superior parietal lobule seizures, inferior parietal lobe seizures, and paracentral parietal lobe seizures, are detailed in Appendix II.

Occipital Lobe Epilepsy

Occipital lobe epilepsy is a relatively rare localization-related, symptomatic epilepsy representing only about 5% of large epilepsy surgery series, although seizures arising from multiple lobes, including occipital cortex, are slightly more common (Williamson et al., 1992). Almost all patients experience some form of visual phenomena as the initial seizure manifestation.

Positive visual symptoms include elementary visual hallucinations often described as bright lights or colored lights, or less frequently as dark rings or spots or simple geometric patterns, usually in the periphery of the contralateral visual field. The visual phenomena may remain stationary or slowly move across the visual field.

Negative visual phenomena may include amaurosis, scotomas, hemianopsias, or quadrantanopsias in the contralateral visual field. Negative phenomena more often follow initial positive symptoms, but occasionally negative symptoms may be the first seizure manifestation. Visual field cuts (e.g., hemianopsia) are of high localizing value.

Other manifestations of seizures confined to the occipital lobes include tonic and clonic eye deviation, head deviation, blinking, a sensation of eye movement, and nystagmoid eye movements. Eyelid flutter or forced blinking and a sensation of eye pulling have also been associated with occipital onset seizures. In occipital epilepsy, the direction of eye and head deviation is usually away from the side of the epileptogenic focus.

Occipital onset seizures typically spread to the temporal, frontal, supplementary motor, or parietal areas and result in seizure semiology typical of these areas. The most common path of EEG discharge spread is infrasylvian to the ipsilateral temporal lobe typically resulting in epigastric distress, unresponsiveness, and automatisms; but seizures may spread to the contralateral hemisphere as well. Occipital seizures spread to the frontal lobe in about one-third of occipital onset patients and to multiple lobes simultaneously in about one-third of patients (Williamson et al., 1992). Different patterns of spread may occur with each seizure within the same individual.

Cryptogenic Localization-related Epilepsies
The focal onset cryptogenic epilepsies are partial seizure disorders that have no genetic or identifiable underlying neuropathological cause. No etiology can be established after thorough medical evaluation, but because of their semiology and EEG characteristics, cryptogenic seizures are presumed to be symptomatic. In practice, this most often turns out to be due to a difficult to diagnose neurodevelopmental brain disorder, such as cortical dysplasia, focal heterotopias, neuronal migrational disorders, or microdysgenesis. As brain magnetic resonance imaging (MRI) has become more prevalent and powerful, the number of cryptogenic epilepsies is decreasing as modern techniques are detecting smaller focal pathologies that were previously not visible. Thus, the prevalence of cryptogenic epilepsies has been shrinking while the number of symptomatic epilepsies has been growing.

Generalized Epilepsies and Syndromes
Generalized epilepsies and syndromes are epileptic disorders in which the initial clinical symptoms indicate involvement of both cerebral hemispheres in conjunction with bilateral EEG changes. The most common generalized seizures types have already been described earlier in Chapter 2. The generalized epilepsy syndromes are classified on the basis of whether the syndrome is idiopathic (unknown cause, thought to be genetic), symptomatic (known cause), or cryptogenic (unknown cause, but presumed to be symptomatic due to some neurodevelopmental etiology).

Idiopathic, Generalized Epilepsy Syndromes
The International League Against Epilepsy has described the idiopathic epilepsies as follows (Commission on Classification and Terminology of the ILAE, 1989):

> Idiopathic generalized epilepsies are forms of generalized epilepsies in which all seizures are initially generalized (absences, myoclonic jerks and generalized tonic clonic seizures), with an

EEG expression that is a generalized, bilateral, synchronous, symmetrical discharge (such as is described in the seizure classification of the corresponding type). The patient usually has a normal interictal state, without neurological or neuroradiological signs. In general, interictal EEGs show normal background activity and generalized discharges, such as spikes, polyspike spike-waves, and polyspike-waves ≥3 Hz. The discharges are increased by slow sleep. The various syndromes of idiopathic generalized epilepsies differ mainly in age of onset. No etiology can be found other than a genetic predisposition towards these disorders.

As mentioned earlier, patients with idiopathic generalized epilepsies are usually normal neurologically and show no major clinical signs or symptoms between seizure episodes. Interictal EEG is typically abnormal in untreated patients and may show generalized discharges of spikes or polyspike-and-wave complexes. Epileptogenic discharges may be precipitated by sleep deprivation, hyperventilation, or photic stimulation.

Research into the genetic origins of the idiopathic generalized epilepsies has strongly implicated abnormalities in the γ-aminobutyric acid $(GABA)_A$ receptor as predisposing patients to generalized seizures. Careful selection of the appropriate antiepileptic drug (AED) regimen is important since drugs effective in the partial epilepsies may be ineffective in idiopathic generalized epilepsy (Panayiotopoulos, 2005). Moreover, AED efficacy even differs within generalized epilepsy subtypes, and thus, it is often worthwhile to make sure patients are taking the best AED for their particular condition.

Most neuropsychologists will never be called upon to evaluate or treat patients with most of the idiopathic generalized epilepsies listed in Table 3.1, especially those seizure disorders occurring in the neonatal period. However, some neuropsychologists may be asked to evaluate children who have had these disorders as infants, and so, these syndromes will be briefly reviewed later with an emphasis on issues of importance to neuropsychologists, such as whether the disorder is associated with developmental delays, learning disabilities, mental retardation, or other psychological issues. For a more thorough description of these rare epilepsy syndromes, the reader is directed to one of the more comprehensive epilepsy textbooks that discuss these childhood syndromes, such as Willie, Gupta, and Lachhwani (2006), Panayiotopoulos (2005), or Wallace and Farrell, (2004). The common or important idiopathic, generalized epilepsy syndromes, such as benign neonatal familial convulsions, benign neonatal convulsions (nonfamilial), benign myoclonic epilepsy of childhood, juvenile myoclonic epilepsy (JME), and epilepsy with tonic-clonic seizures upon awakening, are detailed in Appendix II.

Childhood Absence Epilepsy (Pyknolepsy)
This is the syndrome equivalent of the typical absence seizure type described in Chapter 2. *Pyknolepsy* is a term more commonly used in Europe to mean either a high frequency of absence attacks or as a synonym for the childhood absence epilepsy syndrome (Panayiotopoulos, 2005). It should be recalled, however, that this syndrome is more narrowly defined than absence seizures generally. These seizures are thought to be genetically determined, but the precise mode of inheritance has not yet been identified. These children are normal neurologically and experience normal development. Prognosis for patients with childhood absence epilepsy is excellent, with seizures remitting in most cases by adolescence. Among those with recurrent absence attacks, many will go on to develop generalized tonic-clonic seizure in adolescence.

Juvenile Absence Epilepsy
In the juvenile-onset absence epilepsy syndrome, seizures typically begin around puberty, although the range of age of onset extends between 10 and 17 years old. Juvenile absence seizures are very much like childhood absence seizures, with some distinguishing features. The frequency of absence seizures is less in juvenile absence epilepsy than in the childhood variety. Children may have 20, 50, or even hundreds of absence spells each day, whereas in the juvenile form there are usually only one, or only a few, episodes each day.

In juvenile-onset absence seizures, consciousness is less affected than in the childhood variety even though the paroxysmal discharges may last slightly longer in the juvenile seizures (Panayiotopoulos, 2005). Furthermore, tonic-clonic seizures are much more frequent in juvenile than childhood absence, occurring in 50%–80% of cases (Browne and Holmes, 2004). The tonic-clonic seizures in juvenile absence may take place at any time of day. A tonic-clonic seizure is usually the presenting complaint that brings these adolescents to medical attention. Absence seizures become more difficult to control with AEDs in older age groups.

Juvenile-onset absence seizures are difficult to distinguish from complex partial seizures since lapses of consciousness and automatisms may occur in both seizure types. Hyperventilation is useful in triggering the EEG and clinical signs of absence seizures in patients of all ages, whereas photic stimulation seldom brings on absence seizures. The EEG signs of juvenile absence epilepsy are the well-known 3-Hz spike-and-wave complexes. Absence status epilepticus is much more common in teens and adults than in children. As with most of the idiopathic epilepsy syndromes, juvenile absence epilepsy is thought to be predominantly caused by genetic factors.

Cryptogenic or Symptomatic Generalized Epilepsy Syndromes
These syndromes are primarily seen in infants and younger children. Lennox-Gastaut syndrome is one of the more important generalized cryptogenic encephalopathies that cause epilepsy. The other major disorders, such as West syndrome, epilepsy with myoclonic-astatic seizures, and epilepsy with myoclonic absences, are briefly discussed in Appendix II.

Lennox-Gastaut Syndrome
Lennox-Gastaut syndrome is another severe epileptic encephalopathy of childhood with seizures beginning between ages 1 and 8 years (peak age 3–5 years). These patients may have one, or more often a combination of several different seizure types. Most common are tonic seizures, absence attacks, and atonic seizures; although myoclonic, generalized tonic-clonic, and partial seizures may also be present in some cases. Seizures and status epilepticus are frequent in these children, and seizures are poorly controlled with AEDs (Commission on Classification and Terminology of the ILAE, 1989).

Etiology is similar to that found in West syndrome, with perinatal ischemic birth injuries and malformations of cortical development predominating. About one-third of cases are idiopathic or cryptogenic, and there is no evidence for a genetic cause (Panayiotopoulos, 2005). Development is severely affected, with progressive neuropsychological impairment being the rule. Mental retardation is common, and the overwhelming majority of patients surviving into adulthood continue to have seizures.

Symptomatic Generalized Epilepsy Syndromes
The symptomatic generalized epilepsies occur mostly during infancy and early childhood. They are characterized by generalized seizures with clinical and EEG characteristics that differ from idiopathic generalized seizures. Ictal EEG is less symmetrical and less rhythmic

in symptomatic than idiopathic syndromes, and interictal EEGs differ from idiopathic cases as well (Commission on Classification and Terminology of the ILAE, 1989). The most common seizure types include myoclonic, tonic, atonic, or atypical absence seizures. There are usually neurological, clinical, and neuropsychological signs of a diffuse, nonspecific encephalopathy.

Epilepsies and Syndromes Undetermined Whether Focal or Generalized With Both Generalized and Focal Features

Acquired Epileptic Aphasia (Landau-Kleffner Syndrome)
Landau-Kleffner syndrome is a childhood disorder characterized by progressive language deterioration associated with focal or multifocal EEG spikes. Modal age of onset is 5–7 years old (range 2–8 years), and males are twice as likely to have the disorder. These children are initially neurologically normal, having achieved age-appropriate developmental milestones including speech and language. The core linguistic deficit is an verbal auditory agnosia that gradually worsens to include other signs of language dysfunction, such as reduced fluency, paraphasias, perseverations, and eventually muteness, as well as nonverbal auditory agnosia (i.e., inability to recognize meaningful environmental sounds such as a doorbell ringing in many cases) (Panayiotopoulos, 2005). Some children will display telegraphic speech or will talk in very simple sentences. These children are at first often thought to be suffering from some type of hearing loss, autistic process, or severe psychiatric disturbance. In addition to the linguistic deficits, most children with Landau-Kleffner develop other cognitive and behavioral impairments such as hyperactivity, disinhibition, or attention-deficit disorder. Dementia with a global intellectual decline is not characteristically seen in these children.

All children with Landau-Kleffner syndrome show linguistic dysfunction, but only about three-quarters of these children also have seizures. In those children who have seizures, the most common seizure type is partial motor, with or without loss of awareness, but focal myoclonic jerks, complex absence spells, and generalized tonic-clonic seizure types may also occur (Neville and Cross, 2006). Seizure frequency is rare, and seizures are usually easily controlled with antiepileptic medications. Controlling the seizures with medications does not affect the rate or severity of cognitive deterioration.

Nearly all children with Landau-Kleffner will develop continuous spike waves during sleep with concomitant neuropsychological deterioration at some point during the illness. Etiology is unknown. Most often there are no detectable structural abnormalities on brain MRI. Functional brain imaging studies have shown hypoperfusion within the temporal lobes. Seizures and EEG abnormalities are age-dependent (similar to continuous spike waves during sleep which is discussed later in this chapter) and gradually wind down and finally remit around age 15 years. After maximal recovery has taken place, about half are left with irreversible cognitive impairments, whereas only about 10%–20% are able to live completely normal, independent lives (Panayiotopoulos, 2005).

With regard to treatment of Landau-Kleffner syndrome, among the older AEDs, valproate and ethosuximide are the usual drugs of choice. Recall that there appears to be no linkage between reducing seizure frequency or epileptogenic activity on EEG and altering the course of this syndrome. Among the newer AEDs, levetiracetam monotherapy has reportedly improved both the language and seizure disorder in one case study (Kossoff et al., 2003). If treatment with AEDs is suboptimal, current practice is to treat these children with corticosteroids (adrenocorticotropic hormone [ACTH] or prednisone), which apparently has resulted in some improvement. Finally, some centers will treat medically intractable cases of Landau-Kleffner with multiple subpial transactions (see Chapter 11 for a full description of this surgical procedure) if the epileptogenic tissue can be localized to a fairly circumscribed region of the brain.

Epilepsy with Continuous Spike-Waves During Slow Wave Sleep (ECSWS)

Epilepsy with continuous spike-waves during slow wave sleep (ECSWS) is a rare epilepsy syndrome (but one of interest to neuropsychologists since most cases will eventually come to neuropsychological attention at some point) with onset during childhood; modal age 4–5 years old. It is characterized by continuous spikes and waves during non-REM sleep that are associated with either partial or generalized seizure types while asleep, and atypical absences when awake (Commission on Classification and Terminology of the ILAE, 1989). Since the nocturnal seizures are continuous, they are a type of status epilepticus that will eventually produce significant brain damage and dramatic cognitive deterioration. The seizures usually remit after a period of years (between 2 and 10 years), most commonly in early- to mid-adolescence. The cause of this disorder is unknown, but abnormalities are seen during brain MRI in about one-third of cases including focal or diffuse cortical atrophy, focal porencephaly, or malformations of cortical development. There is no evidence for any genetic cause, and a family history of seizures is uncommon in these patients.

Neuropsychological deterioration usually begins to become evident one to two years after onset of the seizures. It is thought that the repetitive seizures cause gradual, progressive decline in cognitive functions. The cognitive domains affected depend upon the location of the continuous spikes; frontal localization is most common (Panayiotopoulos, 2005).

These frontal lobe seizures may cause executive dysfunction or frontal lobe behavioral syndromes, which worsen over time and progresses to involve language functions (as the brain dysfunction expands into Broca's area). As the condition continues to progress, patients may show disinhibition, hyperkinesias, agitation, aggressiveness, and inattention; and the disorder ultimately results in a frontal lobe dementia syndrome and/or psychosis. Temporal lobe continuous spike waves during sleep (CSWS) may produce memory disorders or aphasias. Motor deterioration is also common in this phase of CSWS, with patient developing ataxia, hemiparesis and apraxia, and in some cases with motor strip involvement, children may develop dysarthria, muteness, weakness of the face or tongue, or drooling.

Following the neuropsychological deterioration phase of CSWS, the seizures gradually begin to diminish in frequency and severity (usually between 2 and 7 years after seizure onset) and then eventually completely remit in all cases regardless of etiology (Neville and Cross, 2006). Electroencephalography slowly improves to normal and some neuropsychological recovery may also occur. However, many children with CSWS will have permanent and severe neuropsychological impairments that are stable after resolution of the seizure disorder, usually by early adolescence.

Special Syndromes
Situation-related Seizures (Gelegenheitsanfälle)
Febrile Convulsions

Febrile seizures are the most common type of seizure disorder in early childhood. These seizures have a relatively benign course and long-term prognosis. Febrile seizures are an age-related disorder in young children (who have a genetic predisposition toward seizures) that occurs during an acute febrile illness. *Febrile* is defined as a body temperature greater than 38°C (100.4°F) measured rectally, although some prefer a higher cut-off at 38.5°C (101.3°F). Although by definition febrile seizures must occur after the age of 3 months but before the age of 5 years, 90% of febrile seizures occur within the first 3 years of life and 50% during the second year (modal incidence between 18 and 24 months) (Duchowny, 2006).

Seizures typically occur within the context of common childhood illnesses, most commonly viral upper respiratory tract infections, and less frequently, otitis media or gastrointestinal infections. Generalized tonic-clonic seizures are the most common type of seizure experienced during febrile spells; occurring in approximately 80% of cases (Panayiotopoulos, 2005). Febrile seizures usually occur early in the course of the infectious illness as the temperature curve is rising. Seizure duration is usually brief; lasting only for a few minutes. Children are at greater risk for the development of a febrile seizure if they have a first- or second-degree relative with a history of febrile seizures. Although the etiology of febrile seizures is thought to be primarily genetic, no specific mode of inheritance has been identified. One-half will go on to experience another febrile seizure in the future. Febrile seizures are classified into one of two categories, namely, simple and complex, based on differences in the initial medical work-up, treatment, and prognosis.

Simple Febrile Seizures
Simple febrile seizures are isolated generalized tonic-clonic seizures that last less than 15 minutes (most are of shorter duration) and do not generally occur in neurologically normal children and do not cause any permanent deficit. Simple febrile seizures are by far the most common type of febrile seizure, occurring in 80%–90% of cases (Verity et al., 1985). Children are excluded from this diagnosis if their seizures are due to a central nervous system (CNS) infection (e.g., meningitis or encephalitis), if the child had a previous nonfebrile seizure, and if a CNS abnormality has been identified.

Complex Febrile Seizures
Complex febrile seizures are diagnosed when the seizures are prolonged, repetitive, or associated with other risk factors for poor prognosis. Specifically, these seizures are considered "complex" if any or some combination of the following are present:

- Seizure duration longer than 15 minutes
- Seizure recurrence within 24 hours
- Seizures occurring in children with focal seizures or preexisting neurological deficits

If the child has a focal, rather than a generalized, type of seizure, the risk for later developing nonfebrile seizures is greatly increased. The overall risk of developing nonfebrile seizures later in life across all severity levels of febrile seizures is 3%. Neuropsychologists who work at epilepsy surgery centers commonly encounter patients with intractable TLE who have a history of febrile seizures. One-third of adult patients with MTS and complex partial seizures of temporal lobe origin have a history of prolonged (i.e., complex) febrile seizures (Cendes, 2004).

The American Academy of Pediatrics (AAP) has published practice parameters and suggested standards for the evaluation and treatment of children with simple febrile seizures (American Academy of Pediatrics, 1996; American Academy of Pediatrics, 1999). There is no need for an expensive, time-consuming medical evaluation in these patients. Specifically, the diagnostic yield on brain MRI, EEG, and lumbar puncture is very low, and prognosis for complete recovery is excellent. Similarly, the AAP advises against treating these children with AEDs, concluding that the benefits of treatment do not outweigh the risks.

No neuropsychological deficits have been identified in children with a history of febrile seizures as a group. When children who had simple and complex febrile seizures in the past were compared with healthy controls at 10 years of age, there were no differences between the groups with regard to academic achievement, intelligence, or prevalence of behavior disorders (Verity et al., 1998). Other studies in the United States and the United Kingdom have reported similar findings; namely, no increased prevalence of cognitive deficits, developmental disorders, or learning disabilities exist later in life among children with a history of febrile convulsions.

Isolated Status Epilepticus

Status epilepticus is either a prolonged (>30 minutes) continuous seizure or a series of brief seizures occurring so rapidly that the patient does not have time to recover from one before another has begun. Status epilepticus is classified under special syndromes because it is not itself a disease, but rather a symptom of some underlying CNS or systemic disorder that produces secondary CNS effects. Thus, the majority of individuals with status epilepticus do not have epilepsy; only about 25% of status epilepticus cases occur in patients with epilepsy. The most common causes of status epilepticus in adults is cerebrovascular disease and medication withdrawal or noncompliance, while the most common etiology in children is fever with infection (see Table 3.4).

Convulsive status epilepticus should be considered a medical emergency, and the underlying cause must be identified and promptly treated to prevent

Table 3.4 Comparison of the etiology of status epilepticus in children and adults (in Richmond, Virginia)

ETIOLOGY	% OF CHILDREN (<16 YEARS OLD)	% OF ADULTS (>16 YEARS OLD)
Fever / infection	35.7%	4.6%
Medication change	19.8%	18.9%
Unknown	9.3%	8.1%
Metabolic	8.2%	8.8%
Congenital	7.0%	0.8%
Anoxia	5.3%	10.7%
Central nervous system infection	4.8%	1.8%
Trauma	3.5%	4.6%
Cerebrovascular	3.3%	25.2%
Ethyl alcohol / drug-related	2.4%	12.2%
Tumor	0.7%	4.3%

Adapted from DeLorenzo RJ, Towne AR, Pellock JM, Ko D. Status epilepticus in children, adults, and the elderly. *Epilepsia* 1992;33 (Suppl 4):15–25.

irreversible CNS injury. The goal of treatment is to stop the seizures as soon as possible and treat the underlying cause. Status epilepticus is one of the most common neurological emergencies; tonic-clonic status has a mortality rate of approximately 20%, and there is a serious risk of cerebral damage among survivors. It is estimated that approximately 55,000 people die each year in the United States from status epilepticus (DeLorenzo et al., 1995).

Although status epilepticus is most frequently encountered in children and the elderly, it can occur at any age. Any type of epileptic seizure can progress to status epilepticus, including generalized tonic-clonic, absence, complex partial, or simple partial, but the most common is generalized tonic-clonic seizures, which also is the most life threatening. Idiopathic generalized status epilepticus is an emergency since irreversible brain damage or death may result if appropriate treatment is not instituted within 30–45 minutes after onset; this is especially true for adults. Although status epilepticus is currently defined by seizure activity lasting longer than 30 minutes, the Epilepsy Foundation of America advises the public to call for emergency assistance when a convulsive seizure has lasted more than 5 minutes, and the Working Group on Status Epilepticus of the Epilepsy Foundation of America (1993) recommends that emergency room physicians initiate treatment if the convulsions continue for more than 10 minutes.

Epidemiology of Epilepsy Syndromes

There have been several reasonably representative epidemiological studies of newly diagnosed epilepsy syndromes in children and adults. More than 90% of adults present with localization-related (62%) or unclassified (~30%) epilepsies. Most of the unclassified adult cases will ultimately be diagnosed as localization-related epilepsies. Children, on the other hand, are more evenly divided; approximately 50%–60% of children have generalized epilepsies and 40%–50% are localization-related cases.

Idiopathic epilepsy is much more common in children that adults. In children, approximately 48% of all newly diagnosed seizures are idiopathic; approximately 9% are localization-related and around 40% are generalized seizures. On the other hand, symptomatic epilepsy is more common in adults than children; between 9% and 20% of epileptic children have an initial identifiable structural lesion, whereas around 30% of adult cases have a symptomatic discovered cause during the initial work-up.

Etiology of Epileptic Syndromes

In the French Coordination Active du Réseau Observatoire Longitudinal de L'Epilepsie (CAROLE) sample, approximately one-third of the cases were idiopathic (i.e., genetic), about one-half were cryptogenic (i.e., developmental), and in the remaining 18%, some identifiable symptomatic etiology was found (Jallon et al., 2001). In 88% of these cases, some remote static disorder was identified; most commonly cerebrovascular accidents, head injuries, and pre- or perinatal insults. The most common etiology of seizure disorder in the remaining 12% of cases was a progressive lesion, such as a tumor.

In the National General Practice Study of Epilepsy conducted in the United Kingdom, the underlying neuropathology of the seizure disorder was discovered in 32%, and a probable cause was identified in an additional 9.4%, of patients with localization-related epilepsy (Manford et al., 1992). In this study of both children and adults, vascular disease (cerebral infarction in 31%, arteriovenous malformations and venous angiomas in 43% of cases) and tumors (glioma, meningioma, and metastasis in 31% of cases) were the most common causes of newly diagnosed localization-related, symptomatic epilepsy. The high number of vascular etiologies in this study was most likely due to the older age of the sample.

Unfortunately, these studies have not accounted for the ability of modern, high-resolution MRI techniques to detect subtle developmental lesions. The developmental etiologies of epilepsy, including cortical dysplasia, heterotopias, dysembryogenic neuroepithelial tumors (DNETs), hamartomas, agyria/pachygyria, and tuberous sclerosis, are now being recognized as important determinants of seizures. As radiological and other medical techniques continue to improve our diagnostic acumen, the ability to detect these once undetectable lesions will continue to improve. This in turn will reduce the number of cryptogenic epilepsy cases and increase the number of localization-related, symptomatic epilepsies.

Nonepileptic Seizures

Nonepileptic seizure is the current preferred term for a paroxysmal behavioral event that looks like an epileptic seizure, but that isn't actually caused by epilepsy. That is, the clinical event mimics an epileptic seizure, but it is not due to electrical neuronal disruption, measurable on EEG, in the cerebral cortex. The older name for these nonepileptic events was *pseudoseizures*, but this term is no longer in use due to what has been viewed as its derogatory implications. These events are difficult to diagnose, and many are misdiagnosed as epilepsy, which can result years of inappropriate treatment with AEDs and of not being properly treated for the underlying disorder. Nonepileptic seizure patients are fairly commonly encountered on epilepsy monitoring units, where 20%–40% of patients undergoing video-EEG recording have nonepileptic seizures.

There are two broad types of nonepileptic seizures: organic or physiologic nonepileptic seizures and psychogenic nonepileptic seizures. *Physiologic nonepileptic seizures* are caused by a sudden disruption of brain function. Medical causes of these events are most often either ischemia from cardiac arrhythmias or metabolic disturbances secondary to severe hypoglycemia or hyponatremia. Other conditions that may mimic epileptic seizures but are actually physiologically based include syncopal episodes (fainting), migraine attacks, and transient ischemic attacks. Furthermore, disorders of sleep, such as night terrors, somnambulism, rapid eye movement (REM) sleep behavior disorder, or restless leg syndrome, may be mistaken for nocturnal seizure events.

Psychogenic nonepileptic seizures are often due to psychological conflicts or significant current life stressors, such as death or divorce. There is frequently a history of past sexual or physical abuse during childhood or early adolescence.

Many of these individuals have somatoform disorders, depression, or histrionic-type personalities, and some consider psychogenic nonepileptic seizures to be a type of conversion disorder. Psychogenic nonepileptic seizures are discussed in more detail in Chapter 8.

Sudden Unexplained Death in Epilepsy

The death rate among individuals with seizure disorders is two to three times that of the general population. Causes of death that are directly related to the fact the individual has epilepsy may include death due to some underlying disease in the symptomatic epilepsies; accidents during an epileptic seizure, such as trauma, drowning, choking, or burning; status epilepticus; suicide; death related to treatment of the epilepsy; and sudden unexplained death in epilepsy (SUDEP).

Sudden unexplained death in epilepsy, is defined as sudden, unexpected, nontraumatic, nondrowning death in a relatively healthy individual with epilepsy in which postmortem examination does not reveal a cause for death. Sudden death means within minutes rather than hours, usually following a seizure. Death is not directly related to a seizure or status epilepticus. Sudden death usually occurs after a seizure is over and often after the individual has regained awareness of his surroundings.

Sudden unexplained death in epilepsy accounts for 10%–17% of deaths in patients with epilepsy. Sudden death in epileptics is more frequent in males, African-Americans, and younger epilepsy patients. Most cases of SUDEP occur in patients who are between 20 and 40 years old (mean age is 28–35 years old). It is very rarely observed in children. Incidence estimates have ranged between 0.35 cases per 1,000 and 5.6 cases per 1,000 person years of follow-up (Ficker et al., 1998; Neuspiel and Kuller, 1985). Thus, the standardized mortality ratio of sudden unexplained deaths is 24–40 times higher in people with epilepsy than in the general population.

The cause of SUDEP is unknown, but death appears to be multifactorial; cardiac factors, pulmonary edema, and central apnea may individually or in combination cause death. Fatal cardiac arrhythmias, tachycardia, and abnormalities of the cardiac conductive system have all been implicated. Pulmonary edema has been documented in the overwhelming majority of SUDEP cases often thought to be due to cessation of spontaneous respiratory drive (i.e., central apnea syndrome) during sleep. Sixty percent of SUDEP cases occur during sleep. Cardiac and respiratory events could be caused by direct propagation of electrical discharges from the brain to the cardiac and respiratory pacemaker centers in the brainstem.

Other risk factors for developing SUDEP include subtherapeutic levels of AEDs, being on multiple AEDs, poorly controlled seizures, alcohol use, and a history of epilepsy surgery for an intractable seizure disorder. Most physicians do not routinely counsel patients with epilepsy about SUDEP unless they are high-risk patients. Measures to reduce the risk of SUDEP include optimizing treatment with AED monotherapy if possible, controlling seizure frequency, avoiding seizure-provoking situations, abstaining from alcohol and drugs, and caregiver training regarding acute seizure management and cardiopulmonary resuscitation. Postictal stimulation is thought to reduce the chances of developing apnea following a seizure. An Epilepsy Foundation Joint Task Force on SUDEP has recently made recommendations about (1) when SUDEP should be discussed with patients, (2) research to identify risk factors, and (3) possible preventive strategies to avoid SUDEP (So et al., 2009).

4

Diagnostic Tests in Epilepsy

The initial evaluation for suspected epilepsy usually consists of electroencephalography (EEG) and magnetic resonance imaging (MRI). The EEG is crucial for establishing whether epilepsy is present and can aid in classifying if the seizure disorder is focal or generalized. Accuracy of initial diagnostic interictal scalp EEG in outpatient settings is only between 29% and 55% but repeating the EEG will eventually show interictal discharges in 80%–90% of patients (Pillai and Sperling, 2006). MRI is the structural imaging method of choice in the initial work-up of patients with epilepsy to determine the presence and nature of the underlying pathology, if any exists (Commission on Neuroimaging of the ILAE, 1997).

Electroencephalography

Electroencephalography is the most important diagnostic tool for the diagnosis of epilepsy. Interictal EEG (in conjunction with seizure semiology) is useful not only in confirming a diagnosis of epilepsy, but also assisting in the identification of seizure type, prognosis, and optimal treatment. After a diagnosis of epilepsy has been established by EEG, subsequent serial EEGs also may be used to monitor response to treatment or progression of disease.

The EEG signals recorded from the scalp are detecting the summated effects of thousands of excitatory postsynaptic potentials (EPSPs) from neurons firing in the cortex nearer to the surface of the skull. Surface (scalp) EEGs do not detect signals well from deeper sources within the brain. In the partial epilepsies, EEG recordings can pick up interictal epileptic discharges from an often large area of cortex with abnormal excitability (Kennett, 2000). This is referred to as the *irritative zone*, which most frequently contains the focal area of the brain where seizures are generated, which in turn is called the *ictal onset zone*.

Interictal Scalp Recordings

Scalp EEG recordings follow a standardized positioning of the electrodes called the *10–20 international system of electrode placement*, as well as standard procedures for seizure detection recording established by the American Clinical Neurophysiology Society (1994). The '10' and '20' in the 10–20 system of electrode placement refers to the 10% or 20% inter-electrode distances (Homan et al., 1987). The international 10–20 electrode placement is shown in Figure 4.1. As can be seen in Figure 4.1, the even numbers refer to electrodes placed on the right and the odd numbers refer to electrodes placed on the left side of the head.

The Fp electrodes cover the frontal polar regions, F3, Fz, and F4 cover the frontal convexities, and F7 and F8 detect signals emerging from the temporal poles. The lateral temporal lobes are covered by T3, T4, T5, and T6. Finally, P3, Pz, and P4 cover the parietal lobes and O1 and O2 cover the occipital lobes. The "z" refers to electrodes placed on the midline.

The EEG recording devices have at least 16 channels to record the electrical activity from both hemispheres, as well as simultaneously recording eye movements (OCG), muscle artifact (EMG), electrocardiography (EKG), and respiration which are needed to identify EEG artifact. Activation procedures,

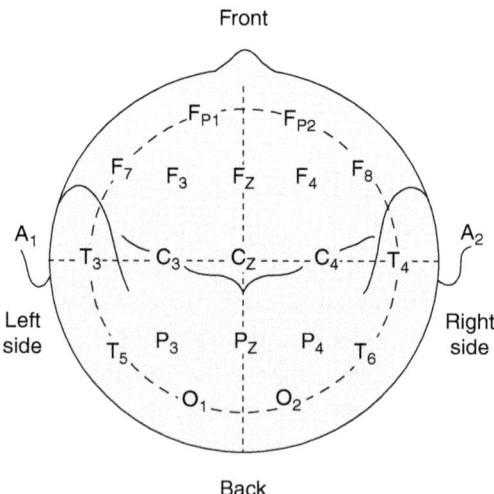

FIGURE 4.1. The 10–20 international system of standardized EEG electrode placement. Courtesy of Grass Technologies, An Astro-Med, Inc. Subsidiary.

including hyperventilation, photic stimulation, or barbiturate drug activation, may be included as part of a standard EEG recording session which typically lasts 45–60 minutes.

Between 50% and 60% of patients with epilepsy may have a normal interictal EEG during their initial EEG evaluation (Salinsky et al., 1987). For this reason, a normal EEG should not be solely relied upon to rule out epilepsy or to diagnose nonepileptic seizures. Most epilepsy patients will show interictal EEG abnormalities after serial examinations using activation procedures, especially sleep; in one study only 8% of patients with epilepsy continued to have a normal EEG after multiple recordings with activation procedures (Binnie and Prior, 1994). Special electrodes, such as sphenoidal electrodes to detect temporal lobe seizures, supraorbital electrodes for orbitofrontal seizures, or nasopharyngeal electrodes for frontal polar seizures, may be used to improve diagnostic yield in appropriate cases.

A variety of factors may interfere with EEGs. Muscle artifact, eye movements, and blinking may obscure the underlying EEG waveforms or be misinterpreted as representing neuronal electrical activity. An attempt to mitigate these interfering influences is made by recording EMG and OCG at the same time as the EEG. Many different drugs may also alter the EEG pattern.

Specific Electroencephalographic Patterns in the Partial Epilepsies

The EEG onset of partial or localization-related epilepsies is characterized by a sudden change of frequency and amplitude from the previous ongoing background electrical activity, and the appearance of a new EEG rhythm. The focal onset of an electrographic seizure will typically evolve through several phases (Chabolla and Cascino, 2006): (a) focal attenuation of EEG activity; (b) focal, rhythmic, low-voltage, fast activity discharge; and (c) progressive increase in amplitude with slowing that spreads to a regional anatomical distribution.

If a complex partial seizure is captured on a scalp EEG recording, most will progress through these typical EEG phases. Standard scalp EEG recordings are most often normal in patients with simple partial seizures.

The main types of EEG abnormalities in patients with partial epilepsies include spikes, spike waves, spike-and-wave complexes, periodic lateralized epileptiform discharges (PLEDs), and intermittent rhythmic δ activity. These transient abnormal EEG waveforms need to be distinguished from other types of transient waveforms that are benign and occur in healthy normal individuals. Spikes and slow waves may occur either in isolation or following a slow wave, which is referred to as a *spike-and-wave complex*. A spike is a predominantly negative waveform with steep ascending and descending limbs that by definition last less than 70 msec (Chabolla and Cascino, 2006). Sharp wave discharges are broader waveforms than spikes with a peak that lasts between 70 and 200 msec. Spike-and-wave complexes are formed by repetitive spikes that are followed by a slow wave. Although these epileptiform EEG discharges may arise from any region of the brain, they are most commonly arise from the anterior temporal, frontal, or centrotemporal or rolandic regions.

Periodic Lateralized Epileptiform Discharges

Periodic lateralized epileptiform discharges (PLEDs) consist of recurring episodic EEG discharges that are localized to a particular region within one hemisphere. They are repetitive spike or sharp wave discharges occurring at intervals between 0.5 and 5 seconds, but usually around 1 per second. When present, these waveforms tend to continue throughout the EEG study. Periodic lateralized epileptiform discharges are most commonly seen in the context of acute focal lesions, typically cerebral infarctions, and usually resolve after the acute phase of the cerebral insult is over. The EEG manifestations of PLEDs are often accompanied by clinical seizures. See Figure 4.2 for an example of the EEG waveforms seen in PLEDs.

Intermittent Rhythmic δ Activity

This EEG pattern is characterized by trains of rhythmic, low- to moderate-amplitude, δ frequency (2.5 Hz) slow waves that are commonly associated with structural brain lesions (especially diencephalic, intraventricular, or

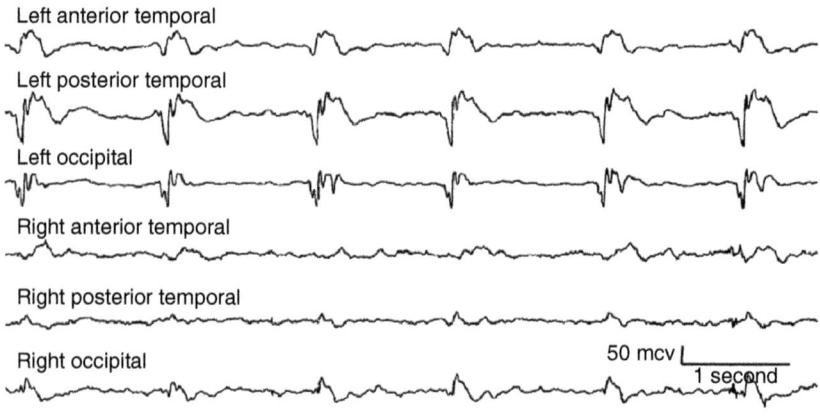

FIGURE 4.2. Example of periodic lateralized epileptiform discharges (PLEDs) during interictal EEG recording. From: http://www.emedicine.com/neuro/topic136.htm with permission.

posterior fossa tumors) or with diffuse encephalopathies. This EEG pattern becomes prominent during non-rapid eye movement (REM) sleep and drowsiness. In adults, the δ activity has a frontal predominance, whereas in children it is more commonly seen in the occipital brain regions.

Specific Electroencephalographic Patterns in Generalized Epilepsies

The most common patterns of EEG abnormality in the primary generalized epilepsies are (Chabolla and Cascino, 2006) 3-Hz generalized spike-and-wave, multiple spike-and-wave pattern, slow spike-and-wave, and generalized paroxysmal fast activity.

3-Hz Spike-and-Wave Pattern

The 3-Hz generalized spike-and-wave pattern is the EEG signature for absence seizures, although similar waveforms may also be seen in idiopathic generalized epilepsy. The EEG pattern is similar during both ictal and interictal recordings and consists of generalized, repetitive, and symmetrical spike-and-wave discharges occurring at approximately 3 cycles per second (see Figure 4.3).

These waveforms often arise more prominently from the anterior regions of the brain. The clinical manifestations of absence seizures, such as eye blinking, staring, or automatisms, may occur when the EEG changes last for more than 3 or 4 seconds. This pattern occurs most frequently in children ages 5–15 years old and is more common among girls.

Multiple Spike-and-Wave Pattern

This EEG pattern is characterized by generalized intermittent brief spikes interspersed with polyspike complexes associated with slow waves of varying frequency (~3.5–6 Hz) (Chabolla and Cascino, 2006). When the bursts of EEG activity are brief (1–3 seconds), there may not be any observable clinical evidence of a seizure. This type of EEG waveform may give rise to several generalized seizure types including generalized tonic-clonic seizures and clonic, myoclonic, or atypical absence seizures.

Slow Spike-and-Wave Pattern

This pattern of EEG activity is most often seen in children with mental retardation or other severe cognitive impairment. The pattern itself consists of generalized, repetitive, bilaterally synchronous sharp waves or spikes occurring at 1.5–2.5 Hz. In many instances, there may be no clear clinical manifestations of the abnormal electrical discharges. The ictal and interictal EEG patterns are

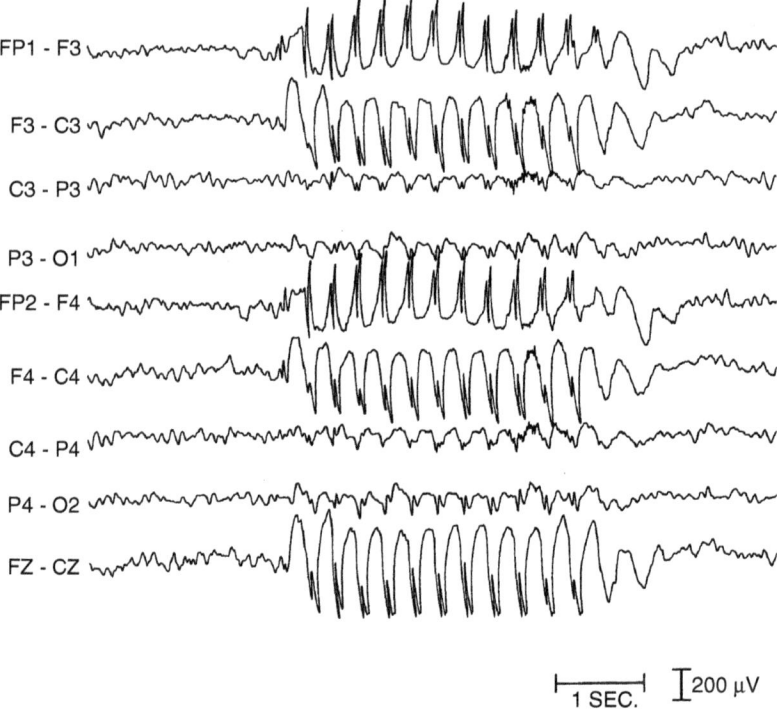

FIGURE 4.3. Example of typical pattern of 3 Hz generalized spike and wave EEG in absence seizures. From: http://brain.fuw.edu.pl/~suffa/Modeling_SW.html, FH Lopes da Silva.

essentially the same whether or not a clinical seizure event is evident. Seizure types may be either tonic-clonic, tonic, atonic, atypical absence, myoclonic, or some combination of these. These seizures may persist into adulthood.

Generalized Paroxysmal Fast Activity
This EEG pattern consists of repetitive spikes occurring around 20 Hz that are seen with tonic seizures, in the sleep recordings of patients with generalized seizures, or at the beginning of a generalized tonic-clonic seizure (Chabolla and Cascino, 2006). Ictal EEG recordings show a generalized, synchronous alterations of low-voltage, fast frequency activity that gradually increases in amplitude.

These descriptions of abnormal EEG patterns are only intended as the most general of introductions. Verbal descriptions do no lend themselves to easy understanding of what the waveforms actually look like. For easier-to-visualize and more detailed descriptions, readers are referred to the graphic displays of abnormal EEG tracings available in EEG atlases and textbooks such as Blume and Kaibara (1995), Blume and Kaibara (1999), Luders and Noachtar (2000), or Neidermeyer and Lopez da Silva (1993).

Structural Neuroimaging in Epilepsy

The majority of epilepsy cases in adulthood, and between 30% and 40% of childhood onset epilepsy, have some underlying structural lesion that is potentially detectable by MRI or computed tomography (CT), and this proportion continues to grow as imaging technology improves. Brain MRI is considerably more sensitive than head CT for detecting the presence and determining the nature of the pathology in epilepsy. The International League Against Epilepsy (Commission on Neuroimaging of the ILAE, 1997) recommends that all patients with intractable localization-related epilepsy undergo MRI to look for epileptic pathologies and to assist in accurate diagnosis of a possible epilepsy syndrome.

The MRI, and in some special instances CT, examination looks for pathological lesions that may cause localization-related epilepsy. These most often include mesial temporal lobe sclerosis (MTS), malformations of cortical development, tumors, vascular abnormalities, cysts and other manifestations of infection, traumatic lesions, and nonspecific gliosis (Anslow and Oxbury, 2000). Computed tomography scanning is the preferred imaging method when there is some type of calcified lesion since lesions that are totally calcified may be undetectable by MRI. Pathological conditions with cerebral calcification that have been associated with epilepsy include low-grade gliomas, oligodendrogliomas, cortical dysplasia, tuberous sclerosis, cavernous angioma, calcified abscesses or aneurysms, and cysticercosis. Although portions of these lesions that are not calcified may be visible on MRI, CT can help clarify the nature of the pathological process in certain cases.

In a fairly representative study of patients with chronic partial epilepsy, MRIs successfully detected lesions in 85% of cases (Cook and Stevens, 1995). Temporal lobe pathology was identified in more than 90% of the patients with temporal lobe epilepsy, whereas only 65% of patients with extratemporal epilepsy showed pathology on MRI. In terms of the frequency of MRI detected pathologies, hippocampal atrophy was the most common pathology (34%)

followed by malformations of cortical development (16%), tumors (16%), and vascular malformations (10%).

Standard brain MRI in the evaluation of epilepsy generally images the whole brain in three planes (i.e., axial/horizontal, sagittal, and coronal) using three basic pulse sequences (namely, T1, T2, and proton density). In addition, other pulse sequences provide certain advantages in evaluating patients with epilepsy. The fluid attenuated inversion recovery (FLAIR) pulse sequence is particularly useful in detecting pathological processes where increased tissue water is an issue; FLAIR is useful in detecting mesial temporal lobe sclerosis and multiple sclerosis plaques.

T1 volume acquisitions (called MPRAGE on Siemens scanners) produce 128 contiguous 1mm images which allow multiplanar reconstruction allowing excellent spatial resolution. An MPRAGE pulse sequence is very valuable when looking for hippocampal atrophy or malformations of cortical development. Unfortunately, cortical malformations are difficult to image in young children (<2 years old) since myelination is incomplete, and this results in poor gray–white matter delineation. The use of contrast enhancement dyes, such as gadolinium, is usually of little benefit in MRI evaluations for epilepsy, but nevertheless can be advantageous in some less common circumstances. Since gadolinium accumulates in brain regions where there has been a breakdown in the blood–brain barrier, it can help identify areas of inflammatory change and some brain tumors.

5

Medical Treatment of Epilepsy

Medical therapies for epilepsy consist of nonsurgical interventions; essentially treatment with either antiepileptic drugs (AEDs) or ketogenic diet. Before treatment is initiated, however, it is important to be certain the patient actually has epilepsy and not another illness (or even some normal paroxysmal phenomena) masquerading as seizures. Between 20% and 30% of patients with nonepileptic events are misdiagnosed as having epilepsy, and these patients are often improperly treatedfor years with AEDs (Gates, 2002). A general discussion of nonepileptic seizures may be found in Chapter 3, and details about psychogenic nonepileptic events are provided in Chapter 8.

Once it has been established that a seizure has actually occurred, the next question the physician faces is whether to pursue further work-up in an attempt to identify the cause of the seizure and whether to treat the seizure with an AED. *Most individuals who experience a single seizure will never have another one.* This is especially true if there is no history of previous seizures, and the neurological examination, including magnetic resonance imaging (MRI) and electroencephalogram (EEG), are normal. In this scenario, most physicians would not begin AED treatment. This is particularly true in children, because the AED side effects pose a greater threat to health and psychosocial adjustment than the risk for another seizure. The same is true for children with febrile seizures; most physicians prefer to withhold AED therapy for children who have experienced a febrile seizure.

The risk that an otherwise normal child who experiences a single generalized tonic-clonic seizure will have a second one is only about 15%. On the other hand, in patients who have a second generalized seizure with EEG or

MRI abnormalities, the risk for recurrent seizures rises to more than 50%, and for patients who have a single partial onset seizure, the risk for recurrent seizures is greater than 75% (Hauseret et al., 1982). Most physicians would choose to treat these higher-risk patients pharmacologically. Nevertheless, the risks of AED treatment must be weighed against the risks of seizures for each individual patient and his or her particular circumstances.

Pharmacologic Therapies

Once a diagnosis of epilepsy has been established by the presence of at least two unprovoked seizures, AED therapy is usually initiated, since the likelihood of having recurrent seizures rises to 80%–90% after two unprovoked seizures (Camfield and Camfield, 1997). The goal of pharmacotherapy is to achieve complete seizure control while eliminating unwanted side effects and maximizing quality of life. The most common side effects of all AEDs include sedation, slurred speech, unsteadiness, clumsiness, dizziness, nausea, and adverse cognitive and behavioral effects. Other fairly common adverse effects associated with particular medications that are not dose-specific include double vision, weight gain, hyperactivity, irritability, sleep disturbances, gum dysplasia, hirsutism, and changes in mood. The common cognitive and motor AED side effects include poor attention and concentration, difficulties with learning, reduced mental processing speed and efficiency, and diminished fine motor speed.

With regard to potentially serious medical adverse events, some AEDs may result in liver damage, blood disorders (aplastic anemia), or severe (even fatal) rashes. The International League Against Epilepsy has issued best-practice guidelines for therapeutic drug monitoring to guide dosage adjustments in order to minimize adverse side effects and maximize seizure control (Patsalos et al., 2008).

Older Antiepileptic Drugs

Before the mid-1990s, six AEDs were in common use: phenytoin (Dilantin), carbamazepine (Tegretol), valproic acid or valproate (Depakote), ethosuximide (Zarontin), primidone (Mysoline), and phenobarbital. When using these older AEDs, treatment decisions were reasonably straightforward. The typical practice in the 1980s and early 1990s was to prescribed carbamazepine, phenytoin, or valproate for focal-onset (partial) seizures and valproate for generalized tonic-clonic and generalized myoclonic seizures. In absence seizures, ethosuximide or valproate were the drugs of choice (Camfield and Camfiled, 1997).

Two randomized, double-blind, prospective studies of the efficacy of these older AEDs (except ethosuximide) were conducted at Veteran's Affairs (VA) medical centers throughout the United States The first study found overall treatment success was best with carbamazepine and phenytoin, intermediate for phenobarbital, and lowest for primidone (Mattson et al., 1985). Seizure control efficacy was similar across all five AEDs investigated for use in generalized tonic-clonic seizures. Carbamazepine provided complete seizure control more often in complex partial epilepsy than phenobarbital or primidone.

In the second VA study, valproate was as effective as carbamazepine for controlling generalized tonic-clonic seizures, but carbamazepine provided better control of complex partial seizures (Mattson et al., 1992). Moreover, carbamazepine had fewer long-term adverse effects (e.g., weight gain, hair loss, tremor) than valproate. Eleven percent of patients on carbamazepine developed a rash, whereas only 1% of those on valproate developed rashes.

The VA Cooperative studies also yielded valuable information regarding seizure control in serial monotherapy and polytherapy. Among patients prescribed their first AED (monotherapy), approximately 60%–70% achieved almost complete seizure control (Mattson and Cramer, 1997). On the flip side, 30%–40% of patients continued to have seizures which were refractory to initial drug monotherapy. When an additional drug was added to the treatment regimen, 40% of those patients who were intractable on monotherapy experienced improved control with the two-drug combination, but only 9% of these individuals achieved complete seizure control. Even fewer patients achieve seizure control after being tried on a third solo drug. Thus, approximately one-third of patients with complex partial epilepsy will have an intractable seizure disorder that is refractory to AED treatment.

Newer Antiepileptic Drugs

During the latter half of the 1990s, eight new AEDs were introduced and approved by the U.S. Food and Drug Administration (FDA). The major advantage of the newer AEDs over the older ones is improved safety and tolerability, with the exception of felbamate, which had a high rate of aplastic anemia and a rate of liver failure similar to valproate. Because of this, the use of felbamate declined considerably soon after its initial introduction, although it is still an effective AED in a few selected patients (e.g., short-term use in Lennox-Gastaut syndrome).

Although not yet thoroughly investigated, it appears that the newer AEDs are about as effective as the old AEDs in controlling seizures. Excluding felbamate (Felbatol), the newer AEDs include gabapentin (Neurontin), lamotrigine (Lamictal), levetiracetam (Keppra), oxcarbazepine (Trileptal), tiagabine (Gabitril), topiramate (Topamax), zonisamide (Zonegran), vigabatrin (Sabril, not commercially available in the United States), and more recently (2005) pregabalin (Lyrica). Considering both the old and new AEDs together, Table 5.1 summarizes Nadkarni, LaJoie, and Devinsky's (2005) recommendations for first- and second-line AED selection for the major seizure types and seizure syndromes.

Table 5.1 Long-term antiepileptic drug treatment by seizure type or epilepsy syndrome based on recommendations of Nadkarni, LaJoie, & Devinsky (2005)

SEIZURE TYPE	FIRST-LINE DRUG	SECOND-LINE DRUG
Primary Generalized Seizures:		
Absence	Ethosuximide, valproic acid	Lamotrigine
Myoclonic	Valproic acid	Clonazepam, lamotrigine, primidone
Tonic-clonic	Valproic acid, carbamazepine, oxcarbazepine, lamotrigine	Phenobarbital, phenytoin, primidone, topiramate
Generalized Epilepsy Syndrome:		
Childhood-onset absence	Ethosuximide	Valproic acid, lamotrigine
Benign rolandic	Carbamazepine, gabapentin, oxcarbazepine	Lamotrigine, levetiracetam, topiramate, valproic acid
Adolescent-onset absence	Valproic acid	Ethosuximide, lamotrigine
Juvenile myoclonic	Valproic acid	Acetazolamide, clonazepam, lamotrigine, levetiracetam, primidone, topiramate

(continued)

Table 5.1 (Continued)

Infantile spasms Lennox-Gastaut	Corticotropin (ACTH) Valproic acid, lamotrigine	Clonazepam, valproic acid Carbamazepine, oxcarbazepine, topiramate, zonisamide
Partial Seizure Types & Epilepsy Syndromes:		
Simple partial, complex partial, secondary generalized tonic-clonic	Carbamazepine, gabapentin, lamotrigine, levetiracetam, oxcarbazepine, topiramate, valproic acid, zonisamide	Phenobarbital, phenytoin, primidone, tiagabine

Adapted from Nadkarni S, LaJoie J, Devinsky O. Current treatments of epilepsy. *Neurology* 2005;64(Suppl 3):S2-S11.

With regard to the newer AEDs, evidence suggests that gabapentin (Neurontin), lamotrigine (Lamictal), topiramate (Topamax), and oxcarbazepine (Trileptal) are effective first-line monotherapies for newly diagnosed partial or mixed seizure disorders (French et al., 2004). The considerations that enter into a physicians decision about which AED to prescribe include the drug's: particular side effect profile, time it takes to reach an effective dose, interactions with other medications, patient's comorbid medical conditions, and cost. It is recommended that patients start at a low dose, and the dose is then slowly increased to avoid many of the adverse side effects and unnecessary discontinuation.

With regard to comorbid disorders, lamotrigine (Lamictal) is a good drug choice in patients with comorbid depression or bipolar disorder and for those patients who have concentration problems or somnolence (Nadkarni et al., 2005). For those who have comorbid migraine headaches or obesity, topiramate can be particularly useful. In patients who are taking other medications, levetiracetam (Keppra) and gabapentin (Neurontin) may be advantageous since they have minimal interactions with other drugs. For example, levetiracetam may be used in patients with brain tumors due to its lack of interaction with chemotherapy agents. With regard to side effects, levetiracetam (Keppra), gabapentin (Neurontin), and lamotrigine (Lamictal) have minimal cognitive side effects compared to other AEDs.

Antiepileptic Drug Adverse Effects

Negative side effects are common in most all AEDs, but less so in the newer AEDs. Adverse effects of the drugs are more often seen soon after initiation of treatment or when drugs reach toxic blood levels because of excessive dosing or other factors, such as impeded drug clearance. The most common side effects across all AEDs include sedation, adverse cognitive and behavioral effects, dizziness, and gastrointestinal upset (Mattson and Cramer, 1997).

Idiosyncratic reactions although rare may develop in some individuals due to individual sensitivity or allergic reaction to a particular drug. The most common idiosyncratic reaction is rash. The rash of Stevens-Johnson syndrome should be taken seriously since it may be fatal. Other serious idiosyncratic reactions include potentially fatal liver damage and aplastic anemia. Aplastic anemia has been reported in patients taking felbamate and valproate. There appears to be an increased risk of severe rash in children taking lamotrigine. Hepatic failure has occurred in young children on valproate polytherapy. Idiosyncratic reactions may, however, occur with any of the standard AEDs with a frequency ranging between 5% and 10%. Idiosyncratic reactions are less frequent in patients taking valproate, gabapentin, and vigabatrin (Mattson and Cramer, 1997). In December 2007, the FDA issued a recommendation that people of Asian ancestry take a genetic test before they are treated with carbamazepine due to an excessive risk (about 10 times higher in some Asian countries) for development of Stevens-Johnson syndrome and other rashes among these individuals who have a specific type of immune-system gene.

After many years of use, some differential effects of particular AEDs have emerged. Long-term use of phenytoin may result in gum hyperplasia, hirsutism, acne, a coarsening of facial features, peripheral neuropathy, and cerebellar atrophy. Long-term use of carbamazepine has been associated with hyponatremia, and valproate may cause significant weight gain, pancreatitis, and reduced hearing.

Reduced bone density and subsequent risk of developing osteoporosis has been linked to the duration of AED use, as well as with the number of AEDs taken. Bone loss affects children and adults of both sexes equally and has been associated with certain enzyme-inducing AEDs (namely, phenytoin, carbamazepine, and barbiturates). To counteract these potential effects, it is recommended that patients taking these medications ensure they get adequate daily amounts of calcium and vitamin D and monitor bone density at 5-year intervals.

Switching Antiepileptic Drugs and Generic Equivalents
In recent years, there has been accumulating evidence that breakthrough seizures and unexpected side effects may occur when patients are switched from their brand-name drug to a generic version of the same drug or vice versa. These problems have also been reported when the switch has occurred between different manufacturers' versions of the same generic drug and when a single manufacturer switches to a new formulation of the same drug. Such changes to epilepsy patients' AEDs may be made by pharmacies without the patients' or their doctors' knowledge, since pharmacies may change generic drug manufacturer suppliers as often as once a month. The American Epilepsy Foundation has called on the FDA to develop new standards to determine the equivalence of all AEDs (both generic and brand-name). Even though the majority of patients can safely switch to different AED formulations, patients and health care professionals should be vigilant against the potential harm of switching among AEDs until there are new FDA standards for equivalence.

Antiepileptic Drugs in Women
Certain AEDs decrease the efficacy of oral contraceptive medications in women by increasing the elimination of the contraceptive drug and causing an insufficient hormonal dose to prevent ovulation or implantation. The so-called enzyme-inducing AEDs, (phenytoin, carbamazepine, phenobarbital, and primidone) are the drugs most likely to cause this effect, although topiramate and oxcarbazepine may do so at certain doses. The other AEDs do not reduce the efficacy of oral contraceptives (Morrell, 2006). Recently, the reverse effect has also been found. Hormonal contraceptives have been shown to produce an 84% higher clearance rate for lamotrigine, thus increasing the risk of seizures (Pennell, 2008). Thus, there is a risk of breakthrough seizures in pregnant women who are taking lamotrigine due to increased clearance rate of the drug during pregnancy. This often necessitates increasing the dose of lamotrigine during pregnancy and reducing the dose after pregnancy to avoid toxic effects.

Some clinicians and researchers have suggested that fertility may be reduced in women with epilepsy for a variety of reasons, including the effects of AEDs and other biological or psychosocial factors. In a recent Finnish population study (Viinikaine et al., 2008), 59% of women with active epilepsy had given birth, and thus, the rate of childlessness among these women was 41%. The corresponding figure for the entire Finnish population is 77% of all women have given birth, and the rate of childlessness is 23%. Therefore, about 18% fewer women with active epilepsy have children as compared to all women, although this is probably not a direct effect of seizures or AED treatment.

In an attempt to determine the reasons for this discrepancy, the authors next examined the prevalence of severe comorbidities (such as mental retardation or significant psychiatric illness restricting social life or requiring hospitalization) among women with epilepsy on AED monotherapy and found that 15% of women on valproate, and 18% of women on carbamazepine, had severe comorbid conditions that may be expected to secondarily affect the ability to have children. After excluding these severe comorbid factors, they concluded that there was no significant difference between the fertility rates in women with epilepsy and the general population. Moreover, they did not find that women wishing to become pregnant while on valproate had reduced fertility compared with women using carbamazepine.

Pregnancy and Teratogenic Effects
Pregnancy registries exist in the United States and the United Kingdom to track the effects of AED treatment on the fetus. To date, the majority of newborns exposed to AEDs in utero do not have major congenital malformations, although there is less data on the newer AEDs. The risk of teratogenic (production of developmental deformities) effects is less with monotherapy and increases with polytherapy.

The FDA classifies AED medications into two categories: category C and category D. Category C drugs are those that have been shown to cause teratogenic effects in animals, but the risk in humans is unknown. All of the new AEDs are classified as category C drugs. Most of the newer AEDs (e.g., gabapentin, lamotrigine and levetiracetam) do not appear to cause congenital malformations in animals, with the exception of topiramate and vigabatrin (Panayiotopoulous, 2005). Category D drugs are those that have been shown to cause teratogenic effects in both animal and human pregnancies. Category D drugs include valproate, phenytoin, and carbamazepine.

The risk of major congenital malformations is 3.4%–7.8% for infants exposed in utero to one AED. This rises to 6.5%–13.5% for infants exposed to AED polytherapy (Kaaja et al., 2003; Morrow et al., 2004). Malformations caused by AEDs are more common in children exposed to high serum concentrations during the first trimester. Data from the United States and the

United Kingdom Registries have yielded the following rates of major congenital malformations for individual drugs: 5.9% for valproate, 3.8% for phenobarbital, 2.3% for carbamazepine, and 2.1% for lamotrigine (Holmes et al., 2004; Morrow et al., 2004). The prevalence of nonsyndromic congenital malformations among newborns not exposed to AEDs is approximately 2%. Orofacial clefts and skeletal abnormalities have been reported with valproate and carbamazepine. Other teratogenic abnormalities across all AEDs studied include cardiac and neural tube defects, facial stigmata, genitourinary malformations, and gastrointestinal tract abnormalities.

When women with epilepsy have children, there is a risk that some AEDs may be transferred to infants during breast-feeding. Levetiracetam and primidone are the AEDs most likely to transfer into breast milk in amounts that may be clinically important, whereas valproate, phenytoin, and carbamazepine probably are not transferred in significant amounts (Harden et al., 2009).

Antiepileptic Drug Treatment in Children
Many of the studies of AED efficacy do not include younger pediatric patients and specific pediatric epilepsy syndromes. This is especially true of the newer AEDs, which have not been available long enough for systematic study in children. Thus, current AED prescribing patterns tend to be guided by less rigorous research and expert opinion. The next sections will briefly review the current AED prescribing practices for the common childhood epilepsy disorders, based upon a survey of U.S. physicians who specialize in pediatric epilepsy.

Absence Seizures in Children. The treatment of choice for typical childhood absence seizures is ethosuximide followed by valproic acid and lamotrigine (Wheless et al., 2005). Multiple AEDs can worsen seizures that involve generalized spike-wave discharges, and these should not be used in absence seizures. These include phenytoin, carbamazepine, gabapentin, tiagabine, oxcarbazepine, and vigabatrin.

The treatment of choice for juvenile absence seizures are valproic acid and lamotrigine (Kaddurah and Moorjani, 2008). These two drugs work well for both absence and generalized tonic-clonic seizures. Ethosuximide is not effective in preventing tonic-clonic seizures, and since childhood absence attacks may evolve into juvenile tonic-clonic epilepsy, ethosuximide is not used in this age group.

Benign Epilepsy of Childhood with Centrotemporal Spikes. As discussed in Appendix II, the *benign* in this seizure syndrome refers to the ultimate remission of seizures typically seen in this condition during adolescence and not to the cognitive, motor, or psychosocial problems sometimes associated with benign epilepsy of childhood with centrotemporal spikes (BECTS). Some children experience a significant decline in language and other cognitive functions which can improve with AED treatment (Berroya et al., 2004). Oxcarbazepine and carbamazepine are the current treatments of choice according to expert opinion (Wheless et al., 2005). Gabapentin and lamotrigine have also been found to be effective and well-tolerated in children with BECTS.

Juvenile Myoclonic Epilepsy. There are no FDA-approved AEDs for monotherapy in juvenile myoclonic epilepsy (JME), although levetiracetam has been approved as an adjunctive therapy. Valproic acid is the drug of first choice in JME for adolescent males, whereas lamotrigine or levetiracetam are the recommended AEDs for treating JME in adolescent females (Wheless et al., 2005). This is due to the potential teratogenic effects of valproic acid on the developing fetus, which may present as subtle neuropsychological deficits in these children later in life (Meador et al., 2009). Thus, valproic acid is not a good AED for women of child-bearing age.

Partial Seizures. Carbamazepine is the most common AED initially prescribed for either complex or simple partial seizures, although oxcarbazepine is often among the first choice of drugs among the newer AEDs since it has a similar mechanism of action as carbamazepine. The Therapeutics and Technology Assessment Subcommittee of the American Academy of Neurology (AAN) and Quality Standards Subcommittee of the American Epilepsy Society (AES) published practice parameters for the use of AEDs, and they also recommend using gabapentin, topiramate (approved for initial monotherapy in children 10 years and older), lamotrigine, and levetiracetam either as first-line or adjunctive treatment, especially if oxcarbazepine or carbamazepine have been found to be ineffective (French, Kanner, Bautista, et al., 2004; Sankar, 2004).

Adverse Antiepileptic Drug Effects in Children

Although there are few studies of the child-specific adverse effects of AEDs, there are several well-established physical side effects. In general, children seem to be more vulnerable to the negative lipid, endocrine, and bone mineral side effects of AEDs than adults. These negative side effect details may be found in the AAN-AES practice parameters for AED therapy in pediatric populations (Sankar, 2004). Valproic acid is more likely than the other AEDs to cause liver disease in children under the age of 2 years, particularly when used in conjunction with the so-called enzyme-inducing AEDs (namely, phenytoin, phenobarbital, carbamazepine, and primidone). Valproic acid has also been associated with weight gain and insulin resistance. The enzyme-inducing AEDs are associated with lower bone mineral density. Weight gain is commonly seen in children taking carbamazepine and gabapentin in addition to valproate (valproic acid).

Phenytoin has been shown to cause gum hyperplasia, coarsening of facies, and hirsutism (male pattern of hair growth) in children. Lamotrigine may cause a rash, such as that associated with Stevens-Johnson syndrome, which is more common in younger children (Sankar, 2004). With regard to psychological adverse effects, topiramate and levetiracetam have been found to result in changes in mood, irritability, and hyperactivity in some children. In contrast, lamotrigine may have positive effects on mood and concentration difficulties. Even fewer studies have been conducted on the cognitive adverse effects of AEDs in children, but similar to adults, phenytoin, topiramate, and valproic acid seem to have the worst cognitive effects, whereas levetiracetam, lamotrigine, and gabapentin appear to have the fewest cognitive side effects (e.g., Pressler et al., 2006).

In children, the use of AEDs has been associated with deficiencies in processing speed, language, and verbal learning and memory, even among children with newly diagnosed seizures who had no epileptiform EEG activity (Fastenau et al., 2009). When examining individual drugs, these authors found neuropsychological deficits in at least one cognitive domain among 31% of children taking valproic acid, 27% of children taking oxcarbazepine, and in 21% of the children taking carbamazepine. Valproic acid was associated with slower processing speed relative to the other AEDs. This suggests that school-aged children may be more vulnerable to the cognitive side effects of AED monotherapy than adults. The cognitive and behavioral side effects of AEDs (primarily in adults) are discussed in more detail later.

Cognitive and Behavioral Effects of Antiepileptic Drugs

In contrast to major physical malformations, more subtle developmental consequences have been reported in children born to mothers with epilepsy. In utero valproate exposure is more likely than other commonly used AEDs to impair cognitive development. For example, valproic acid (also known as valproate) has been reported to cause developmental delays in children under the age of 6 years and result in lower IQs than carbamazepine (Loring et al., 2007). Children on valproate are more often enrolled in special education classes in school than those on other AEDs. Similarly, children who had been exposed to valproate in utero had lower Bayley Developmental Quotients at 2 years compared to children exposed to carbamazepine, lamotrigine, or phenytoin (Meador et al., 2007).

Although there may be some small differential effects of individual AEDs on neuropsychological performance, for the most part they exert a similar pattern of cognitive test impairment that is likely attributable to the general sedating and inhibitory effects of AEDs. These AED sedating effects are due to decreased neuronal excitability, which in turn increases seizure threshold and reduces the likelihood of seizure propagation. Most of these drugs will cause varying degrees of deficit on tests measuring attention, concentration, vigilance, learning, memory, mental processing speed, reaction time, and motor speed. Some of the more common behavioral adverse effects of AEDs include irritability, hyperactivity, aggressive behavior, and emotional lability. In contrast to these negative side effects, many AEDs have positive psychotropic effects on mood, and thus, are frequently used as mood stabilizers in psychiatric patients with bipolar disorder for example.

Most of the cognitive and behavioral AED adverse effects are dose-dependent (fewer adverse cognitive effects at lower doses). The risk of side effects is increased with polytherapy and less likely in patients on monotherapy. Many of the negative side effects can be circumvented by starting therapy on low doses and slowly titrating upward in small dose increments.

With regard to the specific differential effects of individual AEDs, there are really no significant differences in the neuropsychological adverse effects of the older AEDs such as carbamazepine, valproate, or phenytoin in adults. Although no strong differences in seizure control efficacy have been found between the older and newer AEDs, the newer AEDs have fewer negative side effects, including fewer cognitive or behavioral adverse effects. Gabapentin, lamotrigine, and levetiracetam have all been shown to produce fewer neuropsychological test performance deficits in adults relative to the most commonly prescribed AED, which is carbamazepine (Meader et al., 2007). A brief overview of the known cognitive and behavioral negative side effects of the newer AEDs follows.

Gabapentin (Neurontin)
Although gabapentin is better tolerated than carbamazepine and other older AEDs, there is still some risk for adverse neuropsychological effects with this drug. Healthy control subjects taking gabapentin performed worse than no-drug controls on a serial addition task, Rey Complex Figure memory, and verbal selective reminding, and showed EEG slowing as well (Meador et al., 1999). Gabapentin has been reported to produce some negative behavioral side effects in some children. The package insert for the drug reports that gabapentin may result in any of the following adverse behavioral effects in children: emotional lability (primarily behavioral problems); hostility, including aggressive behaviors; thought disorder, including concentration problems and changes in school performance; and hyperkinesias, primarily restlessness and hyperactivity. Similar behavioral problems have not been reported in adult patients with epilepsy.

Lamotrigine (Lamictal)
Similar to gabapentin, lamotrigine produces fewer negative cognitive and behavioral effects than the older AEDs. Nevertheless, lamotrigine can result in reduced finger tapping speed, EEG slowing, and increased subjective feelings of depression when compared to healthy normal subjects not taking any medication (Meador et al., 2001). No significant negative neuropsychological effects of lamotrigine have been reported in children (although few studies exist). In a recent study of new-onset epilepsy in the elderly, lamotrigine was the best tolerated drug overall (Rowan et al., 2005). With regard to its psychotropic effects, lamotrigine has been reported to have beneficial effects on mood in both patients with epilepsy and patients with bipolar disorder. The FDA approved the use of lamotrigine for the treatment of bipolar disorder in 2003.

Levetiracetam (Keppra)
Levetiracetam also has a superior side effect profile compared to the older AEDs. In a study comparing levetiracetam, carbamazepine, and oxcarbazepine in healthy control subjects, levetiracetam had the fewest negative side effects on neuropsychological and EEG measures. Similar to the other new AEDs, even though all the newer AEDs have fewer side effects than the older AEDs, these drugs nonetheless produce some negative cognitive and behavioral effects when compared with nondrug performances in normal subjects. When compared with healthy controls not taking any drug, healthy subjects taking levetiracetam performed worse on a serial addition task, Symbol Digit Modalities Test, and self-rated fatigue or tiredness (Meador et al., 2007).

Although there appear to be few cognitive side effects of levetiracetam, there are some reports of increased irritability or hostility in some individuals. According to levetiracetam's package insert, behavioral side effects have been found in 13.3% of cases. In 2005, the FDA announced a safety warning for levetiracetam concerning increased risk for somnolence, fatigue, and behavioral abnormalities in children using the drug. Children with epilepsy who are taking levetiracetam have been reported to have higher risks of developing depression (3%), hostility (11.9%), nervousness (9.9%), and personality changes (37.6%).

Oxcarbazepine (Trileptal)

Although there are few investigations of oxcarbazepine's cognitive and behavioral side effects, the information that does exist suggests little difference between it and the older AEDs; making oxcarbazepine one of the less beneficial newer medications from a cognitive and behavioral side effects perspective. For example, some have reported oxcarbazepine produces the same neuropsychological effects as phenytoin, which can produce moderate cognitive side effects. There have been similar reports in children taking oxcarbazepine showing no neuropsychological benefits over carbamazepine, valproate, or combination carbamazepine/valproate therapy. Oxcarbazepine's mild to moderate negative neuropsychological effects are accompanied by slight EEG slowing (Loring et al., 2007).

Tiagabine (Gabitril)

In several studies using tiagabine as an add-on drug, no significant cognitive or behavioral adverse effects were found. In comparison with phenytoin, tiagabine (again as an adjunctive agent) treatment resulted in better verbal associative fluency and improved psychomotor speed. Comparing tiagabine monotherapy to carbamazepine monotherapy over the course of 1 year, the neuropsychological effects of tiagabine were similar to those of carbamazepine with the exception of worse associative verbal fluency in the carbamazepine group. Furthermore, tiagabine produced no significant adverse effects on mood (Aikia et al., 2006).

Topiramate (Topamax)

Of all the newer AEDs, topiramate has the greatest potential to produce negative neuropsychological effects, although there may be individual variation in vulnerability to these adverse effects. In many head-to-head trials, topiramate produced worse cognitive side effects than gabapentin, lamotrigine, tiagabine, carbamazepine, and valproate (Loring et al., 2007). Topiramate has reportedly caused somnolence, slowing, memory deficits, and language problems including word-finding difficulties and associative verbal fluency. With regard to deficits on specific neuropsychological tests, topiramate has produced impaired performances on Digit Symbol, Symbol Digit Modalities, delayed recall for stories (WMS, Logical Memory), Selective Reminding, and Controlled Oral Word Association. Many of these cognitive and behavioral problems are seen even when the starting doses of topiramate are low and titration is slow. These negative cognitive effects (namely, difficulty with concentration, attention, memory, psychomotor speed, and speech), which are often quite severe, have also been reported in children (Mula et al., 2003; Thompson et al., 2000).

Zonisamide (Zonegran)
There are fewer studies of the cognitive and behavioral effects of zonisamide than the other new AEDs, probably owing to its relatively later approval (2000) for use in the United States. In one add-on study, zonisamide was found to impair cognition, with mental and motor slowing and memory deficits that tended to become muted with time (over the course of 24 weeks). Others have found slowing of mental activity, word-finding difficulties, depression, irritability, and psychosis. Depression and psychotic behaviors may be especially common in children. In one study, 19% of children treated with zonisamide for several years had some form of episodic psychotic behavior (Miyamoto, et al., 2000).

Pregabalin (Lyrica)
Since pregabalin recently obtained FDA approval for use in the treatment of partial-onset seizures (and peripheral neuropathic pain) in 2005, little is known about the cognitive or behavioral side effects, or even if there are any. Pregabalin monotherapy in healthy controls caused no deficits in reaction time, short-term memory, or vigilance, but did result in subjective sedation and poor divided attention in one investigation (Hindmarch et al., 2005). In his summary of pregabalin, Panayiotopoulos (2005) reported this drug's less common (between >1% and <10% of cases) adverse reactions may include attentional disturbance, memory impairment, euphoric mood, irritability, and significant weight gain, whereas the most common side effects (>10% of cases) were somnolence and dizziness.

Lacosamide (Vimpat)
The newest AED at this time is lacosamide, which has been approved as an adjunctive therapy for the treatment of complex partial seizures in patients who are 17 years and older. Similar to many of the other newer AEDs, the side

effect profile is thought to be better than the older AEDs; dizziness, headache, nausea, and diplopia are the most common adverse effects reported. So far, there seem to be few cognitive or behavioral side effects; there are low rates of somnolence, cognitive impairment, and behavioral abnormalities that are roughly comparable to placebo. Lacosamide is one of the non–enzyme inducing AEDs, and therefore there are no known clinically significant drug–drug interactions (Beyreuther et al., 2007).

Use of Antiepileptic Drugs for the Treatment of Bipolar Disorder

Bipolar disorder is a chronic psychiatric condition characterized by recurrent manic, hypomanic, depressive, or mixed symptomatic episodes. Although pharmacotherapy cannot cure bipolar disease, a combination of antidepressant, antipsychotic, and AED agents is often used to treat the acute symptoms and to maintain remission. Although lithium is still the most commonly prescribed drug for bipolar disorder, AEDs have also become frequently used as adjunctive therapy. In one study, both the older AEDs, valproate (36.8%) and carbamazepine (5.8%), and the newer AEDs, lamotrigine (15.4%), gabapentin (14%), and topiramate (5.2%), were prescribed (Ghaemi et al., 2006). Unfortunately, the efficacy of some of these AEDs has not been demonstrated, and this prompted a consortium of State Medicaid agencies to undertake a drug class review to compare the effectiveness and adverse side effects of AEDs in the treatment of bipolar disorder (Melvin et al., 2008).

After a systematic review of this drug class for treating bipolar disorder, the consortium concluded that AEDs are no better than lithium, which has been established as an effective treatment (Melvin, et al., 2008). The group also came to the following conclusions:

- Current evidence supports the use of three AEDs, namely, carbamazepine, valproic acid, and lamotrigine, to maintain remission in outpatient adults with bipolar I disorder.
- There is no clinical trial evidence to support the use of either gabapentin or topiramate either as a primary or adjunctive treatment for bipolar disorder.

Suicidality and Antiepileptic Drugs

After reviewing data from 199 placebo-controlled clinical trials of 11 different AEDs (N = 43,892), the FDA issued a warning in 2008 to health care professionals alerting them about an increased risk of suicidality in patients taking AEDs (Box 5.1). Suicidality was defined as including both suicidal behavior (completed suicides, suicide attempts, and preparatory acts) and suicidal ideation. Results were consistent across all 11 drugs studied. The 11 AEDs studied included:

- Carbamazepine (Tegretol, Carbatrol)
- Felbamate (Felbatol)
- Gabapentin (Neurontin)
- Lamotrigine (Lamictal)
- Levetiracetam (Keppra)
- Oxcarbazepine (Trileptal)
- Pregabalin (Lyrica)
- Tiagabine (Gabitril)
- Topiramate (Topamax)
- Valproate (Depakote)
- Zonisamide (Zonegran)

The studies reviewed were examining the efficacy of AEDs in epilepsy, psychiatric disorders (e.g., bipolar disorder, depression, anxiety disorders), and other conditions (e.g., migraine headache and neuropathic pain syndromes). The relative risk of suicidality versus placebo was highest in patients receiving treatment for epilepsy (relative risk [RR] = 3.6), than for selected psychiatric patients (RR = 1.6) or other conditions (RR = 2.0). The FDA announcement may be accessed online at: <http://www.fda.gov/medwatch/safety/2008/safety08.htm#Antiepileptic>

BOX 5.1 U.S. Food and Drug Administration Alert

Information for Healthcare Professionals
Suicidality and Antiepileptic Drugs

FDA ALERT [1/31/2009]: The FDA has analyzed reports of suicidality (suicidal behavior or ideation) from placebo-controlled clinical studies of eleven drugs used to treat epilepsy, as well as psychiatric disorders, and other conditions. These drugs are commonly referred to as antiepileptic drugs (see list below). In the FDA's analysis, patients receiving antiepileptic drugs had approximately twice the

(continued)

risk of suicidal behavior and suicidal ideation (0.43%) compared to patients receiving placebo (0.22%). The increased risk of suicidal behavior and suicidal ideation was observed as early as one week after starting the antiepileptic drug and continued through 24 weeks. The results were generally consistent among the eleven drugs. Patients who were treated for epilepsy, psychiatric disorders, and other conditions were all at increased risk for suicidality when compared to placebo, and there did not appear to be a specific demographic subgroup of patients to which the increased risk could be attributed. The relative risk for suicidality was higher in the patients with epilepsy compared to patients who were given one of the drugs in the class for psychiatric or other conditions.

All patients who are currently taking or starting on any antiepileptic drug should be closely monitored for notable changes in behavior that could indicate the emergence or worsening of suicidal thoughts or behavior or depression.

The suicidality risk may extend beyond 24 weeks, but this risk could not be reliably determined because most trials did not last longer than 24 weeks. Although this risk of increased suicidality with AED use is likely real, it should not be forgotten that this represents less than one-half of 1% of the patients studied. Thus, clinical psychologists and psychiatrists who only see patients with psychiatric disorders are probably faced with a considerably higher risk of suicidality in their patients each month than this figure represents.

Nevertheless, patients taking AEDs should be monitored for the development of suicidality or depression, and patients and their families should be warned about this risk and urged to contact their healthcare provider if there are worsening signs of depression or suicidal thinking (U.S. Food and Drug Administration, 2008).

Compliance with Antiepileptic Drug Therapy

Lack of compliance in taking AEDs is a common problem in patients with epilepsy. It has been reported that 30%–50% of patients with epilepsy are nonadherent to their prescribed AED therapy when using large insurance claims databases (Rosenfeld et al., 2004). Although there may be many reasons why patients fail to take their medication as prescribed, some common reasons include cost, unpleasant side effects, and pregnancy. Unfortunately, noncompliance will often result in lack of seizure control which may lead to increased risk for physical injury (e.g., fractures, head injuries, burns) and psychosocial problems (e.g., reduced social interaction, loss of employment, loss of driver's license, depression).

In a recent study of 33,658 patients with epilepsy, 26% of patients were noncompliant with AED therapy, and this nonadherence was associated with a 3-fold increase in mortality risk compared to those who were adherent (Faught et al., 2008). Moreover, AED noncompliance was associated with other serious complications including an 86% higher incidence of hospitalization and a 50% higher incidence of emergency room visits. There were also significantly higher incidences of motor vehicle accident injuries and fractures.

Neuropsychologists are probably among the most sensitive of health care professionals to the adverse cognitive side effects of AEDs, but this study underscores the importance of medication compliance. Neuropsychologists should never advise their patients with epilepsy to reduce dosage or stop taking their AEDs because the threat of death and other problems related to uncontrolled seizures is of greater concern than unwanted medication side effects.

Ketogenic Diet

The ketogenic diet is a treatment option used primarily in infants and children who have difficulties controlling generalized seizures with AED therapy. The diet itself consists of foods high in fat and low in carbohydrates and low in protein. Patients are typically hospitalized and basically starved for several days before the diet proper is initiated. This period of fasting is conducted to create a state of bodily ketosis, and then this ketotic state is maintained by the diet composed of three or four parts fat to one part carbohydrate and protein (3:1 or 4:1 fat-to-nonfat ratio).

Mechanism of Action

The exact mechanism underlying how the ketogenic diet works is unknown, but it likely has to do with how ketone bodies alter brain metabolism. Ketone bodies are the major fuel for brain metabolism during starvation conditions, and ketone bodies alter brain metabolism and increase the brain's energy reserves (Nordli and DeVIvo, 2006). It is thought that maintaining a constant state of ketosis promotes neuronal stability by increasing cerebral energy reserves thereby raising the seizure threshold.

The Diet Itself

Meals must be scrupulously planned and measured out, and ketosis must be constantly monitored. Dietitians and other supporting professionals trained in ketogenic diet management should design, instruct, implement, and monitor the entire process. Continuing on the diet can be very demanding on both the patient and the family. Staying with the diet can be difficult since the child often feels hungry and the foods offered are frequently not ones most children seem to crave.

Sudden cessation or eating or drinking restricted foods may result in breakthrough seizures and even cause status epilepticus. The nutritional well-being of the child is of constant concern. The diet lacks certain required nutrients, and thus, is often supplemented with calcium, iron, folate, and multivitamins.

Adverse Effects of the Ketogenic Diet

Adverse side effects of the diet occur in about 10% of cases and may include hypoproteinemia, anemia, gastritis, colitis, abnormal liver function tests, and kidney stones (Thiele, 2003). The ketogenic diet also results in significantly reduced bone mass density; thus, these children regularly take vitamin D supplements. Children's growth and development must be regularly monitored, and caloric and nutritional adjustments should be made as appropriate.

Certain AEDs are avoided in patients on the diet since they can cause adverse effects. These drugs include the carbonic anhydrase inhibitors, such as topiramate and zonisamide, and valproate. Dose adjustments may need to be made in the child's prescribed AEDs as the diet can affect drug metabolism as well.

Efficacy of the Ketogenic Diet

The ketogenic diet has been shown to be a safe and effective treatment method for children with drug-resistant epilepsy. It is most frequently used to treat idiopathic or cryptogenic generalized epilepsies, and may be particularly effective in children with myoclonic seizures. Overall estimated results are that 20%–33% of patients become seizure free, approximately 30% have a greater than 90% reduction in seizure frequency, and perhaps one-third of patients do not respond or do not continue on the diet (Panayiotopoulos, 2005).

If the ketogenic diet is going to be effective in reducing or eliminating seizures, it should be evident fairly soon (within the first 1 or 2 weeks) after starting (Kossoff et al., 2008). If it is not going to be effective, this should be evident after about 2 months, and the diet can be discontinued. When the diet is successful, patients generally continue on it for 2 or 3 years, and then they are often gradually weaned off of the diet over the course of an additional year.

There have been few studies of the ketogenic diet in adults or among patients with complex partial seizures. From the little data that exist, the diet appears to work reasonably well with these two kinds of epilepsy patients as well. There is some preliminary information that the less radical Atkins diet, which also creates a ketotic state, may be somewhat efficacious in adults with epilepsy.

With regard to treatment of patients with complex partial seizures, Nordli and DeVivo (2006) conclude there is no compelling data to favor the use of the ketogenic diet over the newer AEDs or potentially curative epilepsy surgery. They suggest that children with refractory partial seizures should be evaluated to determine if they are candidates for focal resective epilepsy surgery, and if they are, surgery should not be delayed to wait for a trial of the ketogenic diet to finish. If, on the other hand, AEDs have failed, and the patient is determined to be a poor surgical candidate, then the ketogenic diet should be tried.

6

Neuropsychological Assessment in Epilepsy

Assessment of patients with epilepsy is in essence no different than the evaluation process conducted on patients with other chronic neurological or psychiatric conditions. The purpose of testing is typically guided by the referral question. Although no single pattern of neuropsychological test results can be used to diagnose epilepsy or to differentiate among the various seizure subtypes, there are many uses of the assessment results other than epilepsy diagnosis.

Similar to most assessments, medical records are reviewed, collateral information is gathered from family members and teachers, a clinical interview is conducted with the patient, and neuropsychological tests and questionnaires are administered. The overall goal of neuropsychological assessment in epilepsy is to establish a profile of the patient's strengths and weaknesses across multiple cognitive and behavioral domains and arrive at a neurobehavioral diagnosis that will help explain the patient's presenting complaints and provide information about the course and prognosis of deficits. The results may also be used to monitor the efficacy of medical treatment, signal the need for alternative or supplemental treatments, track deterioration over time, or assist in educational and vocational planning.

Selection of Neuropsychological Tests

Decisions as to which tests to use in the evaluation of general epilepsy patients will depend upon the specific referral question. In general, however, most epilepsy neuropsychologists will typically conduct a comprehensive examination that covers the major cognitive domains including intelligence, attention-concentration, learning and memory, language, visual-spatial and visuoperceptual abilities, executive functions, and sensorimotor skills. Depending upon the age of the patient, referral question, and presenting complaints, other domains may be included in the evaluation such as academic achievement, psychological-emotional functioning, adaptive behavior, or health-related quality of life (QOL). Reason for referral often depends upon the patient's age with academic, behavioral, or vocational planning concerns being prominent in children, and job performance concerns, cognitive deterioration, or disability determination being more common among adults. These commonly assessed domains in adults are presented along with some of the more frequently used tests in Table 6.1.

Table 6.1 Neuropsychological domains and commonly employed tests used in adults with epilepsy

Intelligence:
    Wechsler Adult Intelligence Scales (WAIS-III, WAIS-IV)
    Test of Nonverbal Intelligence (TONI-3)

Attention-Concentration:
    Digit span
    Spatial span
    Digit Symbol
    Paced Auditory Serial Addition task (PASAT)
    Trailmaking Test
    Continuous Performance tests (vigilance)
    Cancellation and visual search tasks

Learning and Memory:
    Wechsler Memory Scale (WMS-R, WMS-III, WMS-IV)
    Selective reminding tests
    California Verbal Learning Test (CVLT-II)
    Rey Auditory Verbal Learning Test
    Rey-Osterrieth Complex Figure Test
    Recognition Memory Test

Language:
    Boston Naming Test
    Boston Diagnostic Aphasia Examination (selective subtests)
    Multilingual Aphasia Examination (selected subtests)
        Controlled Oral Word Association
        Token Test

(continued)

Table 6.1 (Continued)

Visual-spatial:
- Judgment of Line Orientation
- Block design
- Rey-Osterrieth Complex Figure – copy

Visual-perceptual:
- Facial Recognition Test
- Visual Form Discrimination
- Hidden Figures/Overlapping Figures

Executive/Frontal:
- Wisconsin Card Sorting Test
- Delis-Kaplan Executive Function System (DKEFS)
- Stroop interference trial
- Mazes
- Word, animal, or figural fluency
- Trail B
- Behavior Rating Inventory of Executive Function (BRIEF-A)
- Frontal Systems Behavior Scale

Sensorimotor:
- Sensory-Perceptual Examination
- Left-right Orientation
- Tactile Form Perception
- Grooved Pegboard
- Finger tapping

Achievement Tests:
- Wide Range Achievement Test (WRAT-IV)
- Peabody Individual Achievement Test
- Woodcock-Johnson Psychoeducational Battery

Psychological-Emotional:
- Minnesota Multiphasic Personality Inventory
- Personality Assessment Inventory
- Beck Depression Inventory

Health-Related Quality of Life:
- Quality of Life in Epilepsy (QOLIE)
- Washington Psychosocial Seizure Inventory

Although many of the underlying assessment assumptions and basic approaches to neuropsychological testing in children and adults is similar, children present significant challenges primarily because cognitive and psychological maturation is a moving target that changes across developmental stages. Because of this, test selection must be tailored to the individual child based upon considerations such as age, developmental level, educational exposure, emotional stability, and neurological disease characteristics. Children are not simply little adults, and so the brain–behavior relationship assumptions typically applied in adults are often inappropriate for children. Despite this well-accepted dictum, many neuropsychological tests frequently used in pediatric populations are simply "dumbed down" versions of their adult counterparts, based on the localization of higher cortical functions in adults using the lesion method. Therefore, much work is still needed to design developmental stage–appropriate neuropsychological tests, in addition to gathering more normative test data that are appropriate for the variety of subgroups of children we are asked to evaluate. For a detailed overview of the commonly assessed neuropsychological domains, practical issues involved in pediatric assessment, and a comprehensive normative pediatric reference book, the reader is referred to Ida Sue Baron's *Neuropsychological Evaluation of the Child* (2004).

The neuropsychologist's task is to interpret the pattern of deficits across tests to understand the relative contributions of each of the seizure history variables to assist with treatment planning. Deficit patterns across neuropsychological tests can suggest particular sites of cerebral dysfunction, treatment effects, and various neurologic or psychologic processes that underlie the deficit pattern. The neuropsychologist attempts to integrate neuropsychological test data, history, clinical interview, behavioral observations, and available laboratory and radiological evidence into one cohesive summary report that arrives at a neurobehavioral diagnosis or summary, discusses the neurological and psychological implications, and informs professionals about follow-up management, potential educational strategies, and treatment.

Factors Contributing to Cognitive Decline in Epilepsy

The major factors to be considered in how epilepsy impacts test performance include the etiology of the seizure disorder, location and extent of the lesion causing epilepsy, seizure frequency, age of seizure onset, and seizure type. Other factors that also may affect test results, such as medication adverse effects and the effects of subclinical epileptiform discharges on testing, will be covered in the section on confounding factors in test interpretation.

Etiology

The underlying neuropathological cause of epilepsy and its severity and location are probably the most important variables affecting test performance. This depends in large part upon the age of the patient when the seizures began. In early childhood, the most common causes of epilepsy that the neuropsychologist is likely to encounter include birth injury, perinatal anoxia, febrile convulsions, head trauma, and cortical dysgenesis. In later middle age (50–65 years old), the most likely cause of new onset epilepsy will be brain tumors or possibly leakage from some small vascular abnormality, such as an aneurysm or venous angioma. Seizures will most commonly begin after age 60 because of strokes or neurodegenerative conditions. The pattern of neuropsychological test results in these neurologic diseases will be very much the same as when these conditions exist in the absence of epilepsy. In other words, the disease makes more "noise" on neuropsychological testing than does the epilepsy or its treatment.

Location and Extent of Lesion

The location and extent of lesion(s) causing symptomatic epilepsy are also important moderating variables that may affect the type and severity of neuropsychological impairment. The expected pattern of deficits is related directly to what we know about the localization of the higher cortical functions in the cortex. A few of the more common signs and symptoms of focal epileptogenic lesions are described here.

The most common partial epilepsy in adults arises from the mesial temporal lobe structures, and as expected, is frequently associated with impairments in new learning and memory. Lesions in the right parietal lobe will often be associated with constructional difficulties and other deficits of spatial thinking, neglect phenomenon, or visual-perceptual impairment. Left parietal lobe lesions causing epilepsy may also cause higher cognitive impairments in language, reading, writing, calculation, or verbal intellectual functioning. Prefrontal lesions are frequently associated with executive dysfunction and may show deficits in card sorting, associative fluency, planning, mental flexibility, and complex problem solving. Precentral or premotor epileptic lesions may be associated with upper or lower extremity weakness or disruption of the integration of complex motor actions. Occipital lobe lesions are often accompanied by visual-perceptual defects or visual illusory phenomenon.

Seizure Frequency and Severity

The frequency, severity, and lifetime number of seizures has also been shown to influence neuropsychological test results. It is generally accepted that repeated seizures can lead to eventual neuronal death. Long-term temporal lobe epilepsy that has been refractory to conventional medical treatment has been associated with progressive hippocampal damage and consequent memory decline. Thus, the more seizures a patient has suffered over his lifetime, the greater the expected damage and likelihood of having neuropsychological deficits. Patients with primary tonic-clonic generalized seizure who have had more than 100 convulsive attacks during their life were more impaired neuropsychologically on the Halstead-Reitan battery than patients who had fewer seizures (Dodrill, 1986). Similar results have been found for patients who have experienced status epilepticus; the more episodes of status epilepticus, the worse the neuropsychological outcome. Irreversible cerebral damage with repetitive or prolonged seizures may be due to anoxia, lactic acidosis, or excessive action of excitatory neurotransmitters with release of free radicals (Meador, 2002). Cognitive decline may be seen even with less obvious or severe neurological insults. For example, neuropsychological deficits worsen over time in children who have poor seizure control.

Age of Onset and Duration of Seizure Disorder

Taken as a whole, the literature suggests an earlier age of seizure onset and longer duration of active epilepsy is associated with more severe neuropsychological impairments. The influence of these moderator variables are, however, not as strong as etiology, site and extent of lesion, and frequency and severity of seizures. It is difficult to separate the influence of some of these variables since the causes of epilepsy in newborns are among the most devastating insults to the brain, such as intracranial hemorrhage, large brain malformations, or lack of oxygen during birth, which can have enormous influence on future cognitive development and ultimate neuropsychological attainment. Nevertheless, as a general rule of thumb, the earlier in life seizures begin, the worse the neuropsychological outcome will be. Poor neuropsychological outcome may be related to progressive brain damage caused by continuing seizures over a period of years. For example, there is convincing evidence that pharmacoresistant temporal lobe epilepsy will result in progressive cortical atrophy in extratemporal brain regions if not treated surgically (Bernhardt et al., 2009).

Seizure Type

Although there have been some contradictory findings, studies overall suggest that adult patients with primary generalized tonic-clonic seizures show more cognitive deficits than patients with complex partial seizures, who in turn, show more impairment than patients with absence seizures. Moreover, patients who suffer from multiple seizure types are more impaired than those with a single seizure type. The other moderator variables discussed earlier usually have a greater impact on neuropsychological outcome than the type of seizure disorder, but there are some exceptions to this. For example, certain seizure types seen in early childhood, including infantile spasms, severe myoclonic seizures, Lennox-Gastaut syndrome, and epilepsy with continuous spike-waves during slow sleep (ECSWS) have poor prognosis for neuropsychological functioning later in life. In most cases, however, neuropsychologists will get better prognostic information from knowing about etiology, extent of lesion, medications, and seizure frequency than from information about seizure type.

In summary, the strongest predictor of cognitive deficits in epilepsy is the underlying neurological disease causing the seizures, and patients with symptomatic seizures show the most severe cognitive deficits. Epilepsy caused by primary progressive neurodegenerative disorders, such as Alzheimer disease, will show a pattern of global dementia on neuropsychological testing, whereas focal brain lesions will result in a more circumscribed pattern of neuropsychological impairment that depends upon the brain region affected. The majority of patients with idiopathic (i.e., genetic) seizures do not show any significant cognitive impairment. Finally, there is a greater likelihood of cognitive deficits on testing in patients with earlier age of onset, higher seizure frequency, history of multiple episodes of status epilepticus, longer duration of seizure disorder, more interictal electroencephalographic (EEG) abnormalities, and multiple seizure types.

Additional Considerations in Children

Most of the seizure history risk factors for the development of cognitive impairments cited in the previous section have been thought to hold for school-aged children, as well as adults. Thus, early age of seizure onset, higher seizure frequency (incomplete seizure control), more EEG abnormalities, polypharmacy, and generalized symptomatic epilepsy subtypes have been considered risk factors for neuropsychological impairment in children (e.g., Mandelbaum and Burack, 1997). A recent study of 282 children at the time of their first recognized seizure found that 27% of children who had only experienced a single seizure already showed neuropsychological deficits (Fastenau et al., 2009). A second unprovoked seizure was associated with impairments in attention-concentration, visuoconstruction, and executive functioning even in children who were not taking any antiepileptic medications. This suggests cognitive deficits are not limited to children with frequent seizures, generalized symptomatic epilepsy, early onsets, or who are on high doses of antiepileptic drugs (AEDs). Among children who had multiple risk factors (i.e., two or more seizures, currently taking an AED, having symptomatic or cryptogenic etiology, and exhibiting epileptiform activity on initial EEG), 40% showed cognitive deficits in attention-concentration, visuoconstruction, and executive functioning. The high frequency of neuropsychological impairments so early in the course of epilepsy indicates that epilepsy is not a static condition, but rather a chronic condition that continues to evolve over time due to genetic, physiological, and developmental factors.

When examining the seizure type associated with neuropsychological risk in children, the old finding in adults that generalized seizures are worse than localized (focal onset) seizures does not appear to hold. Similar to adults, however, symptomatic/cryptogenic seizure syndromes seem to be predictive of greater cognitive impairment than the idiopathic syndromes. Fastenau and colleagues (2009) found that 33% of children with symptomatic/cryptogenic epilepsy showed some form of cognitive impairment, whereas 22% of children with idiopathic epilepsy displayed deficits. Table 6.2 shows the percentage of

Table 6.2 Percentage (%) of children showing neuropsychological impairment (in at least one cognitive domain) by type of epilepsy syndrome

SEIZURE SYNDROME	% OF CHILDREN WITH DEFICIT
Generalized idiopathic absence	31.6%
Generalized idiopathic tonic-clonic	17.1%
Generalized symptomatic/cryptogenic	28.6%
Localization-related idiopathic	17.0%
Localization-related cryptogenic	31.8%
Localization-related symptomatic	37.5%

Adapted from Fastenau PS, Johnson CS, Perkins SM, Byars AW, deGrauw TJ, Austin JK, Dunn DW. Neuropsychological status at seizure onset in children: risk factors for early cognitive deficits. *Neurology* 2009;73:526–534.

children who exhibited a neuropsychological deficit (≥1.3 SDs below the sibling control group mean) in at least one cognitive domain by type of seizure syndrome. As may be seen in Table 6.2, children with localization-related symptomatic epilepsy showed the highest frequency of impairment, whereas localization-related idiopathic showed the least impairment. It is worth noting that, contrary to clinical lore, absence epilepsy is not a benign childhood condition since almost one-third of these children showed some type of cognitive deficit.

Confounding Factors in Test Interpretation

The evaluation of patients with seizure disorders may be complicated by a variety of factors specific to epilepsy, and the neuropsychologist must take these into account when interpreting results. A standard interpretation of the reason for failure of a specific test may often be incorrect for patients with epilepsy. Specific tests may be failed for any number of reasons and are not necessarily due to the most common cause of that specific test failure that is found in the literature. The major factors that may affect test performance include the underlying cause of the epilepsy in symptomatic cases, the epilepsy itself, adverse antiepileptic medication effects, and the effects of subclinical epileptiform discharges (often called transient cognitive impairment or TCI). Other confounding factors include motivation and secondary psychiatric and psychosocial influences. The psychological disorders in epilepsy will be covered in Chapter 7.

Medical Adverse Effects of Antiepileptic Drugs

Although negative side effects are less common in the newer antiepileptic drugs (AEDs), all AEDs may cause sedation, dizziness, and adverse cognitive and behavioral effects. The more medically oriented effects of AEDs are covered in Chapter 5. This discussion will focus on the impact of these drugs on testing. The common neuropsychological AED drug effects are generalized motor and mental slowing and inattention. Thus, the cognitive test domains most likely to be adversely affected by AEDs include attention, concentration, vigilance, learning, memory, mental processing speed, reaction time, and motor speed of the hands. Moreover, these generalized slowing effects may be expected to affect most all timed tests, such as the Performance subtests on the Wechsler Adult Intelligence Scale (WAIS). These adverse effects become worse with higher doses (which may cause toxic blood concentrations) and when more than one AED is being taken for seizure control (polypharmacy). Some of the neuropsychological measures that have been found to be the most sensitive to the adverse effects AEDs include Finger Tapping, Grooved Pegboard, reaction time tasks, Digit Symbol, Trailmaking A and B, Digit Span, Stroop test, Paced Auditory Serial Addition Task (PASAT), rapid visual scanning and search tests, and continuous performance tasks. The more common negative behavioral effects of AEDs include irritability, hyperactivity, aggressive behavior, and emotional lability. Also recall that the U.S. Food and Drug Administration (FDA) has issued a warning that AEDs can cause an increased risk of suicidality (see Chapter 5 for more details).

Transient Cognitive Impairment
Neuropsychologists need to be aware that subclinical epileptiform discharges are common occurrences in patients with epilepsy and that these brief epileptiform events can adversely affect test performance. Aarts and colleagues (1984) first proposed the term *transient cognitive impairment* (TCI) to denote a momentary disruption of cognitive functioning associated with subclinical epileptiform discharges. These subclinical discharges must last at least 3 seconds and cause short-term sensorimotor or cognitive changes without any overt manifestations of a typical seizure (Aldenkamp and Arends, 2004). Patients and bystanders are usually completely unaware of TCI events.

Tests of short-term memory, such as digit span or mental arithmetic, are among the tests most sensitive to the effects of TCI. Other tests commonly used to detect TCI effects include reaction time, finger tapping, continuous performance tests, and various WAIS subtests, although TCI could potentially affect almost any test performance depending upon the location of the epileptiform activity in the brain. Neuropsychological test performances have been shown to deteriorate as the amount of interictal epileptiform activity increases and as the discharges become more generalized and less focal. Neuropsychologists should not only consider the effects of focal lesions, medications, seizure severity and frequency, age of onset and seizure type, but must also consider the effects of subclinical epileptiform discharges when interpreting the test performances of patients with epilepsy.

Postictal Neuropsychological Assessment
Finally, the issue of how long to postpone neuropsychological test administration after a seizure is one of continuing debate. Some studies suggest testing may be resumed after only a brief, ~10–30 minute, rest period, whereas others

have suggested the postictal effects may last up to 24 hours or more. Seizure type is obviously an important consideration since a 5-second absence attack or simple partial seizure will have less physiological impact on the brain than a major prolonged generalized tonic-clonic convulsive seizure with a prolonged period of postictal confusion and somnolence. This issue most often arises for patients who have complex partial seizures with impairment of consciousness. Patients with recent complex partial seizures can produce patterns of cognitive test impairment that appear much more severe and widespread and may even show false localizing results.

Although this issue is important for all neuropsychological testing of epilepsy patients because recent seizure activity may affect the validity of test results, it is of particular importance in epilepsy surgery settings, where patients are tested in the inpatient epilepsy monitoring unit often after experiencing multiple seizures during AED withdrawal. At the Medical College of Georgia, we no longer conduct preoperative neuropsychological assessments in the epilepsy monitoring unit in an attempt to mitigate recent seizure effects on testing. In practice, a 24-hour interval between testing and the patient's last complex partial seizure is a good rule-of-thumb to follow. If this is not possible (either because testing must be conducted on the monitoring unit or because it's not practical for the patient to return at some future date), the validity of the neuropsychological test results should be interpreted with caution if the patient has experienced a recent seizure.

Cognitive Deficits in Epilepsy

The cognitive deficits seen across the major seizure history risk factor variables have been reviewed by Perrine, Gershengorn, and Brown (1991). These have been updated and slightly adapted and are summarized in Table 6.3.

As may be seen in Table 6.3, repeated generalized tonic-clonic seizures and history of status epilepticus have widespread general cognitive effects on measures of intelligence, attention, memory, executive functions, and visuomotor speed. Higher levels of AEDs have negative effects on Performance IQ, attention, memory, and visuomotor speed measures. Frequent ongoing interictal EEG abnormalities have been associated with similar cognitive deficits as AEDs; namely, poor performance on tests of attention, memory, Performance IQ, and visuomotor speed. The effects of focal brain lesions

Table 6.3 Cognitive deficits across various seizure history risk factor variables

	REPEATED GTCS	STATUS EPILEPTICUS	AED LEVELS	BRAIN LESION	EARLY ONSET	SEIZURE FREQUENCY	EEG ABNORMALITIES
Full-Scale IQ	+	+	+/-	+	+	+	+
Verbal IQ	+	+		+/-	+	+/-	
Performance IQ	+	+	+/-	+/-	+	+	+
Attention	+	+	+				+
Verbal Memory	+	+	+	LT		+	
Visual Memory	+	+		RT			
Anomia				LT			
Aprosodia				RT			
Executive/Frontal	+	+	+/-	F		+	+
Visuomotor Speed	+	+	+	+	+	+	+
Psychosensory		+		+/-	+	+	+

+, present; +/−, may be present; LT, left temporal lobe; RT, right temporal lobe; F, frontal lobe; GTCS, generalized tonic-clonic seizures
Adapted from Perrine K, Gershengorn J, Brown ER. Interictal neuropsychological function in epilepsy. In Devinsky O, Theodore WH (eds.), *Epilepsy and behavior*. New York: Wiley-Liss, 1991:181–193.

on cognitive function depend upon where in the brain they are located. Psychosensory deficits, such as agraphesthesia, astereognosis, errors during double simultaneous stimulation in multiple sensory modalities, and finger agnosia, have been associated with history of status epilepticus, early seizure onset, high seizure frequency, and EEG abnormalities.

Cognitive Effects of Recurrent Seizures

As reflected in the preceding sections, the cognitive deficits seen in patients with epilepsy are multifactorial. The most important variables predicting cognition are underlying pathology causing the seizures; regions of the brain affected; frequency, duration, and type of abnormal EEG activity; effects of AEDs; and emotional consequences of living with epilepsy. The effect of these seizure history risk factor variables on the major cognitive domains is presented in Table 6.3. Attempts to disentangle these various risk factors is particularly difficult when trying to determine whether or not repetitive seizures themselves can cause cognitive deterioration. Nevertheless, taken as a whole, the literature suggests that recurrent seizures probably cause at least some mild damage to the brain which will result in cognitive decline that becomes worse over time (Dodrill, 2004). Recurrent seizures seem to have mild, but cumulative deleterious effects on the brain most likely through neuronal metabolic and structural deterioration. In a recent prospective study, Baker and Taylor (2008) found that patients with newly diagnosed epilepsy showed significant deterioration on measures of psychomotor speed, memory, and mental flexibility after only 1 year. Studies with longer follow-up periods report declines in intellectual and other cognitive functions over time (e.g., Jokeit and Ebner, 1999).

As a corollary, evidence of brain structural abnormalities associated with recurrent seizures suggests that temporal lobe epilepsy may be a progressive neurological disorder (Cascino, 2009). For example, Bernhardt, Worsley, Kim, and colleagues (2009) found that progressive neocortical atrophy occurred in the frontocentral and parietal regions over a 2.5–year period in patients with longstanding (greater than 14 years) intractable temporal lobe epilepsy. The authors concluded this progressive atrophy was most likely due to seizure-induced damage secondary to excitotoxicity.

The remainder of this chapter will discuss the major cognitive deficits that have been found in patients with chronic epilepsy across each of the major cognitive domains typically included in comprehensive neuropsychological evaluations. Specifically, the more important findings for patients with epilepsy with regard to attention, memory, intellectual functions, language, visual-spatial and visual-perceptual functions, and frontal lobe and executive functions will be reviewed.

Attention

A number of different cognitive constructs involving attention reflect the widespread neural underpinnings of the attentional system in the brainstem, subcortical, and cortical regions. Selective or *focused attention* is the capacity to concentrate on a particular subset within a wider range of surrounding stimuli while ignoring the extraneous stimuli. This is probably the most commonly used meaning of attention and is measured by tasks such as digit and spatial span, Trailmaking test, and cancellation and visual search tasks. The attention, or immediate memory, span lasts 30 seconds without rehearsal and can hold approximately 7 (+/− 2) items in storage at one time. *Concentration* refers to a more sustained version of attention that requires more mental effort over time and is commonly measured by digit span backwards, mental arithmetic, or Trails B. *Vigilance* is the ability to sustain attention over longer periods of time (~5–10 minutes) and is typically measured using continuous performance tasks.

Attentional complaints are very common in patients with many different types of epilepsy. Since intact attention is required to adequately perform most cognitive tasks (including many tests of memory, language, intelligence, spatial reasoning, and executive functions), patients with selective attentional problems often complain of other cognitive deficits, such as forgetfulness, word-finding difficulty, or thinking inefficiencies. Attention-concentration and vigilance (along with visuomotor speed) are the most vulnerable cognitive domains to the adverse effects of AEDs (Meador, 2002).

Attention and visuomotor speed appear to worsen over the years in patients with temporal lobe epilepsy, and this deterioration has been associated with history of more tonic-clonic seizures, early age of onset, and longer duration of epilepsy (Paizzini et al., 2006). Patients with symptomatic epilepsy, in general, show greater attentional impairments than those with idiopathic or cryptogenic epilepsy, and children with epilepsy seem to display more attentional deficits than adults with epilepsy (Sanchez-Carpintero and Neville, 2003). Finally, children with benign childhood epilepsy with centrotemporal spikes (BCECTS), and children with so-called benign Rolandic epilepsy, show selective deficits of attention and concentration (Kavros et al., 2008). Thus, it would seem that epilepsy and its treatment with AEDs predispose children to develop attention problems that are likely to interfere with cognitive development and academic achievement.

Attention-deficit Hyperactivity Disorder
Although attention-deficit hyperactivity disorder (ADHD) is commonly seen in children with epilepsy it has not been clear whether it develops as a consequence of epilepsy or simply coexists with epilepsy as an unrelated disorder (Dunn et al., 2003). Regardless of etiology, it has been estimated that 30%–38% of children and adolescents with epilepsy meet diagnostic criteria for ADHD. Hermann and colleagues (2007) studied 75 children with new-onset idiopathic epilepsy and found that 31% of them met DSM-IV diagnostic criteria for ADHD as compared with 6% of healthy controls. In contrast to controls, most of the epileptic ADHD children had the inattentive subtype, and 82% displayed ADHD symptoms before they were diagnosed with epilepsy, which eliminates AED medication or psychosocial stress related to seizures as potential causes of the ADHD. Neuropsychological testing of the epilepsy/ADHD group revealed deficient motor, psychomotor speed, and executive functions, and parental report using the Behavior Rating Inventory of Executive Function indicated abnormalities in impulse control, attentional shifting, planning, initiation, and complex problem solving. These cognitive test results and parental ratings suggested possible frontal lobe dysfunction in these children. Attention-deficit hyperactivity disorder in childhood epilepsy was not associated with demographic factors, clinical epilepsy characteristics, or potential seizure risk factors during gestation and birth.

Quantitative MRI volumetrics found the epilepsy/ADHD group in this study had significantly larger frontal lobes and smaller brainstem volumes, while there were no group differences in any other brain region (Hermann, et al., 2007). More than 50% of children with ADHD and seizures had obtained an individual education plan (IEP) or other academic support service, whereas only 15% of children with seizures who did not have comorbid ADHD required such help. The authors concluded ADHD is a prevalent comorbidity in new onset idiopathic epilepsy that appears to be related to neurodevelopmental abnormalities in brain structure.

Some clinicians may be reluctant to prescribe amphetamine drug therapy for epileptic children with comorbid ADHD fearing that this class of drugs reduces seizure threshold. However, this is not the case as methylphenidate has been shown to be safe and effective for children who have epilepsy and ADHD (Sankar, 2004).

Memory

When neuropsychologists talk about the cognitive domain of memory, most are referring to what neurologists call *recent memory*. Recent memory may involve either episodic ("event memory") or semantic ("fact memory") information that is newly encoded ("learned") and then stored in a hippocampal-dependent manner for many months or a few years. With regard to the time-course of recent memory, it begins when the limits of the immediate memory system (i.e., 7 +/− 2 items over the course of ~30 seconds without rehearsal) have been reached. The boundary between recent and remote memory is unclear, but information is thought to gradually be transferred from the mesial temporal lobe/hippocampal memory structures to more permanent storage in the secondary association cortices of the modality involved, such that visual memories are thought to be stored in occipital secondary association areas, auditory memories in temporal lobe association cortex, olfactory memories in perirhinal regions, and so on.

The tasks used to measure recent memory typically consist of learning various verbal (e.g., word lists, paragraph length stories, word pair associations) or nonverbal (e.g., geometric figures, faces, complex scenes, spatial locations) stimuli and then evaluating the retention of these stimuli through either free recall or recognition after some period of delay. Common memory tests used in evaluating the recent memory functions of adult epilepsy patients are presented in Table 6.1. Recent memory may be impaired in epilepsy for a variety of reasons, which may include the underlying lesion that causes the epilepsy may disrupt memory brain subsystems, abnormal electrical discharges exemplifying the epilepsy itself may transiently interfere with learning or recall, or sedating or other adverse physiologic effects of AEDs may result in deficient memory test performance. Focal cortical excision of the epileptic focus, especially in the temporal lobe, may also cause recent memory deficits.

The specialized function of the hippocampal system for the acquisition of material-specific information into memory storage is well established. Medical conditions that disrupt left hippocampal systems, such as complex partial seizures, produce deficits in the ability to learn new verbal information, and less often, can cause smaller decrements in nonverbal learning capacity as well. Complex partial seizures arising from the right mesial temporal lobe including the hippocampus may less consistently cause mild decrements in visuospatial learning. Unfortunately, most tests of nonverbal visuospatial memory are poor predictors of right mesial temporal lobe dysfunction, and thus, nonverbal memory test failures should be interpreted cautiously with regard to their association with right temporal lobe dysfunction (e.g., Lee et al., 1989).

Although memory deficits in epilepsy patients are worse with earlier age of onset of seizures (<5 years old), longer duration of epilepsy (>30 years), more lifetime seizures ($N = \sim 50-100$), and adverse effects of AEDs (especially with polypharmacy), memory impairments are also found in newly diagnosed temporal lobe epilepsy patients, suggesting that poor memory cannot be solely attributed to medication effects or recurrent seizures. With regard to seizure onset location and seizure type, patients with idiopathic generalized tonic-clonic seizures may also show memory deficits. This suggests that seizures do not necessarily arise from the temporal lobes, or be complex partial in nature, to disrupt memory. Similarly, patients with complex partial seizures of frontal lobe origin also may show memory deficits, but there are qualitative differences in temporal and frontal epilepsy patients similar to that found in patients with other types of pathology in these areas. In general, frontal lobe seizure patients more often display erratic learning and delayed recall, which is thought to be secondary to poor organizational strategies during encoding and inconsistencies in plan of search through memory storage during retrieval. Recognition test paradigms are also thought to aid frontal lobe epilepsy patients with recall, thereby improving performance.

As with cognitive deficits in general, patients with symptomatic epilepsies have memory impairments more often than those with cryptogenic or idiopathic epilepsies. Among patients with symptomatic epilepsy, it has been reported that patients whose seizures arise secondary to traumatic brain injury and herpes simplex encephalitis show more severe impairments in new learning and memory than with other common etiologies (Jokeit and Schachter, 2004). Memory is also worse in epilepsy patients with intractable seizures and among those who have experienced status epilepticus (the more episodes, the worse the expected memory outcome). Thus, declines in progressive memory may be seen in patients with refractory, or prolonged, seizures due to permanent cerebral damage caused by repetitive or severe seizures.

In addition to the influence of these seizure history characteristics on memory performance, neuropsychologists should proceed with caution when inferring temporal lobe dysfunction on the basis of poor memory test performance, since there may be many other possible explanations for memory test failure in these patients. Probably the most common confound in memory test interpretation in localization-related epilepsy occurs in patients who have selective deficits of attention-concentration and vigilance secondary to medication side effects. Attentional deficits frequently interfere with normal memory performance within the context of normal hippocampal systems. Cognitive impairments in other cognitive domains, including language, visual-spatial functions, and frontal-executive abilities, may also disrupt various aspects of memory test performance. Many psychiatric conditions and symptoms may also interfere with new learning and memory, such as mood disorders, anxiety, disorganized or obsessional thinking, fatigue due to inadequate sleep, poor motivation, and secondary gain. Careful consideration should be given to the possible contributions of these potentially confounding factors in memory test interpretation because prognosis and recommendations for treatment differ for most of them.

Intelligence

Adults. Examination of full-scale IQ using the Wechsler scales has shown only limited intellectual decline, if any, over time in adults with epilepsy in contrast to other, more narrowly defined, cognitive constructs. This is not surprising given the multifactorial nature of both the adult and child versions of the Wechsler scales where the full-scale IQ ends up averaging across at least four very different cognitive factors; namely, verbal-conceptual factor, attention-concentration ('working memory,' "freedom from distractibility") factor, perceptual-constructional ("perceptual-organizational") factor, and a visuo-motor ("mental processing") speed factor. The Wechsler factor scores may provide useful information in the individual case. For example, the attention-concentration or working memory factor and many of the timed, novel problem-solving tasks comprising the perceptual-constructional and speed subtests are more vulnerable to high blood concentrations of AEDs, active neuropathological processes, recent flurry of seizures, and the effects of subclinical EEG spiking (transient cognitive impairment) than are the verbal subtests. Using the WAIS-III, patients who have undergone left temporal lobectomy show relative deficits on the verbal-conceptual subtests compared to the perceptual-constructional subtests. The WAIS Performance subtests are more sensitive to the effects of bilateral diffuse cerebral dysfunction and right hemisphere disease than are the verbal-conceptual factor subtests. It may be somewhat surprising then that patients who have undergone right anterior temporal lobectomy do not show any significant differences from healthy controls on any of the Wechsler factors.

The same seizure history variables that have been related to neuropsychological impairments in other cognitive domains have also been found to affect intellectual functioning. These have been detailed earlier in the chapter, but reiterating briefly, epilepsy patients who are most likely to show deficits in IQ have: (1) a known neuropathological cause for their seizures (i.e., symptomatic epilepsy), (2) an earlier age of onset, (3) a higher frequency of seizures, (4) a longer duration of epilepsy, and (5) AED polypharmacy.

Children. Cross-sectional group studies in children have generally reported greater intellectual impairment than prospective longitudinal studies. However, when following children over time, most studies report little or no decline in IQ over time; although there is a subgroup of 10%–25% of children (who possess many of the seizure variable risk factors just detailed) who do show clinically significant cognitive decline (Vingerhoets, 2008). There is a concern that longitudinal studies may not actually be measuring intellectual deterioration, but rather a failure to acquire intellectual skills at the same rate as their peers in IQ standardization samples. Children with newly diagnosed idiopathic and cryptogenic epilepsies do not differ from healthy control children in IQs despite displaying significantly worse scores in other cognitive and behavioral domains. Newly diagnosed children with idiopathic or cryptogenic epilepsy more often have histories of receiving special educational services in school and repeating a grade than control children. The observation that some cognitive and behavioral symptoms predate the onset of overt seizures and an epilepsy diagnosis raises the possibility that epilepsy may represent a more general underlying, progressive brain disorder of which seizures are only one manifestation.

Children with generalized symptomatic epilepsy are especially at risk for intellectual impairments. In the severe childhood-onset generalized epilepsies (either symptomatic or cryptogenic) such as West syndrome (infantile spasms) or Lennox-Gastaut syndrome, mental retardation is usually seen as part of the disorder. As mentioned earlier, about 10% of children with mental retardation also have epilepsy, and in children who have both mental retardation and cerebral palsy, about 50% will also have comorbid epilepsy.

Language

Speech and language disorders in epilepsy may be seen when the language zones in the dominant (usually left) hemisphere have been disrupted by some underlying structural brain lesion or by the effects of repetitive seizures.

Patients with focal-onset (localization-related) seizures in the dominant temporal or frontal lobes may show language deficits that are subtle. Frank aphasia in epilepsy is rare unless the aphasia and seizures are secondary to symptomatic causes, such as stoke, trauma, or rapid- growing tumors. Patients with complex partial seizures that involve the left temporal or frontal lobes may present with mild visual naming deficits (on tests such as Boston Naming or Multilingual Aphasia Examination [MAE] visual confrontation naming), poor verbal associative fluency (on MAE Controlled Oral Word Association or F-A-S tests), and less commonly, minor difficulties with aural comprehension (on tests such as the MAE Token Test). Fluency (i.e., number of words spoken in the typical or average utterance), repetition, and comprehension are usually normal in patients with complex partial seizures arising in the left frontotemporal regions. Subtle language problems in patients with dominant hemisphere seizures can interfere with new verbal learning and memory in some patients, although it seems clear that language deficits are not the sole cause of poor memory in most left temporal lobe epilepsy cases.

Intracarotid amobarbital (Wada) testing over the years has shown that early onset epilepsy, especially in the left temporal lobe, may result in reorganization of language functions in the brain. Repetitive seizures that begin before the age of 5 years have been associated with more frequent atypical language dominance. In these cases, some language functions may relocate in either the homologous regions of the right hemisphere or in areas surrounding the language zones in the left hemisphere. Intraoperative monitoring studies suggest that naming functions may reorganize anteriorly in the left frontal lobe, while phonemic decoding and aural comprehension, when affected, may migrate more posteriorly and superiorly (Ojemann et al., 1989). The details of exactly how, where, and when intracerebral language reorganization occurs is not precisely understood. From a practical perspective, however, the risk of causing an aphasia following left temporal lobectomy is thought to be reduced in patients with atypical language dominance.

When Wada testing or functional imaging suggest that language functions are co-located within the epileptogenic zone in epilepsy surgery patients, language mapping (either intraoperatively or extraoperatively) may be conducted to reduce the risk of permanent postoperative aphasia. Transient aphasia is common in the days immediately following a left anterior temporal lobectomy. We examined language on a bedside screening examination before, immediately after, and 1 year after left or right temporal lobectomy and found left temporal lobectomy patients showed impairments in visual naming, responsive (auditory) naming, associated fluency (letter and animal), reading, and writing within the first several days after surgery. These language problems had all resolved at 1 year post-surgery and some aspects of language even improved beyond preoperative levels (Loring et al., 1994). Improvement over preoperative levels may reflect the beneficial effects of seizure control, since patients generally do not begin to taper off their AEDs until at least 1 year after surgery. Although visual and auditory naming were normal on this bedside screening exam, studies using the Boston or MAE naming tests often show mild but persistent postoperative naming impairment after left temporal lobectomy.

Childhood Syndromes Affecting Language. Seizure activity has been implicated in causing language deficits in children with Landau-Kleffner syndrome (acquired epileptic aphasia) and in children with continuous spike-wave discharges during slow wave sleep (ECSWS). These syndromes have both generalized and focal features, and not surprisingly, it hasn't been determined whether seizure onset is focal or generalized. Landau-Kleffner children typically present with a gradual loss of the ability to use or understand spoken language despite normal hearing and normal prior development. The disorder is most commonly seen in males, modal age of onset is between 5 and 7 years old, and only about three-quarters have comorbid seizures. The cause of the disorder is unknown. The seizures and EEG abnormalities gradually disappear around the age of 15 years, and recovery begins to take place. The majority of these children are left with some residual cognitive or psychosocial impairment.

In children with continuous spike-wave discharges during slow wave sleep (ECSWS), the partial or generalized nocturnal seizures are continuous, and thus are a form of status epilepticus that eventually will produce brain damage and progressive global cognitive decline. Modal onset age is 4–5 years old; the seizures usually continue for 2–10 years and then remit in early or mid-adolescence. Although the etiology is unknown, there is no evidence for a genetic cause, and about one-third of these youngsters have MRI abnormalities, such as cortical atrophy, malformations of cortical development, or porencephaly. Some cognitive recovery may occur as the seizures and EEG abnormalities remit in the teenage years, but most ECSWS children will have permanent and severe neuropsychological deficits for the remainder of their lives.

Visual-Perceptual and Spatial Functions

Although there is little information about the effects of focal onset seizures in the posterior regions of the brain on visuoperceptive or spatial functions, the data that does exist suggests these functions are relatively resilient to seizures and seizure surgery. There is less information about focal surgical resection and perceptual-spatial functions because localization-related onset seizures in the parietal and occipital lobes are rare. Furthermore, the nature of the stimuli and task demands often recruit more widespread brain regions than verbal-linguistic tasks so that focal lesions tend to less severely affect spatial functions. Spatial reasoning tasks (such as copying figures, assembling two- or three-dimensional constructions, or judging the spatial orientation of lines) are often confounded by motor and mental processing speed factors. It is important to separate spatial functions from visual-perceptual ones (usually measured by facial discrimination or matching figures) because they depend upon different brain processing streams, are dissociable on tests, and have different localization implications.

Although some patients with right or left temporal lobe epilepsy may perform worse than normals on visuoperceptive or spatial tests, these poor performances tend to be quite mild. These deficits depend upon the etiology of the seizures and the severity of the epilepsy disorder. As an example, in a case of intractable right posterior parietal lobe seizures following a traumatic brain injury, a patient showed poorer performance on tasks requiring mental rotation after a right temporoparietal cortical resection. Postoperative functional MRI in this patient revealed reduced activation around the resection site and increased activation in the posterior regions of the contralateral hemisphere while performing spatial tasks (Zack et al., 2004). The authors proposed spatial thinking had shifted from the right to the left hemisphere before surgery, analogous to how language may shift to the right hemisphere following early left-hemisphere injury.

In summary, visuoperceptive and spatial functions appear to be only mildly affected in right hemisphere onset epilepsy patients if at all. Spatial impairments are more likely to be seen in patients with right parietal, and less often right frontal, lobe destructive lesions (such as stroke or infiltrative tumors) or as a more general effect of more severe epilepsy or adverse medication effects. Visuoperceptive and spatial deficits are not common in patients with right temporal lobe epilepsy.

Executive Functions

The so-called *executive cognitive functions* consist of a variety of difficult-to-measure skills that are necessary for oversight, or cognitive control, of all other mental functions. Executive functions, such as planning and mental flexibility, are used to ensure the individual's actions are in concert with her goals in a flexible and adaptable manner. Common tasks used to measure the executive functions include the Wisconsin Card Sorting Test, associative fluency or generativity tasks, Tower of London, Trails B and other alternation tasks, Porteus or Wechsler Intelligence Scale for Children (WISC) and other maze tasks, Digits backwards, mental arithmetic and other working memory tasks, and Stroop interference and other inhibitory control tasks. Many of these tasks have been included in a recent battery of frontal-executive tests called the Delis-Kaplan Executive Function System (D-KEFS). Taken together, these tests attempt to measure such constructs as planning, organization, flexibility, divergent thinking, set shifting, complex mental tracking, self-regulation, and suppression of an automatic or difficult to inhibit response. These tasks and abilities are thought to primarily reflect functions within the dorsolateral prefrontal cortex.

As expected, many patients with symptomatic, localization-related frontal lobe seizures affecting the dorsolateral prefrontal regions perform poorly on many of the so-called frontal lobe tests. These patients have been reported to show impaired planning, slowed concept formation and concept switching, reduced response inhibition, decreased resistance to cognitive interference, reduced mental flexibility, and increased impulsivity and perseverative behavior (Vingerhoets, 2008). Unfortunately, patients with other seizure types and onset locations have also been reported to show similar deficits. In addition, many patients with focal frontal lobe seizures originating either medially (e.g., supplementary motor area) or ventrally (e.g., orbitofrontal cortex) do not display executive dysfunction. For example, patients with idiopathic generalized epilepsy and those who have undergone anterior temporal lobectomy can perform deficiently on many of these executive function tests. Older children with idiopathic juvenile myoclonic epilepsy show relative deficits on executive tests measuring concept formation, planning, and organization (Devinsky et al., 1997). In the case of the generalized epilepsies, a good deal of electrophysiological and imaging data suggests that a frontal role in generation and maintenance of epileptic activity in these seizures affecting widespread areas of both hemispheres. In other words, there is an anterior preponderance of

EEG activity in generalized seizures, and over time this may affect executive functioning.

Beyond the variety of seizure types affecting dorsolateral frontal regions directly, the next most common type of seizure disorders that may exhibit executive dysfunction are complex partial seizures of temporal lobe origin. Perseverative responses on card sorting is reportedly fairly common among left temporal lobe seizure patients and very common among right temporal lobe seizure patients; Hermann, Wyler, and Richey (1988) found 39% of left, and 79% of right, temporal lobe seizure patients had 20 or more perseverative responses on the Wisconsin Card Sorting Test (WCST) before surgery. There were no differences, however, in the number of categories achieved on WCST between temporal lobe onset cases and controls. There is some evidence suggesting that patients with mesial temporal lobe sclerosis are more likely to display perseverative responses on card sorting. It has been suggested that executive dysfunction in anterior temporal lobe cases may be due to an enlarged functional deficit zone caused by the damaging effects of repetitive seizures on closely connected frontal regions.

In addition to the seizures themselves, treatment effects on executive function tests also need to be considered in interpretation. Antiepileptic drugs may affect certain executive functions, especially working memory and mental processing speed, which in turn, can have negative effects on tests in other cognitive domains such as learning and memory. For example in one head-to-head comparison, patients taking topiramate performed significantly poorer on tests measuring verbal fluency, planning, and working memory than patients taking lamotrigine (Kockelmann et al., 2004).

Epileptic foci located in other prefrontal regions other than the dorsolateral cortex, such as orbitofrontal and medial frontal areas, may give rise to a variety of behavioral and emotional signs and symptoms that formal neuropsychological assessment of frontal-executive functions cannot capture. Some seizures arising in the frontal lobes therefore may not affect executive functioning at all. Orbitofrontal epileptic lesions may result in disinhibition, poor emotional control, and problems shifting mental set, whereas medial frontal lesions often cause amotivation, apathy, and hypokinesia. Careful history taking, clinical interview with family members, and structured questionnaires for self and significant other ratings, such as the Behavior Rating Inventory of Executive Function (BRIEF or BREIF-A) or Frontal Systems Behavior Scale (FrSBe), may be useful in evaluating for signs of frontal lobe disease in the absence of executive dysfunction.

Childhood Learning Disabilities

Learning disabilities are commonly found in children with epilepsy. Studies have found that between 20% and 55% of children with seizure disorders meet various diagnostic criteria for learning disorders in at least one area of academic achievement. With regard to seizure type, several studies have reported higher rates of learning problems in children with partial seizure disorders compared to generalized epilepsy and in children with symptomatic (lesional) syndromes compared with idiopathic syndromes (e.g., Stores and Hart, 1976; Zelnik et al., 2001). Definitive information on the association of other seizure history variables is not currently available, but in general, children and adolescents with earlier age of onset, polytherapy, and poor seizure control appear to be at greater risk for the development of some form of learning disorder.

In a recent study of academic underachievement in children and adolescents with epilepsy, Fastenau, Shen, Dunn, and colleagues (2008) found almost one-half (48%) met criteria for some type of learning disability using an IQ-achievement test discrepancy score of at least 1 standard deviation (SD). Specific learning disorder subtypes obtained using this criteria was ~13% met criteria for a Reading Disorder, ~20% met criteria for a Mathematics Disorder, and ~37% met criteria for a Disorder of Written Expression. Children with earlier age of onset of epilepsy, generalized (nonabsence) seizures, and comorbid ADHD were at increased risk for reading and math learning disabilities. This and other studies clearly show that children with epilepsy are at increased risk for learning disabilities, and even those with controlled seizures and less severe epilepsy need to be screened for learning problems in the schools.

Quality-of-Life Assessment in Epilepsy

Quality-of-life (QOL) measurement in epilepsy is a relatively new pursuit designed to assess the psychosocial impact of seizures and treatment outcome on important aspects of patients' lives, instead of relying on seizure control as the sole measure of outcome. By assessing the impact of seizures across multiple dimensions of daily life, QOL instruments may be used to alert clinicians to early signs of physical or psychological problems and treatment efficacy, or to assist in the selection of appropriate patients for epilepsy surgery. Most of these instruments evaluate the self-rated effects of seizures on the physical, psychological, and social aspects of life.

Epilepsy can impact lives across multiple dimensions. The physical dimension may include the adverse health effects of epilepsy itself, the neurologic disease underlying it, or medications side effects. The psychological aspects include patients' adjustment to having a chronic disease, loss of control issues, and alterations of self-image in light of the social stigma epilepsy carries. Finally, the broadly defined social effects of epilepsy include altered interpersonal and family relations, educational difficulties, employment problems, and the economic impacts. Chaplin, Yepez, Shorovom, and colleagues (1990) conducted in-depth interviews with epilepsy patients to identify their perceptions of how the disease affected their lives; 21 major problem areas were found that significantly affected QOL. These included such things as impact of seizures, employment, relationships, medications, and feelings about self (Buelow and Ferrans, 2006).

Adult Quality of Life in Epilepsy Measures
The most commonly used QOL measure in epilepsy today is in the United States is probably the *Quality of Life in Epilepsy* (QOLIE) and its abbreviated and age-adjusted versions. The QOLIE is a health-related measure designed to survey patient perceptions of the impact of epilepsy and its treatment on their lives. The original instrument, the QOLIE-89, consisted of 17 QOL scales that used the RAND 36–item Health Survey (SF-36) as a generic core (Devinski et al., 1995). The SF-36 is a widely used general health status questionnaire used in medical outcome studies that assesses physical activities, limitations in social activities, limitation in usual role activities, energy/vitality, pain, general mental health, and general health perception. These core SF-36 health dimensions within the QOLIE-89 are applicable to other diseases, and thus may be analyzed separately to compare various aspects of QOL in epilepsy with ratings in other chronic medical conditions, such as heart disease, diabetes mellitus, multiple sclerosis, or asthma. Forty-eight epilepsy-specific items were added to the SF-36 to evaluate circumstances particular to epilepsy, such as cognitive functioning, driving, and seizure-specific health concerns.

There are four versions of the QOLIE. The original QOLIE is composed of 89 items that evaluate QOL issues in detail, takes about 25 minutes to complete, and is primarily used for research or at tertiary epilepsy surgery centers. The QOLIE-31 is composed of 31 items and takes about 15 minutes to complete; it was designed to be used when an in-depth clinical evaluation is needed. The QOLIE-10 is a 10–item questionnaire that only takes a few minutes to complete. It was developed to be used as a quick clinical screening device that could be filled out by patients in the waiting area. The fourth version (QOLIE-48–AD) is a 48–item, age-adapted questionnaire for use with

Table 6.4 The 17 test battery scales on the Quality of Life in Epilepsy (QOLIE-89) Inventory

QOLIE-89 SCALES
    Health Perception
    Seizure Worry
    Physical Function
    Medication Effects
    Role Limitations – Physical
    Role Limitation – Emotional
    Pain
    Emotional Well-Being
    Energy / Fatigue
    Attention / Concentration
    Memory
    Language
    Social Functioning
    Social Support
    Social Isolation
    Health Discouragement
    Overall Quality of Life

adolescents. It will briefly be discussed in the next section. The 17 QOL areas contained in the QOLIE are presented in Table 6.4.

The QOLIE-89 has been found to be sufficiently reliable for group comparisons, given that 16 of 17 QOLIE test battery scales had internal consistency reliability coefficients between 0.73 and 0.88 (Perrine, 1993). Test-retest reliability was also satisfactory, although Medication Effects varied considerably, thus suggesting adverse effects of AEDs may change fairly quickly over time. The Profile of Mood States (POMS) factor was significantly correlated with the cognitive scales, as well as with all other QOLIE scales except Physical Function and Pain. This suggests that a patient's mood state may be one of the most powerful predictors of QOL perception. Thus, neuropsychologists should take mood state into consideration whenever they are evaluating self-reported QOL on instruments, such as the QOLIE.

Another QOL inventory, the *Epilepsy Surgery Inventory-55* (ESI-55), was designed to examine QOL in patients who undergo epilepsy surgery. Similar to the QOLIE, the ESI-55 is composed of the SF-36 (Rand's generic health questionnaire) and supplemented with 19 items relevant to epilepsy (Vickrey, 1993). There are 12 epilepsy-specific items centering around cognitive functions, role limitations due to memory problems, and epilepsy-related health problems. The final seven items are more generic in nature dealing with health perceptions, role limitations, and overall QOL scales. Internal consistency reliability coefficients are satisfactory; all scales exceeded 0.70 except Social Function which was 0.68. POMS mood factor was associated with several scales and was negatively correlated with the ESI-55 Well-Being scale. Seizure-free patients rated themselves as having better QOL than those who continued to have seizures after surgery.

An older measure that is less commonly used to measure QOL is the *Washington Psychosocial Inventory* (WPSI). The WPSI is a 132–item, self-report questionnaire that included three validity scales (i.e., number of items left blank, Lie scale, and a Rare item scale) and seven psychosocial adjustments scales: (1) family background, (2) emotional adjustment, (3) interpersonal adjustment, (4) vocational adjustment, (5) financial status, (6) adjustment to seizures, and (7) medications/medical management (Dodrill et al., 1980). There is also a global psychosocial functioning score. Reliability has been established on both the adult and adolescent versions using both internal consistency and test-retest methods. Criticisms of the instrument have been that the validity scales seem to be overly sensitive, resulting in many invalid WPSI profiles, and the test tends to overpathologize epilepsy patients as having more mental health or emotional issues than they actual do.

Pediatric Quality of Life in Epilepsy Measures

The *Quality of Life in Childhood Epilepsy Questionnaire* (QOLCE) was originally developed in Australia, but has since been revised and validated for use with North American children. The U.S. version (USQOLCE) contains 79 items that are completed by the child's parents and yields 16 subscales covering five domains of life: (1) physical function, (2) social function, (3) cognition, (4) emotional, and (5) behavioral well-being (Sabaz et al., 2003). The questionnaire is suitable for children ages 4–18 years. The 16 QOL subscales are similar to other epilepsy QOL scales used in adults, with a strong focus on neuropsychological and behavioral/emotional issues. The USQOLCE includes the following subscales: (1) physical restrictions, (2) energy/fatigue, (3) attention/concentration, (4) memory, (5) language, (6) other cognitive, (7) depression, (8) anxiety, (9) control/helplessness, (10) self-esteem, (11) social interactions, (12) social activities, (13) stigma, (14) behavior, (15) general health, (16) QOL, and an overall QOL score. The USQOLCE has been shown to be a reliable instrument, with internal consistencies correlations ranging from 0.76 and 0.97, and validity has been established by showing close associations to the Child Health Questionnaire. The USQOLCE also shows the expected directional changes across seizure severity variables and with seizure control after surgery (Sabaz et al., 2006).

The *Impact of Pediatric Illness Scale* (IPES) is a brief, 11–item QOL questionnaire developed for children ages 2 through 18 years old (Camfield et al., 2001). The 11–item questionnaire is completed by the child's parent and takes about 3 minutes to complete. The content is geared toward the psychosocial impact of epilepsy on the child and his or her family. The 11 items are: (1) overall health, (2) relationship with parents, (3) relationship with siblings, (4) relationship between you and your spouse/partner, (5) relationship with friends/peers, (6) acceptability to others, (7) number of activities, (8) school/academics, (9) child's self-esteem, (10) your loss of original hopes for your child, and (11) family activities. Internal consistency reliability of the items is generally good, ranging from 0.48 to 0.78. Test-retest reliability for the total score was consistent over time; Pearson's correlation was 0.81. Validity of the scale was determined by comparing the IPES with externally relevant measures including parents' ratings on the Family Enrichment Scale and Parenting Stress Index, school-aged children's ratings on the Piers-Harris Children's Self-Concept Scale, Brother-Sister Questionnaire, and teacher rating using the Academic Performance Rating Scales. Internal validation on the IPES yielded a Cronbach's alpha for the scale of 0.92. Significant correlations also indicated QOL was negatively related to impact on each of the eleven items. Children with epilepsy who achieved higher total impact scores on the IPES had significantly more behavioral, cognitive, and neurological problems, as well as an earlier age of onset of epilepsy, higher seizure frequency, more medication use, and more physician visits.

An English QOL instrument was recently developed for use among children and young adults with both epilepsy and learning disabilities (Buck et al., 2007). The *Epilepsy and Learning Disabilities Quality of Life* scale (ELDQOL) consists of 70 items dealing with the following 12 subscales: (1) seizure severity, (2) seizure-related injuries, (3) antiepileptic drug side effects, (4) behavior, (5) mood, (6) physical, (7) cognitive, and (8) social functioning, (9) parental concern, (10) communication, (11) overall QOL, and (12) overall health. The questionnaire takes about 20 minutes to complete and has been validated on children and young adults ages 18 months to 21 years. The internal consistency of the Behavior, Mood, Seizure Severity, and Medication Side Effects subscales was good, ranging from 0.74 to 0.95, and the test-retest reliability of each subscale was also high, ranging from 0.80 to 0.96. Evidence for construct validity was shown by moderate to high correlations between all four subscales and the Emotional Impact and Families Activities scales of the Child Health Questionnaire. Greater perceived severity of epilepsy and its impact was significantly related to higher scores on each subscale, and disability level was significantly correlated with higher scores on the Seizure Severity and Behavior subscales.

Adolescent Quality of Life in Epilepsy Measures
Adolescence is a particularly turbulent transitional time of life where puberty emerges and identity formation is beginning to solidify. It is a difficult period even for healthy teenagers, but adolescents with epilepsy have a higher frequency of behavioral problems than teens without epilepsy. The cognitive, behavioral, and social problems of adolescents stem from multiple causes, including psychosocial, family, biological, and medication factors, and so the *Quality of Life in Epilepsy Inventory for Adolescents* (QOLIE-AD-48) was developed to comprehensively measure these factors in health-related QOL (Devinski et al., 1999). The QOLIE-AD-48 consists of 48 items sampling eight subscales: (1) Epilepsy Impact, (2) Memory/Concentration, (3) Attitudes Toward Epilepsy, (4) Physical Functioning, (5) Stigma, (6) Social Support, (7) School Behavior, and (8) Health Perceptions, as well as a Total Score (Overall HRQOL). It is a self-report, self-administered questionnaire designed for use with adolescents ages 11–17 years.

Internal consistency reliability is reasonable with all subscales above the conventional standard of 0.70, ranging from 0.73 to 0.94, except for Health Perceptions, which had a Cronbach alpha of 0.52 (Cramer et al., 1999). Test-retest reliability was 0.83 for the summary (total) score (Overall HRQOL). QOLIE-AD-48 scores significantly worsened (decreased) as seizure severity increased. Construct validity was evaluated by measuring correlations between the summary score and other external measures of self-esteem and self-efficacy. The Overall HRQOL summary score was significantly associated with both the self-esteem scale ($r = 0.65$) and the self-efficacy scale ($r = 0.54$). Furthermore, there were similar QOL perceptions between a parent rating scale and the QOLIE-AD-48 ($r = 0.67$). Thus, there is evidence the QOLIE-AD-48 is a reliable and valid measure of health-related QOL for adolescents with epilepsy.

Driving Issues in Epilepsy

Driving is vital to leading an independent life in the United States and has significant influence on the social and vocational aspects of epilepsy patients' lives. It is estimated that the risk for any type of car crash is about two times higher for individuals with epilepsy than for the general population; however, only a minority of motor vehicle accidents where the driver has epilepsy are caused by seizures. In one study, only 11% of motor vehicle accidents involving people with epilepsy were due to seizures (Hansotia and Broste, 1991). Thus, although there is little doubt that epilepsy poses some driving risk, that risk appears to be relatively small. Because of this small increased risk, all states have laws governing epilepsy and driving. Unfortunately, there are a variety of state regulations that are difficult to keep track of which are based more on expert opinion and politics than on scientific information.

Seizure-free Period Requirements

The best predictor of risk for driving with epilepsy is the length of seizure-free interval, and thus, the core of most state regulations centers on seizure control. Every state in the United States allows individuals with controlled seizures to drive; they differ, however, with regard to the length of the seizure-free interval required and the exceptions allowed. The seizure-free period required to be legally allowed to drive in the United States ranges from 3 months to 12 months. The consensus statement from the American Academy of Neurology (AAN), American Epilepsy Society (AES), and the Epilepsy Foundation of America (EFA) recommends a 3–month seizure-free interval (AAN, AES, and EF, 1994). Noncompliance with long periods of driving restriction is common, and a recommended shorter seizure-free requirement is expected to improve the likelihood of compliance. Most states currently require either a 6– or 3–month seizure-free period before driving privileges are reinstated. There is evidence that states with 3–month seizure-free intervals do not have significantly higher rates of motor vehicle accidents and deaths from seizures as compared to states that have 6– or 12–month seizure-free requirements.

As of November 2008, only six states require physicians to report the names of patients with seizures to their state motor vehicle departments. These six states are California, Delaware, Nevada, New Jersey, Oregon, and Pennsylvania. Most physician groups are opposed to mandatory physician reporting because they are concerned that such a rule would discourage patients from reporting their seizures to doctors (Krumholz, 2009). There is some evidence that mandatory physician reporting of seizures does not reduce the crash rates of patients with epilepsy. All states within the United States currently require patients with epilepsy to self-report seizures to their state motor vehicle departments, although noncompliance with these rules is probably quite high. The seizure-free period, mandatory physician reporting, and other regulations for all 50 states in the United States may be found online by searching the State Driving Laws Database maintained by the EFA (http://www.epilepsyfoundation.org/living/wellness/transportation/driverlicensing.cfm).

Exceptions to Seizure-free Period Requirements
Although the seizure-free interval is probably the most important consideration in driving with epilepsy, there are other mitigating factors that may contribute to the decision about driving. The AAN, AES, and EFA (1994) consensus statement regarding epilepsy and driving recommended several modifiers or exceptions to the proposed 3–month no driving rule. The suggested modifiers for shortening the 3–month rule include patients who have seizures with consistent and prolonged auras warning them of an impeding event, simple seizures that do not interfere with consciousness or motor function, seizures during medically directed changes in medication, patients with an established pattern of pure nocturnal seizures, and nonepileptic seizures from physiological causes, such as toxic or metabolic illnesses, that are not likely to recur. Changes or discontinuation of AEDs can increase the risk of seizures and most physicians advise patients not to drive or to limit their driving while tapering off their medications.

In addition to the modifiers that may shorten the 3–month rule, the consensus statement also proposed several modifiers that are unfavorable and could lengthen the 3–month period of driving restriction. Some of the more important factors recommended for consideration when advising patients not to drive include history of noncompliance with medications and subsequent breakthrough seizures, structural brain disease, uncontrollable brain functional or metabolic disorder, recent history of alcohol or drug abuse, frequent seizure recurrences after seizure-free intervals, and prior crashes caused by seizures.

Regardless of whether the neuropsychologist works in one of the six states requiring physician reporting, it is suggested that neuropsychologists advise patients with breakthrough seizures to not drive and make certain the physician managing the patient's epilepsy knows about the seizure events. As we learn more about the mitigating factors influencing safety in driving among patients with epilepsy, it is likely that state regulations will become more uniform and fair in balancing the demands of public safety with the rights of epilepsy patients to live independently.

Case Example

Preoperative Neuropsychological Test Results in Left Mesial Temporal Lobe Seizure Onset

Patient History:

The patient is a 28–year-old, right-handed white woman with 11 years of education. She had managed a movie theatre, but recurrent seizures forced her to quit 7 years ago, and she has been receiving disability for intractable seizures since that time. She developed seizures when she was 14 years old. There is a history of febrile seizures during infancy, but no other etiologic factors were identified. Seizures occur in clusters of 6–12 over the course of 1–2 days approximately once per month. They are related to onset of her menstrual period.

Seizures begin with an aura of a "confused" feeling. She then develops a blank stare and is unresponsive to environmental stimuli. She often grunts or moans and has lip smacking, drooling, right arm stiffening, and eyes looking towards the right. Sometimes the head will deviate toward the right side as well. She has no memory for the event. Heat may precipitate seizures. Previous antiepileptic drug therapy, including phenytoin, phenobarbital, carbamazepine, topiramate, gabapentin, and oxcarbazepine, have not controlled the seizures.

Multiple neurological examinations have been essentially normal. Continuous video-EEG recording as an inpatient captured multiple habitual seizures; all ictal activity emerged from the anterior portions of the left temporal lobe.

Neuropsychological Test Results:

The following interpretive guidelines are offered to assist with interpretation of the scores below:

SS	Standard Score ($M=100$, $SD=15$); SS ≤ 75 indicates potential impairment.
ss	scaled score ($M=10$, $SD=3$); ss ≤ 5 indicates potential impairment.
z	score ($M=0$, $SD=1$); $z \leq 1.6$ indicates potential impairment.
T	T score ($M=50$, $SD=10$); $T \leq 34$ indicates potential impairment.
percentile	Percentile ranks (M = 50th, -1 SD = 16th, $+1$ SD = 84th, Impaired $<5^{th}$).

General Intellectual Functioning:

Verbal Comprehension Index (SS) .... 72, 3rd percentile
    Vocabulary (ss) 6
    Similarities (ss) 4
    Information (ss) 5

Working Memory Index (SS) .... 75, 5th percentile
    Arithmetic (ss) 5
    Digit Span (ss) 6
    Letter-Number Sequencing (ss) 7

Perceptual Organization Index (SS) .... 99, 47th percentile
    Picture Arrangement (ss) 10
    Block Design (ss) 10
    Matrix Reasoning (ss) 9
    Picture Completion (ss) 11
Processing Speed Index (SS) .... 86, 18th percentile
    Digit Symbol (ss) 7
    Symbol Search (ss) 8

Language:

Boston Naming Test
    Raw 34/60
    W/ Semantic cueing .... 34/60, 2nd percentile
    W/ Phonemic cueing 49/60
Letter Fluency (Three 60-second trials)
    Corrected raw score .... 16, <1st percentile
MAE Token Test .... 44/44 correct, 82nd percentile
Sentence Repetition Test
    Corrected raw score .... 11/14 correct, 25–43rd percentile
WRAT-4 Reading (SS) .... 73, 4th percentile
WRAT-4 Spelling (SS) .... 69, 2nd percentile

Attention-Concentration:

Digit Forwards .... Forward Span = 5, 3rd percentile
Digit Backwards .... Backward Span = 3, 4th percentile
WMS-III Mental Control .... 21/40, 25th percentile
Letter-Number Sequencing (ss) ..... 7, 16th percentile
WMS-III Spatial Span (ss) .... 5, 5th percentile

Trailmaking Part A . . . . 26 seconds, 0 errors, 47th percentile
Trailmaking Part B . . . . 109 seconds, 1 error, 8th percentile
Gordon Diagnostic
Vigilance Test
    Correct . . . . 30/30 70nd percentile
    Total Commissions . . . . 0, 68th percentile

Verbal Learning and Memory:

Verbal Auditory Memory – Immediate (SS) . . . . 68, 2nd percentile
    Logical Memory I (ss). 7
    Verbal Paired Associates I (ss) 2
Verbal Auditory Memory – Delayed (SS) . . . . 77, 6th percentile
    Logical Memory II (ss) 9
    Verbal Paired Associates II (ss). 3
Verbal Auditory Memory – Del. Recog. (SS).80,. . . . 9th percentile
Selective Reminding (CTLR) . . . . 31/72, 8th percentile
30' Delayed Free Recall . . . . 5/12, <1st percentile
30' Delayed Free Recognition. . . . . 7/12, Abnormal

NonVerbal Learning and Memory:

Visual Memory – Immediate (SS) . . . . 88, 21st percentile
    Faces I (ss) 8
    Family Pictures I (ss) 8
Visual Memory – Delayed (SS) . . . . 97, 42nd percentile
    Faces II (ss) 10
    Family Pictures II (ss) 9
Rey Complex Figure 3' Immediate Recall . . . . . .21/36, 18th percentile
Rey Complex 3' Figure Delayed Recall . . . . .19.5/36, 10th percentile

Visuoperception and Spatial Thinking:

Facial RecognitionTest (WCST) . . . . 49/54, 85th percentile
Complex Figure Copy . . . . 36/36, >16th percentile
WAIS-III Block Design (ss) . . . . 10, 50th percentile
Judgement of Line Orientation . . . . 20/30, 9th percentile

Excecutive Functioning:

Wisconsin Card Sorting Test (WCST) . . . 6 categories, >16th percentile
    % Perseverative Responses . . . . 16, 45th percentile

% Perseverative Errors .... 14, 45th percentile
WISC-III Mazes (RS) .... 27/28, 95th percentile

BRIEF-A Score Summary (Higher scores, T ≥65, indicate worse functioning):

	T-Scores
General Executive Composite	52
Behavioral Regulation Index	60
Inhibitory Control	60
Shifting	46
Emotional Control	63
Self-Monitor	61
Metacognition	46
Initiation	42
Working Memory	58
Plan/Organize	45
Task Monitor	49
Organization of Materials	44

Motor:

Grooved Pegboard
Dominant (Right) .... 70 seconds, 9th percentile
Nondominant (Left) .... 80 seconds, 4th percentile

Finger Tapping
Dominant (Right) .... 45, 49th percentile
Nondominant (Left) .... 40, 50th percentile

Gordon Diagnostic System Vigilance
Reaction Time .... 350 msec, 64th percentile

Beck Depression Inventory – II:
A score of 42 was obtained, which suggests severe symptoms of subjective depression.

Healthy-Related Quality of Life in Epilepsy (Lower scores indicate worse functioning.

	T scores		T scores
Health Perceptions	32*	Attention/Concentration	37
Overall Quality of Life	38	Health Discouragement	36
Physical Function	27*	Seizure Worry	32*

Role Limitations-
   Physical 36
Role Limitations-
   Emotional. 36
Pain 40
Work/Driving/Social
   Function 33*
Energy/Fatigue 31*
Emotional Well-Being. 23*

Memory 58
Language 43
Medication Effects 44
Social Support 37
Social Isolation 39

Impressions:
1. There were significant impairments across most verbal-linguistic functions. Specifically, there were relative impairments in verbal IQ, visual naming, associative verbal fluency, verbal new learning and memory, reading, and spelling. There were also weaknesses of attention-concentration and mental processing speed. Remainder of the cognitive exam was essentially normal within demographic expectations.
2. The patient reported significant symptoms of depression. Health-related quality of life was rated as being deficient across multiple functional areas, particularly in emotional well-being, physical functioning, energy/fatigue, and seizure worry.

Comments:
With regard to localization, this pattern of neuropsychological test results clearly suggests left hemisphere dysfunction. There was a significant asymmetry in memory function characterized by defective verbal, and normal nonverbal, memory which in turn suggests left temporal lobe dysfunction. There were no strong signs of frontal lobe dysfunction.

The attention-concentration and processing speed deficits are most likely due to medication side effects in conjunction with the mood disorder. The patient is being followed and treated for her mood disorder by a psychiatrist (although admittedly it does not appear to have been very successful to date).

7

Psychological and Psychiatric Disorders in Epilepsy

Psychological and behavioral alterations have been associated with epilepsy since antiquity. For most of recorded history, epilepsy was considered a psychological disorder with a functional rather than a neurobiological cause. It wasn't until the late 19th century that Hughlings, Jackson, and Gowers first proposed a physiological etiology for epilepsy. Later, as an unfortunate consequence of the split between neurology and psychiatry, research on the relationship between epilepsy and psychopathology separated along either biological or psychological lines. The biological revolution in psychiatry revitalized research into the neurological underpinnings of the psychological and emotional changes in epilepsy. Because this renewal is relatively recent, much of the research literature in this area is anecdotal and descriptive, and methodological difficulties abound that often prevent firm conclusions being reached. Psychiatric conditions and symptoms may occur *ictally* (during a seizure), *peri-ictally* (directly before or after a seizure), or *interictally* (during periods of no seizure activity). This chapter deals only with the interictal psychological disorders.

Some researchers have suggested epilepsy causes psychopathology by directly altering limbic neural circuitry. Some have concluded psychological problems emerge mainly as a consequence of difficulties adjusting to a chronic illness. Others have linked psychopathology with neurochemical abnormalities caused by the protracted use of antiepileptic drugs. Finally, another group of researchers have argued that epilepsy and psychopathology simply co-occur independently, with no causal relationship whatsoever existing between them. Currently, the etiology of psychiatric disorders in epilepsy is thought to be due to a multifactorial interaction of biological and psychosocial factors that may vary from case to case.

Risk Factors

Hermann and Whitman (1991) proposed three main categories of potential causes, or risk factors, for the development of psychological or emotional disorders in epilepsy: (1) neurobiological or brain-related factors, (2) psychosocial factors, or (3) iatrogenic or medication-related factors.

Examination of these factors in children (ages 6–16 years) with refractory seizures suggested psychosocial factors were far more important predictors of psychopathology and lack of social competence than were neurobiological or medication factors. The psychosocial risk factors with the strongest correlations with psychological problems were increased number of stressful life events, poor adjustment to epilepsy, less adequate financial status, and female gender. Other psychosocial factors that were less strongly, but nevertheless significantly, associated with psychopathology included amount of social support, increased perceived stigma, poor vocational adjustment, and external locus of control.

The neurobiological risk factors significantly associated with development of psychopathology included increased seizure frequency, multiple seizure types, early age of onset of seizures, and longer duration of epilepsy.

Other studies have added bilateral electroencephalogram (EEG) abnormalities and greater neuropsychological impairment (more global deficits) to this list. An increased number of behavior problems in children have been associated with poor seizure control, divorced or separated parents, and epilepsy of symptomatic etiology. The only medication factor that emerged as potentially predictive of psychopathology was monotherapy, which was positively correlated with social competence. Polytherapy was negatively related to social competence.

This chapter reviews the basics regarding the current state of knowledge about the interictal psychological disorders by type of psychopathology, beginning with mood disorders and then turning to anxiety disorders, psychotic disorders, and personality disorders. The diagnosis of these conditions is no different in epilepsy than it is in any other medical context, and they are, for the most part, thought to be comorbid with epilepsy and not caused by epilepsy. The next section of the chapter will present the current proposals regarding classification of epilepsy-specific psychiatric disorder; that is, those disorders thought to reflect the effects of ongoing epileptiform activity on the brain. Psychogenic nonepileptic seizures will be discussed in Chapter 8.

Mood Disorders

Mood disorders include depression and bipolar disorder. The depressive disorders are divided into three types in the *Diagnostic and Statistical Manual of Mental Disorders*, 4th Edition (DSM-IV): major depressive disorder, dysthymic disorder, and depressive disorder not otherwise specified.

Depression
Prevalence

Depression is the most common psychological disorder diagnosed in patients with epilepsy. Prevalence varies widely depending upon the sample studied, but it is estimated that depression occurs in 10%–20% of patients whose seizures are well-controlled and in 20%–60% of patients with intractable epilepsy (Schachter, 2006). Although depression is much more common in epilepsy than in the general population (usual estimates are between 5% and 17%), the rates of depression in epilepsy are no different than that found in most other chronic medical diseases, such as cardiac disease, asthma, or renal disease among others; or in most other neurological conditions, including traumatic brain injury, neuromuscular diseases, or multiple sclerosis. These studies suggest that depression is not directly caused by epilepsy itself, but is more likely a secondary functional adaptation to living with the consequences of a chronic illness. This is not to diminish the importance of recognizing and appropriately treating depression in epilepsy. In support of this, the suicide rate is 10 times higher in patients with epilepsy than in the general population, and it may be 25 times greater in patients with complex partial seizures of temporal lobe origin (Harden and Goldstein, 2002; Harris and Barraclough, 1997).

Symptoms

Most epilepsy patients with depression do not meet DSM-IV criteria for major depressive disorder. Rather they tend to present with a dysthymic form of depression with symptoms of irritability, anhedonia, and feelings of hopelessness that tend to wax and wane even over the course of several days. Dietrich Blumer proposed the term *interictal dysphoric disorder* be used to classify this form of depression, which has both depressive-somatoform and affective symptoms as follows:

Depressive-Somatoform Symptoms
 1. Depressed mood
 2. Anergia
 3. Pain
 4. Insomnia

Affective Symptoms
 1. Irritability
 2. Euphoric mood
 3. Fear
 4. Anxiety

This syndrome has been observed repeatedly in patients with epilepsy over the years. For example, Kanner and colleagues (2000) examined patients with refractory seizures and depression who were being treated with pharmacotherapy and found only 29% met DSM-IV criteria for major depression, whereas the remaining 71% failed to meet criteria for any of the DSM-IV categories. The patients that failed to meet DSM-IV diagnostic criteria had a clinical presentation with prominent anhedonia along with fatigue, anxiety, marked irritability, poor frustration tolerance, and mood lability with bouts of crying. Some also reported poor appetite and sleep and concentration difficulties. The authors concluded this symptom picture most closely resembled a dysthymic disorder even though it did not meet DSM-IV criteria for dysthymia.

It has been suggested that the presence of anhedonia is a superb marker for depression in patients with epilepsy (as well as in other medical illnesses) because it is relatively independent of physical complaints due to drugs or the underlying illness (Barry et al., 2001). They also suggest anhedonia is an excellent barometer of the severity of depression in the medically ill. Therefore, epilepsy patients who display little interest in their environment or in their interactions with others should be further assessed for a potentially treatable depression.

Pharmacologic Treatment

Antidepressant medications (AEDs) are the primary medical treatment for any of the depressive disorders seen in epilepsy, including dysthymia, interictal dysphoric disorder, or major depression. Medications that have been found to reduce seizure threshold in nonepileptics, such as clomipramine, amoxapine, maprotiline, and bupropion, are typically avoided. Pharmacotherapy with the selective serotonin reuptake inhibitor (SSRI) class of drugs has become the most popular form of drug therapy for epilepsy for several reasons.

The SSRI drugs are unlikely to increase seizure frequency or seizure severity, overdoses are unlikely to be fatal, and the SSRIs have a relatively favorable side effect profile compared with other classes of antidepressant drugs (Schachter, 2006). Moreover, the SSRIs are well-suited to the dysthymic-type symptoms commonly seen in epilepsy, including irritability and low frustration tolerance (Barry et al., 2001). The SSRIs also have minimal interactions with antiepileptic drugs (AEDs), although fluoxetine can result in increased blood serum levels of carbamazepine and phenytoin. Finally, it has been widely reported that the dysthymic-like depression seen in epilepsy responds well to low dose antidepressant pharmacotherapy. Sertraline is the most commonly cited SSRI used to treat depression in epilepsy.

Bipolar Disorder

Symptoms and Prevalence

Bipolar disorder is a cyclical illness with both manic and major depressive episodes. For diagnostic purposes, manic episodes need to endure for at least 1 week and consist of agitation, insomnia, hypersexuality, grandiosity, and sometimes, psychosis. Although euphoric mood is one of the symptoms sometimes seen in interictal dysphoric disorder, this often does not meet criteria for sustained mania required for a bipolar diagnosis. Prevalence of bipolar disorder in epilepsy is thought be around 8%; far less common than the depressive disorders. There have been several reports of cases or small samples with mania and epilepsy. Mania seems to be more common in patients with temporal lobe epilepsy than with other seizure types and has been described as being more common after right anterior temporal lobectomy.

Pharmacologic Treatment

The mood-stabilizing properties of the AEDs have been known for some time and the AEDs are often used to treat bipolar disorders in psychiatric patients without epilepsy. The AEDs most frequently used to treat bipolar disorder are valproate (36.8%), lamotrigine (15.4%), and carbamazepine (5.8%) (Ghaemi et al., 2006). There is good clinical evidence that these three AEDs are beneficial in maintaining remission in outpatient adults with a primary diagnosis of bipolar I disorder. These drugs all interfere with cell excitability in a manner similar to lithium, and thus, they are used as efficacious anti-mania drugs.

Anxiety Disorders

Anxiety is the feeling of apprehension that occurs in anticipation of some nonspecific future danger, accompanied by a feeling of dysphoria or somatic symptoms of tension. Although less well studied than depression, there is probably more evidence that both neurobiological and psychosocial factors influence the development of anxiety in epilepsy. Symptoms of anxiety, such as feelings of impending doom, autonomic arousal, and fear, are known to occur as direct physiological manifestations preictally or ictally when EEG changes involve the mesial temporal lobe structures and insula. Fear is a common aura, for example, in amygdala onset seizures. In terms of psychosocial risk factors, interictal anxiety may be a secondary emotional reaction to various aspects of having a seizure disorder, including fears about having a seizure, feeling of loss of control, risk of ictal injury or death, or the negative social consequences of having to live with epilepsy.

Although the DSM-IV defines approximately 14 different anxiety disorders, most are excluded if there is some underlying medical illness, such as epilepsy, causing the anxiety disorder. This presents some difficulty for the classification of anxiety disorders in epilepsy. Although most epilepsy patients do not meet full DSM-IV criteria for panic disorder, obsessive-compulsive disorder, or generalized anxiety disorder, many suffer from various combinations of symptoms of these conditions.

Prevalence

In outpatient and inpatient epilepsy series, prevalence of generally defined anxiety disorders has generally been estimated to be between 15% and 25% in patients with all types of epilepsy; which is considerably higher than in the general population, which is in the 2%–3% range (Scicutella, 2001). Prevalence does not appear to differ significantly across seizure types although comparisons have most often been made between complex partial and generalized seizure disorders. Moreover, interictal anxiety is not associated with seizure frequency, and somewhat surprisingly, new onset anxiety disorders have been reported after epilepsy surgery in which seizures were cured.

Symptoms by Subtype

Panic attacks

Panic attacks consist of a discrete period of intense fear or discomfort that is accompanied by at least 4 of 13 symptoms listed in the DSM-IV. The attack is characterized by a gradual onset of symptoms that typically build to peak intensity within 10 minutes or so and is often accompanied by an urge to escape. Panic disorder may be diagnosed with or without agoraphobia.

A number of symptoms overlap in complex partial seizures and panic attacks including fear, palpitations, sweating, trembling, shortness of breath, derealization, dizziness, fear of either loss of control or death, paresthesias, chills, and hot flashes. These overlapping symptoms of panic disorder and some complex partial seizures suggest the disorders may share similar underlying neural correlates, and the mesial temporal lobes, parahippocampal gyrus, and insular regions have all been implicated (Tucker and McDavid, 1997). Clinically, the two disorders may be distinguished by various features of history, symptoms, course, and response to treatment. These are presented in Table 7.1.

It should be recalled that complex partial seizures and panic disorder may coexist as two distinct conditions within the same individual and require different treatment approaches. Further complicating psychiatric diagnosis,

Table 7.1 Distinguishing features between complex partial seizures of temporal lobe origin and panic disorder (PD)

FEATURE	COMPLEX PARTIAL SEIZURES (CPS)	PANIC DISORDER (PD)
Age of onset	Any age	20–30 years old
Consciousness	Impaired	Preserved
Onset of symptoms	Sudden	Gradual
EEG	Abnormal	Normal
Duration of episode	Brief (30 sec–2 min)	Longer (~5–20 min)
Family history of PD	Uncommon	Common
Family history of CPS	Common	Uncommon
Hallucinations	Occasionally	Rarely
Aphasia	Sometimes	Never
Response to antiepileptic drugs	Good	Poor

some patients with epilepsy may develop agoraphobia or social phobia in the absence of panic attacks due to fears of having a seizure while being in unprotected places or social situations.

Obsessive-Compulsive Disorder

Obsessive-compulsive disorder (OCD) consists of persistent obsessions or compulsions that are excessive, unreasonable, and time consuming (>1 hour per day). *Obsessions* are recurrent and persistent thoughts, impulses, or images that are experienced as intrusive. The obsessions cause marked anxiety, and the person attempts to suppress or ignore the intrusive thoughts or images. *Compulsions* are repetitive behavioral or mental acts that the person feels compelled to perform in response to an obsession. Examples of compulsions include repetitive hand washing, checking, ordering, counting, or praying.

Although there has been speculation about a common underlying brain mechanism in OCD and epilepsy, especially in temporal lobe seizures, there is very little data to support such a relationship. These ideas arose because some symptoms of OCD may occur as seizure manifestations. For example, Penfield described a case in which the seizure aura consisted of forced thinking about a particular subject, and others have described compulsions to walk or move a particular limb or part of the face.

Monaco, Cavanna, Magli, and colleagues (2005) examined the association between epilepsy and OCD and found no increase in OCD among patients with idiopathic generalized epilepsy. In contrast, 14.5% of patients with temporal lobe epilepsy were identified as having clinically significant levels of obsessionality or compulsivity (but not necessarily meeting DSM-IV criteria for OCD) as compared with 1.2% of healthy controls. OCD risk in temporal lobe epilepsy was not associated with lateralization of the seizure focus, type of antiepileptic drug, duration of epilepsy, or seizure frequency. OCD was found to be more common among epilepsy patients with a previous history of depression.

Generalized Anxiety Disorder

The DSM-IV defines a *generalized anxiety disorder* (GAD) as consisting of excessive apprehension and worry about a number of activities that occurs frequently (almost daily for at least 6 months) and which the person has difficulty controlling. The person also experiences at least three of the following six somatic symptoms: (1) restlessness, (2) being easily fatigued, (3) diminished concentration, (4) irritability, (5) muscle tension, or (6) sleep disturbance. Schondienst and Reuber (2008) reported that only 12% of patients with medically refractory epilepsy were formally diagnosed with generalized anxiety disorder using the Structured Diagnostic Interview for the DSM-IV (SCID) and suggested that no neurobiological relationship between GAD and epilepsy has been established. It thus appears likely that when GAD is encountered in patients with epilepsy, it will most likely be secondary to emotional or psychological adjustment difficulties associated with living with a chronic illness.

Pharmacologic Treatment
There are no double-blind, placebo-controlled studies of the antianxiety agents in patients with epilepsy and comorbid anxiety disorders. Thus, current pharmacotherapy is guided by studies in psychiatric patients without concomitant seizure disorders.

The SSRI have been shown to provide good symptom relief in panic disorder and obsessive-compulsive disorder. The SSRIs are often recommended for patients with epilepsy because they are less epileptogenic than other antidepressant drugs. The SSRIs also lack anticholinergic side effects and do not produce cardiac arrhythmias, which make this class of drugs a good option for the treatment of anxiety in patients with epilepsy (Scicutella, 2001). Side effects of SSRIs include diarrhea, insomnia, and headaches. The only tricyclic antidepressant shown to be efficacious for treating panic disorders is imipramine. Tricyclics are not usually recommended in epilepsy because they can cause seizures at high doses, although they do not cause seizures at therapeutic doses.

Benzodiazepine medications, especially clonazepam and alprazolam, are useful for treating panic disorder and general anxiety disorder (Ballenger et al., 1988). These drugs may be particularly useful in epilepsy since the benzodiazepines are antiepileptic drugs, as well as antianxiety agents. Side effects of benzodiazepines include tolerance, dependence, sedation, ataxia, memory loss, and sometimes paradoxical disinhibition. Finally, AEDs have also been used for treating anxiety disorders. Valproic acid has reportedly improved symptoms of panic disorder. Carbamazepine was successful in several case studies with comorbid OCD and epilepsy. Oxcarbazepine has been shown to relieve panic symptoms in a patient who had generalized tonic clonic epilepsy (Schachter, 2008).

Psychotic Disorders

Similar to other psychiatric conditions, interictal psychosis in epilepsy may be caused by a variety of factors acting alone or in combination and may include a genetic predisposition for psychiatric problems, recent psychosocial stress;

long duration of temporal lobe seizures, and recent changes in treatment, seizure frequency, or EEG activity. As with the other psychiatric conditions seen in epilepsy, symptoms of psychosis may occur during a seizure (ictally), immediately before or after a seizure (peri-ictally), or during periods without EEG or clinical seizure manifestations (interictally).

Prevalence

Just as in the other psychiatric disorders, the frequency of psychosis in epilepsy is highest in inpatient psychiatric samples, and lowest in outpatient neurological samples. Interictal psychotic disorders are found in 7%–9% of community-dwelling patients with epilepsy (Mendez et al., 1993). Among psychiatrically hospitalized populations, the prevalence of psychosis in epilepsy rises to 19%–25% (Torta and Keller, 1999).

Symptoms

It has long been noted that there are substantial differences between the symptoms of schizophrenia in patients with and without epilepsy. Epileptic patients with interictal psychosis do not show a blunting or flattening of affect or the same degree of social withdrawal that is typically seen in nonepileptic, process schizophrenia. These symptomatic differences led to the use of the term, "schizophrenia-like psychosis," when psychosis was associated with epilepsy (Slater et al., 1963). Compared to patients with schizophrenia, patients with interictal psychosis display: (a) an absence of negative symptoms, (b) higher premorbid functioning, (c) better response to pharmacotherapy with antipsychotic drugs, (d) less deterioration of personality over time, and (e) better long-term prognosis.

Although many methodological difficulties interfere with accurate knowledge with regard to psychosis in epilepsy, there are several seizure history risk factors that repeatedly come up. With regard to type of epilepsy, interictal psychosis is far more common in complex partial seizures of temporal lobe origin. After combining data across multiple studies, Trimble (1991) found that 76% of all case series with epilepsy and psychosis suffered from temporal lobe epilepsy. The interval between age of epilepsy onset and age of onset of psychosis is generally reported to be between 11 and 15 years; which has caused some to hypothesize that a kindling-like mechanism may lead to psychosis over time in temporal lobe epilepsy.

Interictal psychosis is less frequent in patients with generalized epilepsy and neocortical extratemporal epilepsies than in those of mesial temporal lobe origin. Risk factors suggesting a more severe epileptic process have been linked to psychosis. These include a longer duration of epilepsy, multiple seizure types, history of status epilepticus, and poor response to drug treatment (Schmitz and Trimble, 2008). Interestingly, however, seizure frequency has not been associated with likelihood of psychosis. On the contrary, many have observed an inverse relationship between the frequency of seizures and presence of psychosis. This apparently paradoxical relationship has been called *forced normalization*. Forced normalization refers to the observation that when psychotic symptoms emerge, epileptiform EEG activity becomes more normal. Conversely, psychotic symptoms tend to resolve when the EEG and clinical manifestations of seizures return. Forced normalization has been estimated to operate in approximately 10% of patients with comorbid psychosis and epilepsy.

Pharmacologic Treatment
Most epilepsy patients who develop persistent interictal psychosis are treated with antipsychotic drugs, just as in nonepileptics. Medications that have been found to reduce seizure threshold in nonepileptics, such as clozapine, chlorpromazine, and loxapine, are normally avoided (Schachter, 2006).

The newer atypical antipsychotic drugs are most often recommended. Olanzapine and risperidone have been used most frequently in epilepsy because these drugs are less likely to exacerbate seizures (Schmitz and Trimble, 2008). The major side effects of olanzapine include sedation, somnolence, weight gain, and glucose intolerance, whereas risperidone has been shown to cause weight gain and extrapyramidal side effects. Extrapyramidal side effects are less common in olanzapine. Patient taking AEDs that increase hepatic metabolism, the so-called enzyme-inducing drugs, will show lower serum levels of antipsychotic medications, and thus, may require higher antipsychotic doses to obtain a positive clinical effect.

Personality Disorders

Prevalence
The Axis II personality disorders as defined by the DSM-IV have been much less frequently studied in epilepsy than the mood, anxiety, and psychotic disorders discussed earlier. Estimates of the prevalence of personality disorders in epilepsy generally range between 16% and 21% across all seizure types, which is slightly higher than the prevalence of personality disorders in

the general population (estimates range between 6% and 13%) (Swinkels et al., 2005). Axis II personality disorders are more rare in epilepsy than are Axis I disorders.

Symptoms

The DSM-IV classifies the personality disorders into three broad clusters:

1. Cluster A Personality Disorders: Paranoid, Schizoid, and Schizotypal
2. Cluster B Personality Disorders: Antisocial, Borderline, Histrionic, and Narcissistic
3. Cluster C Personality Disorders: Avoidant, Dependent, and Obsessive-Compulsive.

A predominance of cluster C personality disorders is found in patients with epilepsy as compared with cluster A and cluster B disorders; although this result is not specific to epilepsy since cluster C disorders are also most common in the general population.

There appear to be no significant associations between specific personality disorders and seizure type. For example, Lopez-Rodriquez, Altshuler, Kay, and colleagues (1999) found no difference in rate of Axis II disorders between patients with temporal lobe epilepsy and those with other types of seizure disorders. Moreover, no associations between localization of the epileptogenic zone and personality disorder traits have been found. Swinkels, Duijsens, and Spinhoven (2003) did report a modest relationship between cluster C (avoidant, dependent) personality traits and age of onset and duration of seizure disorder, seizure frequency, and number of AEDs. With regard to epilepsy surgery, patients with cluster A personality disorders appear to be at some risk for psychosis during video-EEG monitoring or immediately after surgery. There may be some intensification of symptoms in cluster B patients for a period of time after surgery, although there does not appear to be any increased risk for symptom exacerbation among cluster C patients. This is discussed in more detail in Chapter 11.

Similar to all forms of psychopathology in epilepsy, some hypothesize the psychiatric problems arise secondary to disrupted neural functions (neurobiological explanation), although others think they are related to the psychosocial consequences of living with epilepsy, a maladaptive reaction to a chronic medical condition (psychosocial explanations), or some combination of all of these. In some of the best-designed studies, the neurobiological consequences of epilepsy itself seem to carry little weight in causing psychiatric disorders. In a study of the prevalence of psychopathology in eight chronic medical conditions (i.e., chronic lung disease, diabetes mellitus, heart disease, hypertension, arthritis, physical handicap, cancer, and brain disorders including epilepsy and stroke) and healthy controls, Wells, Golding, and Burnam (1988) found recent and lifetime psychiatric disorders were more common in the medical conditions, but there were no significant differences when comparing any of the eight chronic illnesses. Thus, it appears that although there may be some contribution of neurobiological risk factors to the development of personality disorders, the strongest data suggest that the contributions of psychosocial and adjustment to chronic disease factors are probably more important.

Interictal Behavior Syndrome of Temporal Lobe Epilepsy

Waxman and Geschwind (1975) revitalized interest in the neurological underpinnings of personality traits and behavioral alterations in epilepsy by describing an interictal temporal lobe epilepsy syndrome consisting of hypergraphia, hypergraphia, deepened emotions, circumstantiality, religiosity, and alterations in sexual concerns. Bear and Fedio (1977) reviewed the interictal syndrome and expanded it to include 18 traits in all. Research into the 18-trait behavioral syndrome of temporal lobe epilepsy (TLE) has yielded mixed results. Although many of these traits have been noted for centuries (such as religiosity, irritability, viscosity, hyposexuality, and anger/aggression), the traits are not specific to temporal lobe epilepsy. All the so-called TLE traits have been found among patients with psychiatric disorders and other neurologic disorders, as well as in normals. This led Mungas (1982) to conclude the 18 traits represented nonspecific psychopathology, and as such, did not constitute a specific behavioral syndrome. Furthermore, the Bear-Fedio personality inventory is unable to distinguish TLE from other types of seizure disorders. Thus, although many health care professionals who have spent time with epilepsy patients will have encountered individuals with many of these personality traits, the traits are not specifically pathognomonic for TLE.

Epilepsy-specific Psychological Disorders

The ILAE Commission on Psychobiology of Epilepsy published its proposed classification of neuropsychiatric disorders in epilepsy in 2007 (Krishnamoorthy et. al, 2007). The need for some type of classification arose because evidence accumulated suggesting that psychiatric disorders within the context of epilepsy may be clinically distinct, and yet they are not found in any of the major psychiatric diagnostic classification systems, such as the DSM-IV or the ICD-10. It was hoped that by defining these disorders, future research could focus on learning more about etiology and developing specific therapeutic measures.

The majority of psychiatric diagnoses appear to simply coexist with epilepsy, and the commission suggested ignoring the presence of epilepsy when making these diagnoses. This would be the case for minor and major depression, anxiety and phobic disorders, OCD, bipolar disorder, and undifferentiated forms of schizophrenia. These comorbid psychological disorders should be classified using standardized criteria, without the implication that any neurobiological etiology is suggested by the presence of comorbid epilepsy.

Psychoses of Epilepsy
The ILAE Commission described three types of psychosis in epilepsy: (1) interictal psychosis of epilepsy, (2) alternative psychosis, and (3) postictal psychosis (Krishnamoorthy, et al., 2007). Since this chapter is devoted exclusively to interictal psychological disorders, the directly seizure-related ictal and postictal disorders will not be discussed.

Interictal Psychosis of Epilepsy
This condition was described in more detail earlier in the section Psychotic Disorders. This schizophrenic-like psychosis of epilepsy is a paranoid psychosis with strong affective components that may include auditory hallucinations (including command or third-person hallucinations) and preoccupation with religious themes. There is typically no affective flattening, and personality tends to be well-preserved with an absence of a deteriorating course.

Alternative Psychosis
Alternative psychosis is a psychosis with paranoid and affective features that alternates between periods with seizures and normal behavior, and times when seizures are absent and psychotic behaviors emerge. The behavioral disturbance is often accompanied by paradoxical normalization of the EEG (so-called "forced normalization"). The suggested diagnosis is made without reference to EEG findings. However, if the EEG findings are available and normal, the authors state the diagnosis should be further qualified as "with forced normalization of the EEG."

Affective-Somatoform (Dysphoric) Disorders of Epilepsy

Affective and somatoform symptoms may occur intermittently in chronic epilepsy. The ILAE Commission listed eight symptoms that occur at various intervals and typically last between several hours and several days (Krishnamoorthy et al., 2007). The symptoms include irritability, depressive moods, anergia, insomnia, atypical pains, anxiety, phobic fears, and euphoric moods.

The authors listed four conditions under this classification heading: (1) interictal dysphoric disorder, (2) prodromal dysphoric disorder, (3) postictal dysphoric disorder, and (4) alternative affective-somatoform syndromes.

Interictal Dysphoric Disorder

This disorder should be diagnosed when at least three of the eight symptoms listed earlier are present intermittently and cause impairment in social or occupational functioning. In women, the dysphoric disorder is evident in the premenstrual phase.

Prodromal Dysphoric Disorder

This disorder is characterized by irritability or other dysphoric symptoms that precede a seizure by hours or days and cause significant impairment.

Postictal Dysphoric Disorder

This disorder consists of anergia and headaches in conjunction with depressed mood, irritability, and anxiety which develop after a seizure. The disorder may be prolonged or usually severe.

Alternative Affective-Somatoform Syndromes

In this disorder, dysphoric symptoms occur as a manifestation of forced normalization of the EEG. Typical symptoms include depression, anxiety, depersonalization, derealization, and sometimes psychogenic nonepileptic events. These episodes are characterized by brief, but disabling, changes in affect. The authors indicate patients who fulfill DSM-IV or ICD-10 criteria for major depression, dysthymia, or cyclothymia should be excluded from this epilepsy-specific diagnostic system.

Personality Disorders

The commission recognized three types of distinct, but subtle, personality changes in patients with chronic epilepsy: (1) hyperethical or hyperreligious, (2) viscous, and (3) labile groups (Krishnamoorthy et al., 2007); also refer to the *interictal behavior syndrome of temporal lobe epilepsy* described earlier under the non–epilepsy-specific personality disorders. These are to be coded in a manner similar to the Axis II diagnoses in DSM-IV as either: (1) No personality trait accentuation or disorder, (2) Personality trait accentuation, but not disorder, or (3) Personality disorder specific to epilepsy. The personality disorder diagnosis is made when one of the three types of personality changes (or a mixture of two or more of them) interferes significantly with social or occupational functioning. Moreover, it is not diagnosed if the patient meets criteria for one of the DSM-IV or ICD-10 personality disorders. The three types of personality changes in epilepsy were defined as follows:

- *Hyperethical or hyperreligious type*: This epileptic personality type is characterized by a deepening of emotionality with serious, highly ethical, and spiritual demeanor.
- *Viscous type*: This type is seen as a tendency to be particularly detailed, orderly, and persistent in speech and action, which characterizes viscosity.
- *Labile type*: This epilepsy personality type consists of the presence of a labile affect with suggestibility and immaturity (referred to as "eternal adolescence").

Anxiety/Phobias

These epilepsy-related conditions refer to specific phobias, such as fear of seizures, agoraphobia, and social phobia, that may occur within the context of recurrent seizures. These phobias are distinguished from nonepilepsy phobias because they revolve around epilepsy. The fear of the situation or place, and subsequent avoidance, is due to the fear of having a seizure in that particular situation or place.

Anticonvulsant-induced Psychiatric Disorders

The ILAE Commission included this epilepsy-related psychiatric diagnosis since both AED induction and AED withdrawal have been shown to precipitate behavioral changes (Krishnamoorthy et al., 2007). The onset of the psychological disorder must be within 30 days of AED induction, or 7 days after AED withdrawal for the drug to be considered as being possibly etiologically related to the behavioral changes. The specific drug should be identified by name.

The authors decided to develop a classification system of psychiatric disorders specific to epilepsy that is separate from the current diagnostic systems because they believed that trying to integrate them would cause significant confusion to the user. The ILAE Commission's ultimate goal is to have these epilepsy-specific diagnoses included in future editions of the DSM (DSM-V) and ICD (ICD-11).

8

Psychogenic Nonepileptic Seizures

Nonepileptic seizures are episodes of alerted movement, sensation, or experience that resemble epileptic seizures but are not associated with ictal electroencephalographic (EEG) changes. Nonepileptic events may be due to either physiologic (i.e., identifiable medical etiology) or psychologic causes. Nonepileptic seizures from physiological causes are usually secondary to cardiogenic syncope, due to a drop in cardiac output, decreased heart rate, a drop in systemic blood pressure, or cardiac arrhythmias; but other causes such as metabolic disorders, migraine, or sleep disorders may also be implicated (Reuber and Elger, 2008). *Psychogenic nonepileptic seizures* (PNES) are spells in which no physical etiology for the seizure-like episodes can be found and some psychiatric condition, such as a somatoform or conversion disorder, is thought to be etiologically related (Binder and Salinsky, 2007). When nonepileptic or functional seizures are intentionally produced (or are under volitional control of the patient), malingering or a factitious disorder may be causing the seizures. However, PNES is generally considered an involuntary expression of psychological distress. The ICD-10 refers to PNES as *dissociative seizures,* which are classified within the family of conversion disorders. The major physiologic and psychologic etiologies of nonepileptic seizures are given in Table 8.1.

Diagnosis
Diagnosis of PNES is based on a thorough history and observation of the seizure event, and by ruling out the major possible physical causes for the seizures. Differentiating PNES from other medical causes of nonepileptic seizures is often

Table 8.1 Major causes of nonepileptic seizures

Nonepileptic Seizures with Physiological Causes
  Cardiac syncope
  Metabolic disorders
  Sleep disorders
  Migraine
  Endocrine disorders
  Paroxysmal dyskinesias

Psychogenic Nonepileptic Seizures (PNES)

  Unintentional (produced without conscious awareness):
    Somatoform disorders
    Conversion disorders
    Panic disorders
    Dissociative disorders
    Depersonalization disorders
    Posttraumatic stress disorders

  Intentional (produced with conscious awareness):
    Malingering
    Factitious disorder

---

not a very difficult task for the physician, whereas distinguishing PNES from epileptic seizures usually requires referral to an epileptologist or a specialized epilepsy surgery center for video-EEG monitoring. Nonepileptic seizures may be clinically electrographically indistinguishable from epileptic events, and hence, can be quite difficult to diagnose. Once all other possible causes of the seizures have been eliminated, diagnosis essentially consists of determining whether the patient has the prototypical historical, clinical, seizure semiology, EEG, and psychological features that are characteristic of PNES.

The interval from the first emergence of PNES and diagnosis is unacceptably long; estimates range from 7 to 16 years. Unfortunately the majority of PNES patients are misdiagnosed with seizures, and approximately three-quarters are inappropriately treated with antiepileptic drugs (AEDs). Early diagnosis of PNES is important because prognosis for successful treatment is best when the diagnosis is made within the first 6 months after onset. Moreover, misdiagnosis exposes PNES patients to adverse drug effects, emergency interventions, and the potential for long lasting psychosocial and vocational effects. The potential teratogenic effects of AED therapy are particularly important in this population since the majority of PNES patients are young women of childbearing age. Many patients continue to be inappropriately treated with AEDs even after an accurate PNES diagnosis has been made. With regard to vocational impact, in one study for example, 69% of PNES patients were working at the onset of seizures, whereas only 20% were still working by the time an accurate PNES diagnosis was made, usually several years later; four years after diagnosis, more than 41% of PNES patients (with a mean age of 38 years) had retired on health grounds (Reuber and Elger, 2008).

Diagnosis is complicated by the fact that a number (approximately 10%) of patients with PNES also have bone fide epileptic seizures. Fortunately, a variety of differentiating factors can be gleaned from the history, as well as many different clues from observing the seizure itself that can help distinguish PNES from epilepsy. Clues may also be obtained from postictal blood tests, ictal EEG recordings, provocation techniques, and neuropsychological testing. These will be reviewed later. Cragar, Berry, Fakhoury and colleagues (2002) have detailed many of the tools that may be beneficial in the PNES diagnostic process.

Prevalence

Since there are no studies of PNES in community-dwelling individuals, it is not possible to precisely know the prevalence of PNES in the general population. Among referrals to neurology centers, incidence estimates have generally been from 1.5/100,000 to 3/100,000 per year. PNES accounts for 5%–10% of neurology outpatient visits (Syed et al., 2009). Prevalence of PNES among patients with refractory seizures referred for epilepsy surgery evaluation has been estimated at approximately 20%–40% (Reuber and Elger, 2008). A recent sample of 181 inpatients at the Cleveland Clinic epilepsy monitoring unit found 27% met criteria for PNES, 64% had epilepsy, and 9% had coexisting PNES and epilepsy.

Among patients diagnosed with nonepileptic seizures (NES) at inpatient epilepsy monitoring units with simultaneous video-EEG recordings, 76%–80% had PNES alone, whereas 11%–13% had PNES and active epilepsy (Locke et al., 2006; Langfitt, personal communication, 2008). Between 7% and 13% of NES patients were found to have physiological nonepileptic seizures with the most common causes being cardiogenic syncope, complicated migraine, paroxysmal dyskinesia, and sleep disorder.

Etiology

Although a clear-cut cause of PNES cannot always be established, a variety of predisposing factors have been identified similar to that found in patients with other medically unexplained conditions. The overwhelming majority of PNES cases are in younger women. Across various samples, 70%–80% of PNES cases are female with a typical age of onset around 25 years old. The social histories of PNES patients often include past sexual or physical abuse, neglect, or other psychological or physical trauma. Furthermore, the families of PNES patients are more troubled than epilepsy patients' families and may contribute to the development of nonepileptic seizures through family distress, criticism, and tendency to somatize. Along these lines, LaFrance and Devinski (2004) conceptualized two types of PNES: developmental PNES and posttraumatic PNES. *Developmental PNES* stems from an individual's difficulties coping with the variety of psychosocial tasks encountered during development, whereas *posttraumatic PNES* is thought to result from an individual's maladaptive responses to ongoing physical, sexual, or psychological trauma.

Psychiatric disorders are probably the most significant identifiable etiology of PNES. For example, in one series, 101 of 102 (99%) patients with PNES who underwent inpatient video-EEG monitoring had an identifiable psychiatric condition (Westbrook et al., 1998). By far the most common psychiatric diagnoses in PNES are somatoform disorders, including conversion disorders and dissociative disorders. In most series, approximately 80% of PNES patients have some type of somatoform condition. Anxiety disorders, especially posttraumatic stress disorder, are the next most common diagnosis followed by mood disorders. In the study by Westbrook and colleagues (1998), many patients met criteria for more than one psychiatric diagnosis; 85% had conversion disorders, 52% anxiety disorders, 39% mood disorders, 12% malingering or factitious disorder, and 42% were also identified as having some type of personality disorder.

These demographic, social history, developmental, familial, and psychological factors may interact with one another to produce PNES. In the slightly more than 10% of patients who have both PNES and active epilepsy, the epilepsy itself may serve as a stressor contributing to recurrent PNES. Finally, secondary gain and other environmental reinforcements may help to perpetuate PNES.

Psychological Etiology

Although there are many different psychological theories to account for somatoform and dissociative disorders, one of the more favored explanations in recent years is a cognitive behavioral approach based on the fear/avoidance model of Neil Miller. Goldstein, Deale, Mitchell-O'Malley, and colleagues (2004) have described this model as follows. Although PNES patients rarely describe feelings of anxiety during their spells, some notice symptoms of autonomic arousal and hyperventilation. PNES may represent a dissociative response to unpleasant signs of arousal which enable the patient to avoid unpleasant emotions or thoughts. Thus PNES may be viewed as dissociative responses to unwanted arousal which develops when the patient is confronted with intolerable or fearful circumstances.

Once the psychogenic seizures begin to emerge, they are maintained by a host of psychological, physiological, and social factors that form a vicious cycle. Fear and avoidance are thought to be primary among them. As seizures occur, certain activities or behaviors are modified or avoided in case they trigger another seizure. Avoiding these things keeps attention focused on the possibility of seizures, while fear, worry, and the seizures continue to occur in a progressively more restricted lifestyle. Other factors may contribute to maintaining this cycle and worsening of the overall condition. Some of these may include disagreements among doctors about the correct diagnosis and subsequent uncertainty as to optimal treatment(s), fears about the consequences of repeated spells, secondarily caused mood disorders, reassurance seeking ("doctor shopping"), and secondary gains associated with the benefits of assuming the sick role. This psychological model of symptom maintenance suggests a variety of possible areas for cognitive-behavioral intervention (see the Treatment section for details).

Symptoms

A thorough history is one of the most important aspects of PNES diagnosis. A detailed description of the seizure events should be obtained. If the patient is able to give a full description of the seizure events, this may suggest a nonepileptic etiology since only patients with simple partial and psychogenic seizures should retain this ability. How does the patient feel after a seizure? Abnormal postictal states are the rule in complex partial or generalized tonic-clonic seizures, but many PNES patients experience a rapid return to a normal mental state. A thorough history will usually suggest whether the patient has some type of psychopathology that requires further diagnostic investigation, and most PNES patients will carry one or more psychiatric diagnosis. Detailed questioning about the patient's family relationships and educational and vocational situation is also important. Family, work, or school stressors are common in PNES patients. In contrast, the families of patient's with epilepsy are more often than not psychologically normal.

Beyond the clues one can glean from the history, there are many signs evident during the seizure itself that can help distinguish it as a nonepileptic event. These signs include seizure semiology, EEG, and some biological tests. Psychogenic nonepileptic events may be precipitated by emotional events such as arguments or other stressful emotional situations. PNES are frequently witnessed by others and are much more likely to occur in a doctor's office, in contrast to epileptic events which are usually not witnessed in outpatient settings. The onset of seizure symptoms is often more gradual in PNES; beginning with slower and more mild motor symptoms and then escalating to become faster and larger in amplitude. Another important distinguishing feature between epileptic and PNES events involves the clonic activity of the upper and lower extremities. In epilepsy, the clonic movements of the limbs are in-phase, whereas many psychogenic movements are out-of-phase, where movements may be nonsynchronous or oriented in multiple directions. In PNES, movements often begin bilaterally, are asynchronous, and can alternate between sides (e.g., alternating kicking of the legs); all of which is rare in epileptic events. In addition, motor activity that starts and stops during the seizure is suggestive of PNES.

Other clinical signs that have been associated with PNES include pelvic thrusting, eye closure during the attack, ictal crying, maintained pupillary light reflex, awareness of one's surroundings, and longer duration of the spell. When the patient is lying supine, the presence of rhythmic pelvic thrusting has long been considered a common sign of PNES. Unfortunately, pelvic thrusting is not specific to PNES, since it is also seen in complex partial seizures of mesial frontal lobe origin.

A very clear distinguishing clinical sign consists of eye closure during the PNES event, which almost ever occurs in generalized tonic-clonic seizures. Eyes are reportedly open in more than 90% of generalized tonic-clonic seizures, whereas they are closed in 80%–90% of PNES patients. Moreover, PNES patients often display a forceful, sustained eye closure with active opposition to examiner attempts to open them. Others have documented that weeping, crying, yelling, and vulgar language is typically seen in PNES but not in epilepsy. Next, it has been suggested that the persistence of normal pupillary reactivity to light during an unresponsive attack is helpful for identifying PNES. Finally, most epileptic seizures last only seconds to minutes (with the notable exception of status epilepticus), whereas PNES events often last longer than 2 or 3 minutes. On the other hand, some authors have emphasized the duration of both epileptic and nonepileptic events can by highly variable and suggest duration is not a reliable discriminator.

When differentiating epilepsy from PNES as an outpatient is not possible, inpatient video-EEG monitoring of the spells may be indicated. In the inpatient setting, ictal EEG, postictal EEG, and the relationship between AED dose and seizure frequency may all be monitored and carefully recorded. Video-EEG recording of the attacks is the gold standard in PNES diagnosis. If the EEG does not show epileptiform discharges during the typical spells, a suspicion of PNES can be raised since EEGs are almost always abnormal during epileptic seizures. The exception to this is that epilepsy arising from deep within the brain, usually in the mesial frontal or mesial parietal lobes, may show normal scalp EEG recordings. Suspicion of PNES is also higher when there is no change in seizure frequency after AED withdrawal.

Assessing patient's responsiveness and memory for events during seizures may also help to distinguish PNES from epilepsy. Some preserved ability to respond to questions or commands during a seizure is more common in PNES than epileptic seizures. Memory for events that occur during seizures is often preserved in PNES, but patients with complex partial seizures are almost always amnestic for seizure events. In one study, items presented during the spells were recalled after 63% of PNES events as compared with only 4% recall in complex partial seizures (Bell et al., 1998).

The Cleveland Clinic epilepsy group, in conjunction with University Hospitals Case Medical Center, developed a 209-item study questionnaire composed of demographic, clinical and seizure-related information, and psychosocial variables which the literature suggested was associated with PNES such as those in Table 8.2 (Syed, et al., 2009). After screening the items for variable predictability in diagnosing PNES, the self-report questionnaire was stream-lined down to 53 items summarizing 10 variables. Several important predictive questions having to do with history of abuse, psychiatric comorbidity, and feigning of illness were eliminated because patients did not answer these questions reliably. The shortened questionnaire was then validated using independent samples from two epilepsy surgery monitoring units with good results. They were able to accurately identify PNES with 94% sensitivity at the first center (and 85% sensitivity at the second center) and 83% specificity at the first center (and 85% specificity at the second center). The questionnaire and

Table 8.2 Signs occurring during seizures that help differentiate psychogenic nonepileptic seizures (PNES) from epileptic seizures

SIGN	PNES	EPILEPTIC SEIZURES
Seizure witnessed	Common	Less Common
Seizure provoked by stress	Common	Less Common
Asynchronous limb movement	Common	Uncommon
Motor symptoms start & stop	Common	Uncommon
Pelvic thrusting	Occasional	Rare
Weeping, crying	Occasional	Very Rare
Eyes closed	Very Common	Rare
Pupillary light reflex	Preserved	Often Abolished
Responsive to surroundings	Partially Preserved	Rarely Preserved
Long duration of spells	Common	Uncommon
Ictal electroencephalogram	Normal	Abnormal
Postictal recovery	Often Rapid	Slower

related documents may be downloaded online from www.neurology.org under the "supplemental data" heading affiliated with the article.

Summarizing, the most helpful ictal clinical signs suggesting a diagnosis of PNES are seizures that may be provoked by stress; occur in front of a witness; involve excessive movement of the limbs, truck, and head; involve movements that may start and stop; include eye closure during tonic-clonic activity; and have long duration of spells. The most important clinical features distinguishing PNES from epilepsy are presented in Table 8.2.

Psychological and Neuropsychological Assessment

Most studies of the ability of neuropsychological assessment to distinguish between PNES and true epilepsy have not shown particularly large or specific effects. Both groups tend to perform below normative expectations, and there are usually no significant differences in pattern of test failure, although several studies report PNES patients perform worse than epileptics on cognitive testing. This may be due to the nonspecific effects that conversion or somatization disorders produce on cognitive testing, typically disrupting tests of attention, visuomotor speed, and indirectly, memory, which are very similar to the nonspecific effects produced by AEDs and recent epileptic seizures. Secondary gain and exaggeration of degree of deficit may also contribute to poor test performances in PNES.

Personality Testing in PNES

Some differences have been found between patients with PNES and patients with epilepsy on personality tests, such as the Minnesota Multiphasic Personality Inventory (MMPI-2), Personality Assessment Inventory (PAI), and NEO Personality Inventory (NEO-PI-R) as well as on some of the more common symptom validity measures, such as the Portland Digit Recognition Test, Word Memory Test, and a negative response bias on the California Verbal Learning Test (Dodrill et al, 1993; Binder et al., 1998; Drane, et al., 2006).

With regard to objective personality testing, most have found that patients with PNES show significant MMPI clinical scale elevations on the so-called neurotic scales reflecting somatization tendencies, namely, Hypochondriasis (Scale 1, Hs), Hysteria (Scale 3, Hy) and sometimes Depression (Scale 2, D). Thus, PNES patients often produce MMPI code types such as, 1-3, 3-1, 2-1, 2-3, 3-1-2, or 3-2-1. Using a discriminant function analysis, Binder and colleagues (1998) found that, using MMPI scales 1 and 3 and the Portland Digit Recognition Test, they could correctly distinguish 80% of PNES from epilepsy patients. On the PAI, PNES patients are much more likely to show higher T-scores on the Conversion subscale relative to lower T-scores on the Health Concerns subscale. On the NEO-PI-R, PNES patients have been reported to obtain higher scores on the Neuroticism scale. Cragar, Berry, Schmitt, and colleagues (2005) examined the MMPI-2 in conjunction with the NEO-PI-R using cluster analysis and reported three distinct PNES cluster groups:

1. *Depressed Neurotics*: Negative affect, low energy, low sociability, depression, MMPI-2 elevations on scales 2 and 3,
2. *Somatic Defenders*: Somatic tendencies when dealing with psychosocial stress, MMPI-2 elevations on scales 1 and 3 with a lower score on scale 2 (so-called "conversion V" code type), and
3. *Activated Neurotic Group*: Negative affect similar to the first cluster, but more socially engaged and anxious, MMPI-2 elevations on scales 1 and 8.

Cognitive Testing in PNES

No typical pattern of cognitive test results can be used to diagnose PNES, just as there isn't a common pattern of neuropsychological findings that can identify specific seizure types in true epilepsy. Nevertheless, in addition to the clues from history, seizure observation, and psychological testing described earlier, some information may be gleaned from neuropsychological testing. PNES patients reportedly perform more poorly on symptom validity tests (i.e., Word Memory Test, Portland Digit Recognition Test) than epilepsy patients. Drane and colleagues (2006) found the rate of symptom validity test failure in PNES was approximately 28% of patients, whereas the comparable failure rate for epilepsy patients was 8%. PNES patients have also been shown to have a negative response bias in that they rarely made false positive errors on the recognition trial of the California Verbal Learning Test (Bortz et al., 1995). A negative response bias index (using a cut-off score of <0 on the CVLT Recognition Trial) yielded a PNES diagnostic sensitivity of 61% and a specificity of 91% when compared with true seizure patients with left temporal lobe foci (who showed a positive response bias). The failure to explicitly recognize words after repeated exposure on the CVLT may reflect certain aspects of psychological denial in PNES.

Finally, neuropsychologists should be aware that many PNES patients may be referred to them for evaluation of cognitive status following a variety of ostensible neurological insults. Probably the most common scenario is referral for assessment following a concussion with brief or no loss of consciousness and no medical evidence of any neurologic damage upon neurological examination or neuroimaging. In one series of PNES patients diagnosed with video-EEG monitoring, approximately one-quarter attributed the onset of their seizures to a mild head injury. Other forms of neurological injury may also be commonly encountered in PNES patients, such as possible birth trauma, high fevers, possible CNS infection, or chronic pain; although there is usually no tissue damage detected despite thorough examination. In fact, PNES patients generally report neurological injury or insult much more often than true epilepsy patients do.

Treatment

Once a diagnosis of PNES has been clearly established, the diagnosis and its presumable causes should be unambiguously communicated to the patient and their family. It is important the patient and family understand the cause of the events are psychological or stress-induced and not epileptic or neurologically caused. Some patients may react with disbelief or anger if they feel they are being accused of faking or as being labeled "crazy." Some health care workers and families find the term PNES burdensome or offensive, and other terms such as "functional seizures," "stress-induced seizures," or "dissociative seizures" may be substituted. Dissociation may also be a difficult concept for patients to grasp. Patients can be told that they have attacks in which a portion of their brain "switches off," and they subsequently lose awareness and control over their actions. It is vital that the patient and family understand and accept the PNES diagnosis; otherwise, they will not follow treatment recommendations. Another good reason for clear communication at this time is that, in many cases, PNES simply resolves (estimates range from 10% to 50% of cases) after an acceptable explanation has been provided. Of course, in PNES patients who do not have concomitant epileptic seizures, AEDs should be stopped.

Following communication of the PNES diagnosis, patients should be referred for psychiatric or psychological assessment to determine the appropriate psychiatric diagnosis. In addition to the somatoform or dissociative psychopathology, comorbid disorders such as depression, anxiety, phobias, or posttraumatic stress disorder should be treated with appropriate pharmacotherapy and psychotherapy as needed. Adjunctive therapies targeted toward stress management using biofeedback or relaxation techniques have also been recommended and used with some success. No randomized clinical psychotherapy treatment trials in PNES have been conducted to date. Many have outlined some general management principles and there are a number of successful case studies. Two treatment approaches have been described in detail: the brief psychodynamic-interpersonal therapy model of Reuber, Howlett, and Kemp (2005) and a 12-session cognitive-behavioral treatment program outlined by Goldstein and colleagues (2004). Although the cognitive-behavioral therapy described by Goldstein's group (2004) was highly successful (i.e., 81% of those who completed treatment either became seizure-free or had at least a 50% reduction in PNES spells at 6-month follow-up), it should be emphasized that the overall psychosocial outcome of PNES is dismal; approximately one-third will return to work, half will continue to receive disability benefits, and psychosocial outcome has been rated as poor or very poor in more than one-half of PNES patients. PNES is clearly a condition in dire need of competent psychological and psychiatric management and treatment, and as such represents a fertile opportunity for psychologists and other mental health professionals interested in developing the treatment segment of their clinical practice.

PART TWO

SURGICAL TREATMENT OF EPILEPSY

9

Neuropsychological Assessment in Epilepsy Surgery

The duties of epilepsy neuropsychologists vary somewhat across surgery centers, but always include performing pre- and postoperative neuropsychological assessment of epilepsy surgery candidates (Helmstaedter, 2004). In addition, most epilepsy surgery neuropsychologists also conduct the cognitive assessment portion of the intracarotid amobarbital (Wada) procedure; design, monitor, and help interpret the results of cognitive testing performed during functional magnetic resonance imaging (fMRI); and plan and conduct the sensorimotor and cognitive evaluations during electrocortical stimulation mapping both extraoperatively and intraoperatively. Neuropsychologists have been designated as essential personnel, and neuropsychological assessment and Wada testing have been designated as essential neuropsychological services, for all epilepsy surgery centers by the National Association of Epilepsy Centers (2001). This chapter will review neuropsychological assessment; the more specialized procedures used in epilepsy surgery that may involve the neuropsychologist will be discussed in Chapter 10.

Before getting into the details of the cognitive assessment procedures, it is important to see where these procedures fit within the overall scheme of the epilepsy surgery decision making process (Fig. 9.1). Before a patient with epilepsy is admitted to the hospital for continuous video-electroencephalographic (EEG) monitoring, it must be determined that the seizures are localization-related (focal onset), intractable or refractory to drug therapy (failure to achieve seizure control after adequate trials of two or more antiepileptic medications), and frequent or severe enough to significantly interfere with normal activities of daily life.

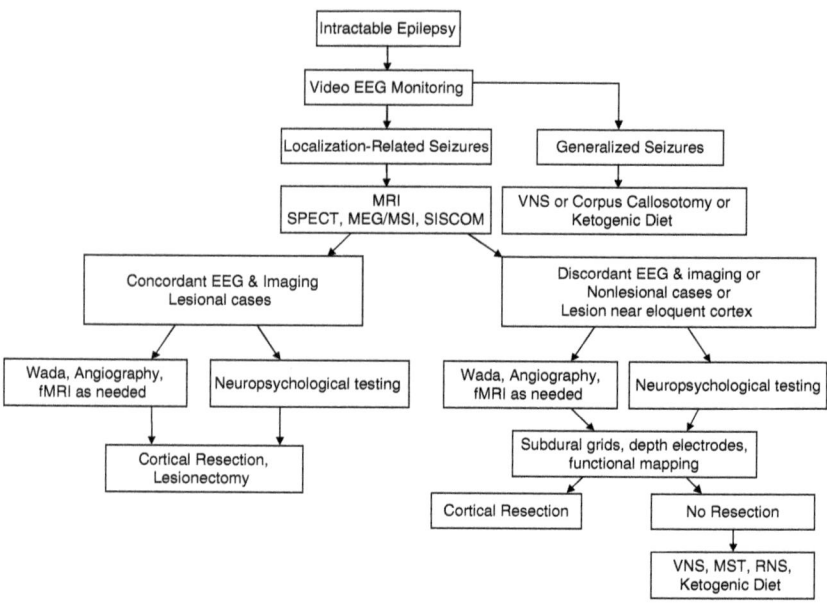

FIGURE 9.1. Overview of the epilepsy surgical decision-making process (Adapated from: Go & Snead, 2008).

After video-EEG monitoring has been successfully accomplished, and the seizures are thought to arise from a focal region of brain that could be surgically resected, structural neuroimaging (MRI) is conducted, and functional imaging (single photon emission computed tomography [SPECT], magnetic source imaging [MSI], or subtracted ictal SPECT co-registered on MRI [SISCOM]) may be conducted. When the seizures have been adequately characterized with regard to seizure semiology, EEG, and/or structural and functional neuroimaging, patients are then usually referred for preoperative neuropsychological assessment, intracarotid amobarbital (Wada) procedure, and fMRI as necessary.

When all of these different sources of data have been gathered on a patient, the results are reviewed by a multidisciplinary epilepsy surgery team (usually consisting of the neurosurgeon, epileptologist, neuropsychologist, and epilepsy nurses) to form a decision about whether to proceed to the next stage in the process. As may be seen in Figure 9.1, neuropsychological assessment and Wada testing take place fairly well along in the epilepsy surgery process, and thus, patients are generally considered good candidates for surgery by the time they are seen by the neuropsychologist.

Preoperative Neuropsychological Assessment

The primary purposes of neuropsychological assessment within the epilepsy surgery context are to help lateralize and localize the seizure focus, predict risk for postoperative cognitive impairment, establish a baseline against which to measure change, help predict seizure relief outcome, and diagnose psychiatric disorders and consider potential impact on ability to cooperate with epilepsy surgery process and ultimate postoperative adjustment. Answers to these questions will assist all involved in preoperative decision-making by weighing the risks of continued intractable epilepsy against the physical and cognitive risks inherent in the surgery.

Purposes of Preoperative Neuropsychological Assessment
Lateralization and Localization

One of the major reasons to obtain preoperative neuropsychological testing is to assist in the lateralization and localization of the seizure focus (Kneebone, 2001). When a focal deficit pattern emerges on testing that is concordant with EEG identification of the epileptogenic zone and with structural brain abnormalities on MRI, there is increased confidence in localization of the seizure focus, as well as an increased likelihood of seizure relief after surgery. Although the pattern of neuropsychological deficits is more sensitive to the brain damage that causes the epilepsy than it is to the presence of the seizure focus itself, the lesion and epileptogenic focus often (but not always) overlap. It is much more likely to obtain a focal deficit pattern on neuropsychological testing when there is an epileptogenic lesion than when the etiology is unknown (i.e., in idiopathic cases). In cases where no discernible brain abnormality is apparent, the neuropsychological effects of epilepsy are more subtle and may be more difficult to localize due to interference from the nonspecific effects of antiepileptic drugs (AEDs), cognitive development factors, psychosocial issues, or subclinical epileptic spikes.

When data from neuropsychological testing is at odds with seizure localization using EEG, the reasons for this discrepancy should be thoroughly investigated since it could have implications for seizure relief and neuropsychological outcome. A variety of factors can confound interpretation of test results. For example, brain lesions early in life may result in atypical functional organization of higher cortical functions, and this can result in "false localization" information on neuropsychological testing. Certain diffuse brain pathologies, such as traumatic brain injury or encephalitis, often present with diffuse patterns of cognitive impairment which may obscure focal patterns of cognitive impairment. In addition, the earlier seizures begin in life, the lower and more generalized the cognitive deficit performance patterns will be. This is thought to be due to early intractable seizures disrupting the normal acquisition of a wide range of cognitive functions regardless of the seizure onset location (Chelune, 1994). Similarly, specific developmental learning disabilities are common in childhood onset epileptic disorders, and these may conceal certain focal deficit patterns on testing.

Clinical lore suggests neuropsychological testing is not overly helpful for lateralizing or localizing the epileptogenic zone when compared with the more "medical" tests, but this is not really the case (Fargo et al., 1980). Although it is true that EEG, MRI, and Wada testing generally have higher predictive capabilities than cognitive testing in lateralizing the side of seizure onset, this does not mean neuropsychological testing has nothing to add in this regard. In fact, cognitive testing can add localization information in epilepsy surgery patients more often than not. The EEG has been reported to accurately lateralize seizures in 89% of temporal lobe cases, structural MRI in 86%, and Wada testing in 80%, whereas single and various combinations of cognitive tests accurately lateralize seizure onset in 66%–73% of patients (Moser et al., 2000; Loring et al., 1993; Kneebone et al., 1997; Kim et al., 2004; Ogden-Epker and Cullum, 2001).

Memory and language tests have been the most commonly used tests to detect the side of temporal lobe seizure onset. The Boston Naming Test has accurately lateralized the temporal lobe seizure focus in approximately 69% of cases across different patient samples and has been used to formulate a multiple regression prediction equation in conjunction with seizure duration, age of onset, and Full-Scale IQ (Busch et al., 2005; Buschet al., 2009). The memory tests that have been found useful in lateralization include Logical Memory, Paired-Associate Learning, and Visual Reproduction from the WMS-R, Auditory Delay Memory Index and Visual Delayed Memory Index from the WMS-III, Rey Auditory Verbal Learning Test, and the Rey Complex Figure Test (Kneebone et al., 1997; Loring et al., 2008; Keary et al., 2007). On the Wisconsin Card Sorting Test, Categories Achieved has been found to be significantly lower in patients with right temporal lobe epilepsy (TLE). Others have reported that configurational interpretation of larger test batteries was useful for localizing seizure onset. Ogden-Epker and Cullum (2001) found WAIS-R Performance IQ, WCST-Categories, and Rey Complex Figure-delayed recall predicted right TLE, and verbal associative (animal) fluency to be predictive of left TLE. Finally, Keary and colleagues (2007) reported the WMS-III verbal and nonverbal delayed memory subtests, WCST-Categories, Wide Range Achievement Test-III Reading, and the Boston Naming Test contributed significantly to the prediction of seizure lateralization in patients with TLE.

Risk for Postoperative Cognitive Impairment

The risks that have been studied the most and occupy the majority of the neuropsychologist's time center around memory and language in temporal lobectomy candidates. After it was discovered that a unilateral temporal lobectomy could result in a dense global amnesia if the contralateral mesial temporal lobe structures (especially the hippocampus) were dysfunctional, neuropsychological testing and Wada memory assessment have been used to determine the functional status of the hippocampus ipsilateral to the seizure focus (its so-called *functional adequacy*), as well as the functional capacity of the hippocampus contralateral to the seizure focus (the *functional reserve*) which will be responsible for the formation of new memories after temporal lobectomy (Loring and Chelune, 2001). Thus, the neuropsychological test patterns that suggest possible risk for global amnesia include impairments in both verbal and nonverbal memory, which imply bilateral hippocampal dysfunction, and material-specific memory impairments in the opposite direction to that expected on the basis of the seizure focus (e.g., verbal memory deficits in patients with right mesial temporal lobe seizures). In both these cases, the preoperative memory test results suggest the possibility of inadequate functional reserve on the side opposite to the proposed surgery.

Prediction of Memory Loss. With regard to predicting material-specific memory losses after unilateral (especially left) temporal lobectomy, many studies have verified the higher the preoperative level of verbal memory capability, the greater the postoperative verbal memory loss will be. This has been well-established for verbal memory following left temporal lobectomy, but not for nonverbal, visual-spatial memory after right temporal lobectomy. The prediction of material-specific memory decline in the individual case is enhanced when the results of baseline neuropsychological memory tests are considered in conjunction with the results of other sources of information. Several mathematical risk models have been developed to predict postoperative memory decline. These risk models have incorporated variables found to be most critical in predicting postoperative memory outcome including side of surgery, presence of unilateral mesial temporal sclerosis on MRI, hippocampal volumetric analysis, age of epilepsy onset, memory ability during the intracarotid amobarbital (Wada) procedure, and level of preoperative memory performance. For example, Stroup, Langfit, Berg, and colleagues (2003) empirically developed a multiple regression equation for predicting memory decline which clinicians may use to estimate an individual patient's risk for verbal memory decline following unilateral temporal lobectomy. Patients with the greatest risk had: dominant temporal lobectomy (left ATL), absence of ipsilateral (to seizure focus) mesial temporal sclerosis, normal preoperative performances on two tests of immediate verbal memory, normal preoperative performances on two tests of delayed verbal memory, and normal Wada memory after contralateral (to seizure focus) amobarbital injection.

Others consider the presence of mesial temporal lobe sclerosis on the side contralateral to the planned surgery to be an additional risk factor for cognitive decline after anterior temporal lobectomy.

Risk of Postoperative Language Impairment. Preoperative neuropsychological assessment of language typically evaluates fluency, naming, aural comprehension, and reading at a minimum. If there are specific concerns about language and other related domains a broader assessment may be conducted. For example, if language deficits are already present, then a more detailed assessment of language is warranted. This usually consists of selected subtests from one or more of the comprehensive aphasia batteries, such as the Boston Diagnostic Aphasia Examination (BDAE), Multilingual Aphasia Examination (MAE), or Western Aphasia Battery (WAB).

As with memory functions, higher preoperative language capacity has been associated with larger postoperative declines. Although acute aphasia is common in the acute postoperative period (usually lasting only a few days or weeks), significant language disorders are not often seen after dominant temporal lobectomy. Nonetheless, subtle deficits in naming are fairly common; although many patients do not spontaneously complain about their word-finding difficulties. Language lateralization is determined using the intracarotid amobarbital (Wada) procedure, and with ever increasing frequency, fMRI. If the epileptogenic zone overlaps with areas of suspected language functions, patients may require more detailed electrocortical stimulation mapping either intraoperatively or extraoperatively using subdural grids to identify eloquent language areas to be avoided during the resection.

Establish a Baseline

Although refractory patients who are candidates for epilepsy surgery almost universally have experienced significant disruption to daily life as a result of their seizure disorder, one of the purposes of a preoperative neuropsychological assessment is to help evaluate and quantify this impact. In addition to a broad based cognitive assessment, the preoperative evaluation also uses psychological-emotional tests, symptom checklists, self-report inventories, and health-related quality of life measures (see Chapter 6 for details on these). These tests, in conjunction with the cognitive measures, attempt to characterize the impact of the seizures on the patient's life to assist in determining if the epileptic disorder is of sufficient severity to warrant surgery.

Baseline assessment may also be used to track cognitive deterioration over time. Although somewhat controversial as to the cause of decline, there is strong evidence that repeated seizures have a deleterious effect on neural tissue. Seizure history variables that may influence cognitive decline include frequency of seizures, number of episodes of status epilepticus, early age of onset of seizures, total number of lifetime seizures, etiology, duration of seizure disorder, multiple seizure types, and use of antiepileptic medications (Chelune, 1994). If a patient's condition is not severe enough to warrant epilepsy surgery, serial neuropsychological assessment may help to establish the eventual need for surgery. Finally, some epilepsy surgery patients experience cognitive deficits or develop new cognitive complaints after surgery and having documented a cognitive baseline allows us to estimate degree of change, if any, with greater precision. Postoperative neuropsychological evaluation will assist in identifying the cause(s) of the changes after surgery, which in turn, will determine the appropriate treatment recommendations.

Baseline assessment may also be used to track improvements following epilepsy surgery. In addition to localized cognitive deficits associated with the epileptogenic region of cortex, more generalized dysfunction may also be seen across many cognitive domains including attention, memory, language, spatial thinking, executive function, and visuomotor speed. This more generalized dysfunction may be due to the widespread structural and physiological effects of the repeated seizures on the brain. These progressive cognitive changes can be arrested, or even reversed, with successful seizure surgery (Helmstaedter et al., 2003).

Prediction of Seizure Control
In addition to seizure history variables (e.g., age of onset, duration of seizures, etiology), EEG characteristics, MRI findings, and neuropsychological test results also have been shown to hold prognostic significance for post-resection seizure control. Results over the past 20 years or so have consistently found that patients who have cognitive deficits restricted to the area of the proposed surgery are most likely to benefit from surgery (Rausch, 1987). There is also evidence that patients whose neuropsychological results suggest that dysfunction is confined to a temporal lobe have better seizure relief outcome than those with test findings indicating extratemporal lobe dysfunction. Neuropsychological test results suggesting widespread cortical dysfunction in both hemispheres have been associated with poor postoperative seizure control. Consistent with this, studies across epilepsy surgery centers have found patients with low preoperative Full-Scale IQs have a worse prognosis for seizure relief than those with higher IQs. This is thought to be due to the fact that epilepsy surgery candidates with low IQs (FSIQ <65) more often have early and more diffuse cerebral involvement. It is important to remember that neuropsychological test results do not have as much predictive power for seizure relief as some other test results, such as multifocal EEG abnormalities. For example, patients with low IQs are not routinely excluded from surgery at most institutions, but this neuropsychological predictive information is often considered as one part of the entire puzzle of pre-surgical test results and could assist with surgical decision-making.

Postoperative Neuropsychological Assessment
A repeat neuropsychological examination is frequently conducted approximately 1 year after epilepsy surgery. Patients, families, and the involved health care professionals should all be made aware of the type and severity of any postoperative cognitive deficits. Patients often need to know what, if anything, has changed, how these cognitive changes may affect their day-to-day lives, and what can be done to help. As noted earlier in the "Risk for Postoperative Cognitive Impairment" section earlier in this chapter, and under the specific cognitive domains in Chapter 6, most attention has been devoted to measuring change in language and memory following temporal lobectomy. However, a variety of postoperative cognitive deficits may be seen depending

upon the site of cortical excision and the functional adequacy of the resected tissue.

Persistent subtle deficits in visual naming and verbal new learning and memory are quite common 1 year after left anterior temporal lobectomy. After left temporal lobectomy, between 25% and 40% of patients will be left with a permanent anomia, and about 60% of patients will show verbal memory decline. Left temporal lobectomies may also result in a small decline (~4–5 standard score points) on some verbal intellectual measures. Most patients rarely spontaneously complain about these cognitive changes, especially if they are seizure-free.

Cognitive changes are much less common after right, than left, temporal lobectomy. Nonetheless, a mild degree of permanent nonverbal memory impairment may be seen in approximately 20%–25% of right temporal lobectomy patients. Other cognitive changes have been reported after non-temporal lobe focal excisions, depending upon the area cortex resected. Improvements may also be seen after surgery. The most common performance gains usually involve attention-concentration and psychomotor speed, which have been associated with seizure-freedom and antiepileptic drug withdrawal.

The majority of patients who undergo corpus callosotomy will not show any permanent cognitive deficits from the procedure. In the acute postoperative period, some patients may display mild signs of callosal disconnection, such as the alien hand phenomena, or attentional deficits. Most patients, however, will either improve or stay the same across most cognitive domains upon re-testing. The alien hand phenomena will typically resolve in the weeks after surgery. Attentional deficits may persist in some patients 1 year after either partial or complete corpus callosotomy.

Case Example
Neuropsychological Results Pre- and Post-Right Anterior Temporal Lobectomy

Patient History:
The patient is an 18 year-old, right-handed white male currently in the 12th grade. He suffered an intrauterine stroke in the right hemisphere, but was born at full-term without complications. Developmental milestones were all attained within the expected time frames. Seizures were first experienced at age 2 years. Spells begin with an aura of hot flashes and a feeling of dizziness. This is followed by a staring spell and lip smacking, with eye and head deviation toward the left side and jerking of the left arm that evolves into

bilateral convulsions. Seizure frequency before surgery was typically 2 or 3 episodes per week. Seizures have been refractory to treatment with multiple antiepileptic drugs (>5 different AEDs). There are no known etiological factors. Specifically, there is no history of birth complications, childhood febrile seizure, head trauma, encephalitis, meningitis, brain tumor, cerebral hemorrhage, brain malformation, cerebral palsy, brain surgery, premature birth, or stroke. There is a history of several episodes of status epilepticus lasting 1–2 hours. Preoperative brain MRI revealed increased T2 and FLAIR signal in the right hippocampus consistent with MTS and periventricular encephalomalacia on the right. Inpatient simultaneous video-EEG monitoring recorded three typical seizures, all of which had their onset in the anterior temporal lobe on the right.

Wada Testing.
Intracarotid amobarbital (Wada) evaluation was conducted prior to surgery. The patient demonstrated exclusive left hemisphere dominance for speech and language. There was a significant asymmetry in object memory performance, with superior performance following the right injection, suggesting right-sided seizure onset and adequate contralateral (left) memory support should the patient undergo a right anterior temporal lobectomy (ATL). Wada results were as follows

	Left Injection	*Right Injection*
Expressive Speech:	Arrest of counting	Normal
Receptive Speech:	Severe impairment	Normal
Visual Naming:	Severe impairment	Normal
Repetition:	Severe impairment	Normal
Reading:	Severe impairment	Normal
Paraphasic Errors:	Semantic & phonemic errors	None
Object Recognition Memory:	0 of 8, 0 False Positives (FPs)	8 of 8, 0 FPs

Surgery:
Surgery was performed 3 weeks after Wada testing. A right frontal temporal craniotomy was performed. An anterior lateral temporal lobectomy was performed with resection being 5 cm superior and 7 cm inferior. Mesial

structures were resected utilizing microscope. Post-resection corticography was performed and no epileptiform activity noted.

Abbreviated description of surgical procedure:
After the bone flap was removed and the dura was opened, attention was directed to the lateral temporal cortex. The pia over the lateral temporal lobe was coagulated and divided using microscissors beginning at the superior temporal gyrus. Attention was then returned to the superior temporal gyrus which was dissected in a subpial manner using microdissection technique. This was carried to the temporal tip. Next, attention was turned to the lateral temporal lobe where corticectomy was performed using bipolar coagulation and suction, carrying this through the white matter until the temporal horn was encountered. Subtemporal cortex was dissected in a subpial manner, carrying this to the temporal tip. The remaining temporal stem white matter was divided and the remaining pia arachnoid was divided using bipolar coagulation and microscissors. The anterior lateral temporal lobe was then removed en bloc.

The microscope was brought in and the remainder of the procedure was performed under microscopic illumination. Residual subtemporal cortex was resected in a subpial manner using microdissection technique and bipolar coagulation with suction. Attention was turned to the hippocampus, which was dissected at the choroidal fissure using microdissection technique. The tail of the hippocampus was divided using bipolar coagulation and suction. The hippocampus and hippocampal gyrus were dissected in a subpial manner using microdissection technique, carrying this posterior to anterior. The head of the hippocampus was disconnected from residual mesial temporal cortex. The hippocampus and uncus were then further dissected in a subpial manner and the hippocampus was removed en bloc.

Attention was then turned to the residual anterior mesial temporal cortex and amygdala, which were dissected in a subpial manner using microdissection technique. Once the amygdala was completely isolated the dome of the amygdala was amputated and removed en bloc. Post-resection corticography was performed with the epileptologist, and no epileptiform activity was noted.

Neuropsychological test results:
The scores here were obtained 1 month before right anterior temporal lobectomy, and post-surgery testing was conducted 18 months after the right ATL. The patient has been seizure-free since surgery. Current medications include Keppra (1500mg, b.i.d.) and Lamictal which was being tapered at the time of testing.

The following interpretive guidelines are offered to assist with interpretation of the scores:

SS Standard Score ($M = 100$, $SD = 15$); SS ≤ 75 indicates potential impairment.
ss scaled score ($M = 10$, $SD = 3$); ss ≤ 5 indicates potential impairment.
z z score ($M = 0$, $SD = 1$); $z - \leq 1.6$ indicates potential impairment.
T T score ($M = 50$, $SD = 10$); $T \leq 34$ indicates potential impairment,
percentile Percentile ranks (M = 50th, −1 SD = 16th, +1 SD = 84th, Impaired <5th).

General Intellectual Functioning:

	Pre-Surgery	Post-Surgery
VIQ (SS)	78, 7th percentile	80, 9th percentile
Verbal Comprehension Index (SS)	82, 12th percentile	84, 14th percentile
Vocabulary (ss)	9	7
Similarities (ss)	5	7
Information (ss)	6	7
Working Memory Index (SS)	78, 7th percentile	78, 7th percentile
Arithmetic (ss)	5	6
Digit Span (ss)	7	6
Letter-Number Sequencing (ss)	7	7
PIQ (SS)	76, 5th percentile	78, 7th percentile
Perceptual Organization Index (SS)	76, 5th percentile	80, 9th percentile
Picture Completion (ss)	8	6
Block Design (ss)	4	7
Matrix Reasoning (ss)	6	7
Picture Arrangement (ss)	6	7
Processing Speed Index (SS)	84, 14th percentile	86, 18th percentile
Digit Symbol (ss)	7	6
Symbol Search (ss)	7	9

Language:

	Pre-Surgery	Post-Surgery
Boston Naming Test		
Raw	42/60	46/60
w/ Semantic cueing	42/60, 6th percentile	47/60, 23rd percentile

Letter Fluency
(Three 60-second trials)
  Corrected .... 24, 6th percentile   33, 33rd percentile
MAE Token Test .... 41 correct, 33rd percentile   43 correct, 67th percentile

MAE Sentence Repetition
Test
  Corrected .... 11, 25-43rd percentile   12, 25-43rd percentile

Attention-Concentration:

	Pre-Surgery	Post-Surgery
Digits Forwards	span = 5, 3rd percentile	span = 6, 23rd percentile
Digits Backwards	span = 4, 14th percentile	span = 3, 3rd percentile
WMS-III Mental Control	21/40, 25th percentile	30/40, 75th percentile
Letter-Number Sequencing (ss)	7, 16th percentile	7, 16th percentile
WMS-III Spatial Span (ss)..	9, 37th percentile	8, 25th percentile
Trailmaking Part A	35 secs, 16th percentile	27 secs, 47th percentile
Trailmaking Part B.........	90 secs, 21st percentile	82 secs, 38th percentile
Gordon Diagnostic Vigilance Test		
Correct	30/30, 72nd percentile	30/30, 72nd percentile
Total Commissions	0, 68th percentile	0, 68th

Verbal Learning and Memory:

	Pre-Surgery	Post-Surgery
Information and Orientation.......................	14/14, 63rd percentile	14/14, 64th percentile
Verbal Auditory Memory – Immediate (SS).................	86, 18th percentile	89, 23rd percentile
Logical Memory I (ss) ...	6	6
Verbal Paired Associates I (ss)	9	9

Verbal Auditory Memory – Delayed (SS)....................	99, 47th percentile	94, 34th percentile
Logical Memory II (ss).......	8	6
Verbal Paired Associates II (ss).	12	12
Verbal Auditory Memory – Del. Recog (SS)..................	85, 16th percentile	90, 25th percentile
Selective Reminding (CLTR)	29/72, 6th percentile	45/72, 49th percentile
30' Delayed Free Recall	12/12, 72nd percentile	11/12, 48th percentile
30' Delayed Recognition	12/12, Normal	12/12, Normal

Nonverbal Learning and Memory:

	Pre-Surgery	Post-Surgery
Visual Memory – Immediate (SS)	61, <1st percentile	75, 5th percentile
Faces I (ss).............................	7	8
Family Pictures I (ss)	1	4
Visual Memory – Delayed (SS)...	70, 2nd percentile	75, 5th percentile
Faces II (ss)...........................	9	8
Family Pictures II (ss)	1	4
Complex Figure 3' Immediate Recall....................................	14/36, <1st percentile	12.5/36, <1st percentile
Complex Figure 30' Delayed Recall.....................................	16/36, <1st percentile	15/36, <1st percentile

Visuoperception and Spatial Thinking:

	Pre-Surgery	Post-Surgery
Facial Recognition Test..	47/54, 72nd percentile	47/54, 72nd percentile
Complex Figure Copy ...	35/36, >16th percentile	36/36, >16th percentile
WAIS-III Block Design (ss)	4, 2nd percentile	7, 16th percentile
Judgment of Line Orientation.....................	10/30, <1.5th percentile	13/30, <1.5th percentile

Executive Functions:

	Pre-Surgery	Post-Surgery
Wisconsin Card Sorting Test (WCST)		
Categories completed	5, 11-16th percentile	6, >16th percentile
% Perseverative responses	20, 18th percentile	15, 32nd percentile
% Perseverative errors	20, 12th percentile	12, 42nd percentile
WAIS-III Similarities (ss)	5, 5th percentile	7, 16th percentile
WISC-III Mazes (RS)	27/28, 95th percentile	26/28, 75th percentile
Verbal Fluency	24, 6th percentile	33, 33rd percentile
Figural Fluency	15 unique, <1st percentile	21 unique, <1st percentile

Academic Achievement:

	Pre-Surgery	Post-Surgery
WIAT Reading Comprehension (SS)	89, 23rd percentile	89, 23rd percentile
WRAT-3 Reading (SS)	91, 27th percentile	93, 32nd percentile
WRAT-3 Spelling (SS)	97, 42nd percentile	94, 34th percentile
WRAT-3 Arithmetic (SS)	78, 7th percentile	85, 12th percentile

Motor:

	Pre-Surgery	Post-Surgery
Grooved Pegboard		
Dominant (Right)	75 secs, 4th percentile	64 secs, 42nd percentile
Nondominant (Left)	112 secs, <1st percentile	122 secs, <1st percentile
Gordon Diagnostic Vigilance Reaction Time	320 msec, 76th percentile	340 msec, 68th percentile

Beck Depression Inventory:

	Pre-Surgery	Post-Surgery
BDI-I	12, minimal	0, none

Health-related Quality of Life (QOLIE-48-AD) (Lower scores reflect worse functioning:

	Pre-Surgery T-scores	Post-Surgery T-scores
Epilepsy Impact	40	59
Memory/Concentration	47	65
Attitudes toward Epilepsy	4	76
Physical Functioning	49	62
Stigma	40	63
Social Support	61	61
School Behavior	56	56
Health Perceptions	46	64
Overall Score	43	66

Impressions:

1. Postoperative cognitive testing reveals persistent deficits in spatial thinking, figural fluency, and visual-spatial new learning and memory with reduced motor speed and dexterity of the left (non-dominant) hand.
2. In comparison with preoperative testing, there has been no significant cognitive decline on any cognitive measure 18 months status-post right anterior temporal lobectomy. Most domains remained the same or showed slight improvement.
3. Self-rated quality of life showed overall significant improvements from pre-surgical ratings.

Comment:

Seizure frequency prior to surgery was approximately 3–4 per month. He has been seizure-free since the right anterior temporal lobectomy. Wada testing suggested there should be no substantial risk for global amnesia or significant visual-spatial memory decline after surgery, and there was no amnesia or material-specific memory decline observed post-surgery.

The patient was taking Keppra (1,500 mg, b.i.d.) and was being tapered off Lamictal at the time of postoperative neuropsychological testing. The patient and his family had no complaints of new onset cognitive problems after surgery and reported improvement in thinking efficiency, memory, attention, and mood at the time of follow-up assessment.

10

Other Neuropsychological Procedures in Epilepsy Surgery

Neuropsychologists design, conduct, and interpret the results of the Wada (intracarotid amobarbital) procedure, cognitive assessment during functional magnetic resonance imaging, and electrocortical stimulation mapping of language and other higher cortical functions. The rationale and methodologies of these procedures are discussed in detail in this chapter.

Intracarotid Amobarbital (Wada) Procedure

The intracarotid amobarbital procedure (IAP), or so-called *Wada* test, was developed by the Japanese epileptologist, John Wada, in the late 1940s early 1950s to establish cerebral language laterality (Wada and Rasmussen, 1960). It has since become a routine part of the pre-surgical work-up for epilepsy surgery. Over time, the purpose of Wada testing evolved beyond its initial role in identifying language laterality to include assessing risk for global postoperative amnesia, gauging the likelihood of postoperative material-specific (verbal) memory loss, and helping to confirm side of seizure onset.

Although the specific procedures vary somewhat from center to center, Wada testing usually consists of injecting 100–200 mg of sodium amobarbital (or some other fast-acting barbiturate drug) into the internal carotid artery. The amobarbital pharmacologically inactivates brain tissue served by the anterior and middle cerebral arteries on one side of the brain for several minutes, during which time the patient's motor, language, memory, and sometimes other cognitive functions are assessed. The idea behind the procedure is that once one hemisphere is anesthetized, its functions will become impaired. Thus, if language is mediated by the anterolateral portions of a hemisphere, the patient will develop aphasia after drug injection (Loring et al., 1992).

Description of the Procedure

Wada testing is typically conducted after cerebral angiography in the hemisphere of interest. Angiography is performed in the angiographic suite with the patient lying in a supine position. Access to the arterial system is obtained by the radiologist making an incision in the right groin area and inserting a catheter into the femoral artery. The catheter is then directed upward through the arterial system until it reaches the internal carotid artery just after the carotid bifurcation on the side of interest. For the purposes of Wada testing, angiography may be used to evaluate the degree of cross-flow from one hemisphere to the other, assess perfusion in the ipsilateral posterior circulation, and make certain there is no fetal origin persistent trigeminal artery, which may supply vital brainstem areas. Wada testing is usually not conducted in cases with a persistent trigeminal artery since the amobarbital could be shunted to cardiac and respiratory centers in the brainstem and harm the patient (Trenerry and Loring, 2006).

The drug is mixed shortly before testing begins. This should usually be within 30 minutes of testing since the potency of amobarbital declines with time. Amobarbital arrives packaged in a small bottle containing 500 mg (0.5 grams) of the drug in powder form. The drug is mixed with either 10 or 20 mL of saline or sterile water, and then stored in either one large (500 mg) or two smaller (250 mg each) syringes. Testing often begins with the patient holding both hands straight up into the air with palms turned backward (rostrally) toward the examiner who is standing behind the patient's head. The patient begins to count aloud and a bolus injection of amobarbital is then delivered into the internal carotid artery by the radiologist over the course of 4–5 seconds. Amobarbital dose is initially between 75 and 125 mg depending on a variety of factors, such as age, body size, gender, type and dosage of antiepileptic drug, and degree of initial drug effect. There has been a trend to use lower doses over time. Children are usually administered 75 or 80 mg, and adults are usually administered 100 or 125 mg. Drug administration is immediately followed by flushing out the catheter line with saline or sterile water. Sometimes a second injection of 25 or 50 mg will be administered if the neurological effects of the first injection are too light or absent.

After the drug and secondary "flush" have been administered, the maximum neurological effects of the drug are usually seen. A contralateral hemiparesis (weakness) or a flaccid hemiplegia (absent tone in the upper extremity) will occur, frequently accompanied by contralateral lower facial droop and less frequently an eye gaze deviation toward the side of injection. If the language-dominant hemisphere has been injected, there will be an immediate arrest of speech; that is, the patient will stop counting and usually become mute for a minute or two. Basic neurologic and language functions are assessed, and the items to be remembered are presented, during the period of hemiparesis. Memory for the items shown while the drug effect was evident is obtained after return to baseline neurologic status. This is usually 8–10 minutes after initial injection. An essentially identical procedure is followed for the opposite hemisphere. Many epilepsy surgery centers wait a minimum of 30 minutes between left and right hemisphere injections to help ensure the effects of the first injection have completely evaporated. At a few centers, the second injection is performed on a different (usually the next) day. Electroencephalogragphy (EEG) is obtained during the period of hemispheric anesthesia at some centers, but not at others.

Wada Language Assessment
Although the language tasks, specific test items, scoring, and criteria for defining left, mixed, or right hemisphere language dominance differ from center to center, most Wada evaluations include evaluation of speech arrest and visual confrontation naming. Speech arrest is defined by cessation of some form of ongoing automatic speech, such as counting or reciting the alphabet, immediately following amobarbital injection. Patients will usually display speech arrest after amobarbital injection only on the side of language dominance, but will continue to speak after the nondominant injection. This is what happens in the vast majority of unselected cases. Patients with mixed language representation are more difficult to verify and classify. Comprehensive language assessment is often necessary to establish bilateral or mixed language since these cases may be missed if all important language domains are not assessed. At the Medical College of Georgia, Loring, Meador, and colleagues recommend a comprehensive evaluation of speech and language across multiple domains to include:

- Oral fluency
- Naming
- Aural comprehension

- Sentence repetition
- Oral sentence reading
- Examination for positive signs of aphasia

In addition, they recommend that explicit criteria be used to define language representation (Loring et al., 1992). Refer to Appendix III for details of the Wada language assessment procedures and rating criteria at the Medical College of Georgia.

Recovery of Language After Amobarbital Injection
When the language-dominant hemisphere has been anesthetized with amobarbital, patients with unilateral language representation usually show severe impairments in fluency, comprehension, naming, and repetition shortly after injection. Language functions begin to recover fairly quickly after approximately 3–3.5 minutes of a dense aphasia depending upon amobarbital dose (Morris et al., 1998). The order of language recovery across language domains usually occurs in a predictable fashion. It has long been observed that reading usually recovers before the other language domains. Naming is usually the next language function to recover, and this is followed by comprehension and repetition. Ravdin, Perrine, Haywood and colleagues (1997) timed this stereotypical recovery pattern and reported the mean time for recovery of naming ability occurred 8 minutes and 29 seconds after injection, comprehension recovered at 9 minutes and 58 seconds post-injection, and repetition returned to baseline at 12 minutes and 30 seconds after injection. These recovery times are slightly longer than those typically seen at the Medical College of Georgia, probably due to higher amobarbital doses in the New York sample. This stereotypic order of language recovery occurred in 72% of their patients.

Mixed or Atypical Language Representation
The prevalence of atypical language representation varies greatly from center to center depending upon the specific tests of language employed, number of language domains sampled, and the criteria used for defining left, mixed, or right hemisphere language dominance. *Mixed language* is bilateral language representation where there is evidence of Wada language disruption after both left and right injections. *Atypical language* refers to both mixed and exclusive right hemisphere representation. Synder, Novelly, and Harris (1990) surveyed

the frequency of mixed language based upon Wada testing across 47 epilepsy surgery centers and found a bimodal distribution; slightly less than half of the centers rarely found mixed language (in 0%–6% of cases) and the other half reported observing mixed language in between 10% and 20% of cases. As might be expected, those centers reporting a lower prevalence of mixed language relied on fewer language domains for defining atypical, or non-left, language; specifically, fewer cases of mixed language were associated with procedures employing serial rote speech, expression of familiar words, or production of partial phonemes.

Atypical language representation has been linked to age of injury, handedness, and region of the left hemisphere affected by a lesion. In general, the earlier the age of injury to the left hemisphere, the more likely language reorganization is to occur. Patients who sustain damage to the left cerebral hemisphere prior to age 5 or 6 years of age are the most likely to show language reorganization. Some studies have suggested a gender effect, in which where males are unable to transfer language to other brain regions at a younger age than are females who retain the ability to reorganize longer (perhaps up to age 8 or 9 years).

Handedness (motor dominance) has been associated with language dominance since the early work of Brenda Milner and her colleagues at Montreal Neurologic Institute (MNI), and these MNI numbers have been relied upon for years to reflect this relationship in the normal population. Of course, since only certain patients were selected from among all epilepsy surgery candidates, they clearly are not representative of the healthy normal population. After 50 years of various Wada samples reported from other centers using more comprehensive evaluation methods, exclusive right hemisphere language dominance is less common than the 15% of left-handers usually attributed to the MNI sample. Furthermore, mixed language is far more common than these early studies suggest among patients who are being considered for epilepsy surgery. Examination of the Medical College of Georgia epilepsy surgery candidates reveals the following relationship between handedness and language representation (using the methods described earlier and without regard to age of injury or age of onset of epilepsy):

Right-handed: 81% show left language dominance
18% show mixed language representation
1% show right language dominance

Left- and mix-handed: 50% show left language dominance
42% show mixed language representation
8% show right hemisphere dominance

Lesion localization has also been associated with likelihood of atypical language. Patients with computed tomography (CT) or magnetic resonance imaging (MRI) confirmed lesions that affect the left frontal and parietal regions but spare the left temporal frequently show atypical language; in one sample from the University of Washington, this occurred in 34% of patients. Woods, Dodrill, and Ojemann (1988) reported that among patients with left hemisphere injury sufficient to cause right hemiparesis, 90% had atypical (either bilateral or right) language representation.

Clinical Implications of Wada Language Representation
The purpose of Wada language assessment is to determine relative laterality of language representation in the brain. Clinically, this information is used to help guide the extent of cortical resection, determine whether finer-grained language localization is required via electrocortical stimulation mapping, functional MRI (fMRI), or magnetoencephalogram (MEG), and to assist with decisions about the type of surgical procedure performed in some cases. The extent of resection most often involves decisions as to the amount of lateral neocortex to be resected in temporal lobe epilepsy patients. In cases in which the language zones are thought to intermingle or coexist with the epileptogenic areas, stimulation mapping of language, either extraoperatively from subdural grid or strip electrodes or intraoperatively using a hand-held electrical stimulation probe, may be conducted. The extent of cortical ablation is then tailored to circumvent eloquent language areas if possible.

Although functional measures, such as fMRI and MEG, show great promise for localizing language areas, they are not yet sufficiently developed to use as reliable guides for extent of resection. They are used, however, as rough guides at some centers. These activation methods typically show much more activity in multiple bilateral cortical regions than is suggested by disruption methods, such as Wada testing or electrocortical stimulation mapping. Moreover, pre- and postsurgical language testing indicates that fMRI activation sites may be included in the resection without causing a postoperative aphasia. This suggests fMRI language sites may participate in some language tasks, but are not be vital for their ongoing normal functioning. Future studies are clearly required before these functional imaging methods can reliably replace stimulation mapping as the gold standard for language localization.

Wada language and memory asymmetries may also be useful when making decisions about whether the seizure surgery should spare lateral temporal neocortex, in conjunction with the rest of the diagnostic evaluation. This may be accomplished using an amygdalohippocampectomy surgical procedure. In this surgery, the amygdala, parahippocampal gyrus, and the anterior portions of the hippocampus are usually resected sparing the lateral and inferior temporal cortex. Although amygdalohippocampectomy is thought to cause little disruption to language, it does produce some negative effects on memory in most cases, especially following left sided excisions. The procedure is described in detail in Chapter 11.

Limitations of Clinical Interpretation with Atypical Language
When Wada test results suggest bilateral or right hemispheric language representation, standard assumptions about usual localization and lateralization of cognitive functions cannot be made with confidence. Atypical language organization may be acquired secondary to early left hemisphere injury, or it may be developmental due to genetic or in utero factors. The etiology of atypical language may influence how nonlinguistic cognitive functions reorganize or develop around the speech and language zones. Unfortunately, it has been difficult to predict lateralization and localization of other functions, such as verbal memory or visual-spatial thinking, in the context of atypical language (Loring et al., 1990). Most conceivable localization scenarios have been observed in the clinic and have been reported in the literature.

If a mirror cortical representation (situs inversus) exists, then all verbal functions should be represented in the right cerebral hemisphere and all nonverbal and visual-spatial functions would be in the left hemisphere. However, this is rarely the case. More commonly, right hemisphere language seems to take over or "crowd out" the normal functions of the right hemisphere so that atypical language patients show poor visual-spatial functioning.

Predictions about location of memory functions also cannot be made with certainty in the context of atypical language. Verbal memory may be mediated by either the left or right hippocampus (or by both to varying degrees) in patients with mixed or exclusive right hemisphere speech and language. Thus, the typical inferences about location and lateralization made using neuropsychological test results can not be relied upon in patients with mixed or right hemisphere language; Wada testing, functional neuroimaging, and electrocortical stimulation mapping must be relied upon to make these determinations.

Excluding Patients from Preoperative Wada Language Evaluation
Wada testing is not performed on every seizure surgery candidate at all centers. The practice of testing all patients appears to be most common in the United States and least common in Europe and Australia (Baxendale et al., 2008). Patients scheduled for right hemisphere resection who are right-handed, have no evidence of atypical language representation, have no history of early left hemisphere insult, and show no ictal or postictal aphasia may not always undergo Wada testing. Other factors that may contribute to the likelihood of avoiding Wada testing include no family history of left-handedness, planned amygdalohippocampectomy, absence of EEG abnormalities in the left hemisphere, absence of neuroimaging abnormalities in the left hemisphere, and normal language functions on neuropsychological testing.

Some epilepsy centers are beginning to forego Wada testing (or any preoperative evaluation of language lateralization) unless the patient has some risk factors for non-normal language representation by history or testing. Unfortunately, many of these risk factors are absent in patients with refractory epilepsy who have atypical language. For example, because 12%–19% of right-handed epilepsy patients have atypical language representation, which is substantially higher than the 2% expected in the normal population, handedness may not be a sufficient predictor of atypical language (Rosenbaum et al., 1989). Rausch and Walsh (1984) reported finding right hemisphere language dominance in right-handed patients with seizures originating from the left hemisphere. Unexpected abnormal patterns of performance on neuropsychological testing may also not show up in cases of atypical language. The same may be said for side of seizure onset. These observations suggest that great caution should be exercised when excluding epilepsy surgery candidates from Wada or other language lateralization studies preoperatively.

Wada Memory Assessment
Purpose of Wada Memory Assessment
Most epilepsy surgery centers evaluate memory during Wada testing by presenting real objects, photographs, or line drawings during the period of hemispheric anesthesia, and then asking for recall after the drug has worn off and the patient has returned to neurologic baseline. The drug effect usually

lasts between 7 and 10 minutes, and so memory recall is usually evaluated around 10 minutes after injection. Over time, the purposes of Wada memory assessment have broadened to predict material-specific (usually verbal) memory loss following temporal lobectomy, aid in confirming side of seizure onset, and assist in predicting degree of postoperative seizure control.

Testing the functionality of the hippocampus for new learning and memory on both the side of proposed surgery and the contralateral (usually healthy) side has become useful for several different reasons. Originally, the purpose of evaluating memory after amobarbital injection ipsilateral to the side of seizure focus was to predict the rare possibility of developing a global amnesia after temporal lobectomy. If the contralateral hippocampus is significantly dysfunctional, then ipsilateral amobarbital injection is expected to produce a transient amnesia due to the bilateral functional inactivation of the hippocampi. Refer to Appendix III for a detailed description of the Wada memory assessment procedures and scoring and interpretation guidelines at the Medical College of Georgia.

A factor to be considered in interpreting Wada memory performance is the timing of stimulus presentation. How soon items to be remembered are presented after amobarbital injection influences recall. Memory for items presented shortly after injection is better for lateralizing the side of seizure onset than are items that are presented after partial return of language function (Loring et al., 1997).

Prediction of Memory Loss. The purpose of Wada memory testing, as originally proposed by Milner and her colleagues at the MNI, was to identify patients with significant damage to the hippocampus on the side opposite of proposed temporal lobectomy. In this discussion (as in most of this book), *ipsilateral* refers to the side of seizure onset and *contralateral* refers to the nonepileptogenic side. Patients failing Wada memory testing after ipsilateral injection are thought to have contralateral hippocampal dysfunction sufficient to interfere with the formation of new memories, and are at risk for development of a postsurgical amnesia. Chelune (1995) and colleagues at the Cleveland Clinic have conceptualized this use of Wada memory testing as evaluating the patient's *functional memory reserve* of the contralateral hippocampus.

Wada memory assessment evolved to predict risk for degree of material-specific memory decline after ipsilateral temporal lobectomy. This prediction relates to the functional status of the hippocampus to be included in the anterior temporal lobe resection and has been termed *functional adequacy* by

Kneebone and colleagues (1995). There have since been numerous studies to support this notion; that the more functional (or "healthy") the ipsilateral hippocampus (as measured by adequacy of memory functioning during Wada testing), the greater the postoperative memory decline after anterior temporal lobectomy. Wada memory assessment allows both the *functional reserve* and the *functional adequacy* to be measured since the ipsilateral injection evaluates the reserve and contralateral injection the adequacy. Furthermore, both scores may be represented simultaneously by a single number; a Wada memory asymmetry score, which is the ratio or proportion of ipsilateral versus contralateral memory items recognized. More about Wada memory asymmetries will be discussed later.

Confirming Side of Seizure Onset. It has been known for a long time that Wada memory testing may assist in confirming lateralization of the epileptogenic zone, particularly in patients with temporal lobe epilepsy. In the initial paper on Wada memory assessment, Milner and colleagues (1962) at the MNI reported that 92% (11 of 12) of patients displayed a memory disturbance after amobarbital injection contralateral to the side of seizure onset. Since this early report, others have confirmed that the majority of patients have worse memory after contralateral injection and most show relatively normal memory after ipsilateral injection. For example, Alpherts, Vemeulen, and van Veelen (2000) obtained correct lateralization of ictal seizure onset in 85% of patients with various neuropathologies. These numbers remain roughly the same regardless of whether the epileptogenic focus is in the left or right hemisphere. Similar (but slightly less robust) results have also been obtained in patients with seizure foci in regions other than the temporal lobes.

Prediction of Seizure Control. Wada memory testing may also be used to predict seizure-relief outcome. Patients with significant neuronal loss in the hippocampus who have evidence of mesial temporal lobe sclerosis on MRI are more likely fail Wada memory testing after injection contralateral to the epileptic focus. Such patients are also more likely to have better surgical outcomes than patients without these signs of neuropathology.

A potentially more powerful method in which to use Wada memory testing in both clinical and research applications is by summarizing both left and right memory scores in a Wada memory asymmetry index. This may be calculated in a number of ways. Wada memory asymmetry scores in their most simply form may be calculated by taking the number correctly recalled after ipsilateral injection and subtracting it from number correct after contralateral injection. In this calculation, a positive score would indicate better Wada memory performance with the nonepileptogenic hemisphere and tend to suggest ipsilateral seizure onset and little risk for amnesia. In contrast, a negative score would represent better memory performance when relying on the epileptogenic hemisphere and tend to suggest possible contralateral dysfunction and risk for poor seizure control after surgery. Another frequently used method to determine Wada memory asymmetries is to subtract the corrected memory score after left injection from the corrected memory score after right injection. In this calculation, positive scores reflect relative left hippocampal dysfunction. This method is discussed in detail in the "Wada Assessment Procedures and Rating Criteria at the Medical College of Georgia" in Appendix III.

Studies from various epilepsy centers have found patients with Wada memory asymmetries that differed by two or three points in the predicted positive direction were significantly more likely to become seizure-free after anterior temporal lobectomy. Between 80% and 89% of adult patients with positive Wada memory asymmetries were seizure-free after left or right temporal lobectomy, whereas only 57%–63% of patients who did not show such an asymmetry were seizure-free at 1-year follow-up (Loring et al., 1994; Perrine et al., 1995). Similarly in children, we have found that Wada memory asymmetries predict seizure control 1 year after temporal and nontemporal focal resections, regardless of the magnitude of the asymmetry (Lee et al., 2003).

Which Drug to Select for Wada: Amobarbital (Amytal) or Methohexital (Brevital)?

Most epilepsy surgery centers use sodium amobarbital for Wada testing since amobarbital has been used the longest, the most data accumulated and the best validation about its effects, has low toxicity, and is relatively brief in duration of action (Buchtel et al., 2002). However, in the late 1990s, a shortage of amobarbital occurred due to a change in drug manufacturer, and when the new manufacturer began resupplying the drug, there was little availability outside the United States. This caused epilepsy centers, mostly outside the United States, to experiment with alternatives to amobarbital, primarily methohexital (Brevital), etomidate, and propofol. Methohexital (Brevital) has become the most widely used alternative to amobarbital for Wada testing.

Because methohexital is faster-acting and more potent than amobarbital, repeated administration is required at lower doses to achieve a similar effect to amobarbital. A total of 3 mg of methohexital is usually administered at first, and this is followed 1–2 minutes later by a second injection of 2 mg. In some cases, if the drug effect is judged to be too weak, additional injections of 2 mg each are administered until the desired drug effect has been achieved. Unfortunately, the multiple doses usually necessary with methohexital require the catheter to stay in the internal carotid artery for a longer period of time than with amobarbital, and this slightly increases morbidity risk (primarily of vasospasm in the younger-aged Wada population). Nevertheless, some examiners prefer the briefer duration of action and less associated sedation of methohexital relative to amobarbital, and the language results yielded by the two drugs have been roughly equivalent.

Several different surgery centers have raised concerns, however, that methohexital does a poor job of detecting Wada memory asymmetries and is, therefore, a poor predictor of postsurgical memory change. For reasons that are unclear, methohexital appears to overestimate the memory capabilities of the hippocampus ipsilateral to the side of seizure onset in temporal lobe epilepsy cases (Andelman et al., 2006). Thus, methohexital seems to be an inaccurate predictor of the *functional adequacy* of the to-be-resected hippocampus.

Wada Testing in Children
Although Wada testing has been validated for use in children, there are several important factors to consider in selecting the stimuli, conducting the test, and interpreting the results. All children are not suitable candidates for Wada testing. Because the angiography necessary for Wada can be a physically uncomfortable and emotionally frightening experience for children, some children may lack the maturity to fully cooperate in all aspects of the testing. Some important considerations in trying to decide whether a particular child is a suitable candidate for Wada testing include age, emotional maturity, tolerance for pain, and intellectual level.

In general, children younger than 7 or 8 years old are considered unsuitable candidates for the procedure, and the results of both Wada language and memory assessments are less reliable in children under 10 years old (Schevon et al., 2007). Emotional maturity and pain tolerance may be gleaned from history, observations, and nursing reports of behavior while in the hospital. For children who are overly intolerant to pain but require Wada testing, sedation may be administered by an anesthesiologist before the radiologist makes the incision for catheter placement. At our institution, propofol is currently favored for this purpose since it is a short-acting agent, and the child can be ready to cooperate with Wada testing 15–30 minutes after the anesthesia is stopped. In some cases, however, a child may awaken from anesthesia in a disoriented or highly charged emotional state and require a good deal of soothing before he is capable of proceeding. A parent can be quite helpful in calming the child in this immediate postanesthesia period, and afterwards, the parent may be escorted out of the angiography suite prior to amobarbital injection.

Level of general cognitive development has also been a predictor of ability to adequately participate in Wada testing. For example, in one series of children (ages 5–12 years), language testing was successful in all children with IQs of at least 70, whereas only 57% of children with IQs less than 70 satisfactorily completed Wada language testing (Szabo and Wyllie, 1993). In this same series of children, those with IQs of 70 and above had good memory retention scores whereas children with IQs of less than 70 were much less likely to show a lateralized memory asymmetry.

If a particular child is thought to be a suitable candidate for Wada testing, the next step is to select appropriate stimulus materials that are tailored to the child's developmental and cognitive level. For language evaluation, a wide-range selection of naming, comprehension, repetition, and reading materials should be gathered. Materials of an appropriate difficulty level for the child are chosen at the pre-Wada baseline assessment, and these will serve as the language items during the Wada. Reading is often not included in our Wada assessment of children because many youngsters cannot read or read very poorly. The Wada memory test stimuli at the Medical College of Georgia consist of real objects that are very common, colorful, and salient in nature (see end of Appendix III for specific memory objects), which are generally suitable for use with children. In children who need it, we sometimes reduced the number of memory stimuli to accommodate those with lower memory abilities.

The symptoms of language impairment seen in children after amobarbital injection differ from that usually observed in adults, and this can make interpretation more difficult. More often than not, children will simply become mute after injection, and thus, there may be no positive signs of aphasia, such as phonemic or semantic paraphasic errors or circumlocutions, to help confirm that language has been affected. Many children display a prolonged period of severe aphasia, with an absence of ability to speak, name, follow commands, repeat phrases, or read, and then suddenly be able to perform all language tests normally. This usually poses no serious impediments to interpretation when language dominance is clearly unilateral, but may complicate interpretation in cases of mixed language representation.

Finally, interpretation of Wada memory asymmetries is usually more difficult in children than adults. The major factors influencing interpretation of memory performance in children appears to be age, IQ, and region of seizure onset. As noted earlier, there is evidence that well-lateralized Wada memory asymmetries are less common in children who are less than 10 years old and in those with full-scale IQs of less than 70. Compiling Wada results from three different surgery centers, we found Wada memory testing was able to accurately lateralize seizure onset in 69% of children, a rate that is somewhat lower than the 70%–88% classification rates reported in adult surgical candidates (Lee et al., 2002). This slightly worse rate of seizure lateralization in children may be due to some combination of age, IQ, and the lower prevalence of temporal lobe seizures found in pediatric populations compared with adults.

Functional Magnetic Resonance Imaging

Functional MRI is a noninvasive procedure carried out in an MRI scanner, and as such, is readily available throughout the country and carries little risk to the patient (e.g., no radiation exposure is associated with the test). The test's primary purpose in the evaluation for epilepsy surgery, at present, is sensorimotor and language mapping; and fMRI is well on its way toward replacing the intracarotid amobarbital (Wada) procedure for these purposes (Swanson et al., 2007).

An fMRI is a measure of increased capillary blood flow within the brain caused by a rise in the metabolic demands of neurons. It is used in epilepsy surgery to identify eloquent cortex which may then be spared during the surgical resection. Specifically, fMRI computes the blood oxygenation level-dependent (BOLD) response, which is determined by measuring relative MRI signal changes in capillary hemoglobin from a deoxygenated to an oxygenated state. The MRI signal changes are determined by measuring the difference between some form of activation task (e.g., motor or cognitive) and a control condition, which must be carefully matched to the activation task in every detail except the variable of interest (e.g., language processing). In other words, the control task should not elicit a BOLD response from the brain regions of interest. Designing cognitive tasks to activate specific brain regions (and matching them to effective control conditions) is more difficult than most would suppose (Tracy and Shah, 2008). An often overlooked problem with fMRI interpretation is that it is unknown whether cortical BOLD activation represents activation of excitatory or inhibitory neuronal populations and networks; it is almost universally assumed activation represents excitatory neuronal activity.

Sensorimotor Mapping Using Functional Magnetic Resonance Imaging
Much evidence has accumulated that shows fMRI can localize cortical areas that are involved in sensory, motor, and a variety of language tasks with reasonable accuracy. Areas of eloquent cortex supporting these functions often need to be identified in patients undergoing surgery in the frontal, parietal, and temporal lobes, so that the neurosurgeons can plan the resection ahead of time. Most epilepsy patients who have undergone fMRI sensorimotor mapping were being evaluated for surgical resections of tumor, vascular malformation, or cortical dysplasia in the primary motor and somatosensory areas (Gaillard, 2006). The primary motor cortical representations of the upper and lower extremities, face, and mouth can be identified through toe wiggling, finger tapping, and tongue movements, whereas the equivalent sensory areas are often mapped by light touch or a puff of air. These fMRI methods have been validated through good correlations with electrocortical stimulation mapping and sensory evoked-potentials at the time of surgery.

Language Mapping Using Functional Magnetic Resonance Imaging
The fMRI has also been shown to reliably determine hemispheric dominance for language with reasonably good agreement with intracarotid amobarbital (Wada) procedure language results. An advantage of fMRI over Wada testing, however, is fMRI has greater spatial resolution, and thus, has the potential to further localize language functions contained in particular lobes and even sulci within both cerebral hemispheres. The disparity between fMRI and Wada for lateralizing language may be reduced to 5%–8% when multiple measures of language are used in both procedures (Gaillard et al. 2004). In general, fMRI is more likely to suggest widespread cortical language representation than Wada testing, including bilateral language. Because of this, fMRI may show multiple areas of activation that may be involved in a language task, but that are not necessarily required to perform the task. This has been shown when some of

these "nonessential" areas have been included in the surgical resection without any demonstrable postsurgical language deficits.

The fMRI tasks that have been used to lateralize language functions have mainly involved visual naming, phonemic fluency, semantic fluency, generating a rhyming word, verbal generation from nouns, semantic decision making, auditory processing tasks, and reading of single words, sentences, or stories. Gaillard (2006) has summarized results of the studies using these paradigms as follows. Verbal fluency paradigms have been shown to reliably activate inferior frontal cortex (Brodmann's areas 44 and 45) and midfrontal and dorsolateral prefrontal cortices (Brodmann's areas 9 and 46). Semantic decision tasks (e.g., determining whether a word represents something alive or not, or whether a word pair is abstract or concrete) also activate mid-dorsolateral frontal regions (Brodmann's areas 9, 46, and 47). Reading and some auditory processing tasks most often produce fMRI activation in the posterior temporal lobe (superior and middle temporal gyri), as well as in the dorsolateral prefrontal cortex of the language-dominant hemisphere. Many of these tasks also result in less potent activation of homologous regions in the nondominant (right) hemisphere.

As these results suggest, activation of the anterior language areas has been accomplished with relative ease, whereas selective activation of posterior language cortex has been more difficult. Control conditions designed to nullify the nonlinguistic aspects of these language activation tasks have included listening to unfamiliar languages and reverse speech (Ahmad et al., 2003).

There is little doubt that fMRI will replace Wada testing to lateralize language functions in the near future. Unfortunately, most of the fMRI language studies present on normalized group averages using standardized structural MRIs, and so it is difficult to apply these results to individual epilepsy cases with confidence. Further development and standardization of tasks and fMRI statistical decision making methods are still needed before fMRI language mapping becomes a reliable clinical procedure.

Functional Magnetic Resonance Imaging and Memory
In contrast to using fMRI to lateralize language, the ability of fMRI to lateralize memory functions is much less developed and less well validated at this time. The primary goals of fMRI memory activation in epilepsy surgery is the same as Wada memory assessment; namely, to predict whether the mesial temporal lobe structures contralateral to the side of surgery will be able to support

memory functions (no postsurgical global amnesia) and to gauge the degree of risk for material-specific memory decline after a unilateral lobectomy. Much work needs to be accomplished before the predictive correlates of relative activation patterns across various brain regions is understood in terms of postoperative memory outcome in the individual case. Although imperfect, Wada testing is currently used as the standard for prediction of postsurgical memory impairment.

Early work in fMRI memory lateralization has nevertheless produced some valuable results, and it is expected that fMRI could eventually replace the IAP for prediction of memory impairment risk. For example, studies from two epilepsy centers showed left memory dominance of preoperative fMRI activation predicted verbal memory decline after left temporal lobectomy (Binder et al., 2008; Richardson et al., 2006). Unfortunately however, it has been difficult to find tasks that selectively and consistently activate the left and right hippocampi and associated mesial temporal lobe structures. Complex visual scenes, human faces, and spatial mental navigation encoding tasks have been shown to activate the parahippocampal regions bilaterally in healthy controls, but reduced activation on the side of seizure onset in patients with temporal lobe epilepsy (Kilgore et al., 1999; Jokeit et al., 2001). Verbal encoding tasks have shown a predilection for left, and facial encoding tasks for right, mesial temporal lobe activation, and some have found greater activation in left hippocampal regions during verbal fMRI encoding and predicted a greater postoperative decline in verbal memory. Again although promising, such results have been inconsistent and cannot yet be relied upon for reliable clinical prediction.

Electrocortical Stimulation Mapping

The cortical localization of language and other cognitive functions varies from person to person, and this makes it impossible to rely solely on anatomical landmarks to identify eloquent areas of cortex in the individual patient. In patients who are candidates for epilepsy surgery in the language-dominant hemisphere, electrocortical stimulation mapping may be used to identify eloquent language areas that are in and around the epileptogenic zone. Cortical sites where stimulation produces language impairment are considered vital for normal language function and are, therefore, generally not included in the surgical resection to preserve language postoperatively.

Electrocortical stimulation mapping is the gold standard for identifying essential language cortex. This is considered different from sites that participate in (but are not essential for) language identified using functional imaging techniques, such as fMRI or PET, which in some instances can be surgically removed without causing a permanent aphasia (Silbergeld, 2001). Language mapping may be accomplished with either chronically implanted subdural electrodes at bedside or intraoperatively during the craniotomy with local anesthesia just prior to surgical resection to remove the epileptic focus.

Extraoperative Stimulation Mapping

When the seizure focus has not been precisely localized using noninvasive measures, such as scalp EEG or MEG, or when the epileptogenic zone is near eloquent (language or sensorimotor areas) cortex, patients may undergo phase II monitoring in which invasive grid, strip, or depth electrodes are placed on or within the brain. When the results of noninvasive tests for seizure localization are discordant or have not localized the exact area of seizure onset, subdural electrodes may be implanted over extended periods of time (usually days or up to one week) to record interictal and ictal EEG abnormalities while the patient is in the epilepsy monitoring unit. Precise localization of the epileptic zone is important because it must be resected as completely as possible to obtain good seizure relief. In addition to recording EEG abnormalities, these implanted electrodes may be electrically stimulated to identify eloquent cortex surrounding the epileptogenic zone. Language mapping is usually conducted in cases where the planned cortical resection is to include the posterior temporal lobe or posterior inferior frontal lobe of the language-dominant hemisphere.

Stimulation Parameters and Methodology

The stimulation intensity used in mapping should be strong enough to cause depolarization of the neurons underlying the stimulating electrode without spreading to adjacent cortex or causing harm to the brain. Stimulation parameters differ slightly from center to center but most commonly consist of 3–20 second trains of constant current 50–60 Hz biphasic square wave pulses of 0.2–0.5 msec duration (Luders et al., 1994; Ojemann, 1994). Amperage is usually higher during when mapping from grids and lower when stimulating intraoperatively. Extraoperative mapping amperages range from 2 to 17mA, whereas intraoperative amperages typically range from 1 to 10 mA. The two major constant current stimulators in current use are the Grass Instruments (models S-12 or S-212, Burlington, MA) or the Ojemann Cortical Stimulator (Radionics, Inc.). Electrical current is typically delivered through platinum contacts (3 mm in diameter) embedded within flexible polyurethane grid or strips placed in the subdural space on the cerebral cortex. The distance between electrode contacts is generally 1 cm. Stimulation is usually bipolar between two adjacent contacts. Cortical stimulation begins before the test items are presented and continues throughout the preselected stimulation period (usually ~10 seconds). The test stimuli are withdrawn after a positive error or when the stimulus period has ended. Cognitive testing and EEG are conducted and monitored throughout the stimulation session.

There are two primary methods used during extraoperative stimulation language mapping. One method is to begin stimulation at low current intensities (~1 or 2 mA) during language testing and gradually increase intensities (up to around 17 mA) at *each and every* electrode site until one of the following occurs (1) an afterdischarge, (2) language disruption, or (3) maximum safe current level is reached (~15–17 mA). This method takes hours to complete.

The other cortical stimulation mapping procedure is to establish the afterdischarge threshold in nonepileptogenic cortex (usually in sensory or motor areas). An afterdischarge (AD) is an electrographic seizure which is usually subclinical and not consciously experienced by the patients. An AD may spread to other areas of cortex beyond the bipolar pair of contacts that initiated the AD. Once the AD threshold has been established, language testing is then conducted at a current level that is 0.5–1 mA below the AD threshold, and the same current level is used to evaluate all contacts. This method takes less time to complete than gradually increasing the current level at each contact method.

When an AD occurs during mapping, a break of one to two minutes is usually taken before beginning stimulation again to ensure the neuronal refractory period has ended. If a clinical seizure occurs, then stimulation mapping is typically stopped for the rest of the day and resumed on a subsequent day. Extraoperative mapping should ideally take place after the necessary video-EEG seizures have been captured so that the patient's serum antiepileptic drug concentration is at therapeutic levels. The nursing staff will have a short-acting intravenous antiepileptic drug, such as Versed (midazolam) or Ativan (lorazepam), readily available in case seizures are elicited during electrical stimulation mapping.

Language and Related Tests Used During Mapping
The role of the neuropsychologist is to select, administer, score, and interpret the results of cognitive testing. The single most common language task used to detect cortical language sites during stimulation mapping is visual

confrontational naming. Naming is an excellent task choice for mapping language because naming deficits occur in *all* the classical aphasia types, stimulation and lesions in most peri-Sylvian regions can cause naming deficits, and naming errors may be either "positive" (i.e., paraphasic or circumlocution responses, intrusions, perseverations) or "negative" (i.e., an inability to come up with the name of the object, speech arrest). Stimuli are usually presented on cards or on a computer monitor and consist of line drawings of simple common objects (such as animals, clothing, furniture, food, tools), which may be selected from commercially available naming tests from standardized aphasia batteries, such as the Boston Naming Test.

Other frequently used tasks to evaluate language during cortical stimulation mapping include repetition of simple phrases and sentences, reading aloud, reading comprehension, comprehensive of aural commands, writing to dictation, and automatic speech tasks (e.g., counting, reciting nursery rhymes, months of the year, pledges, prayers, or the alphabet). The tasks to be used are designed to assess the specific cortical areas to be mapped and difficulty levels of the stimuli are tailored at baseline testing for each patient. Test items that are not answered correctly at baseline cannot be used to identify stimulation errors during mapping. Less common tasks that have been used at times include responsive or "auditory naming," for example, "What is a type of boat that travels under the water and can fire torpedoes?" (Hamberger et al., 2001) in temporal lobe cases; or mental calculations, finger localization, or right-left discrimination for mapping inferior parietal lobule regions (Morris et al., 1984). Neuropsychologists may devise any number of tasks to evaluate the functional capacity of cortical tissue underlying electrode contacts being considered for inclusion in a cortical resection based upon our knowledge of the behavioral geography of the brain.

Interpretation of Results
It is generally thought that language errors must be reliably reproducible at the stimulation site through repeated trials before inferring the site's importance in the function tested. Ojemann (1987) requires consistent errors on three intraoperative trials, and Luders and colleagues (1994) require failure on two extraoperative trials, and verification that the patient can perform the task normally without stimulation, before concluding there is a cause and effect relationship between the stimulation and the observed language deficit.

Stimulation of primary sensory and motor areas usually produce "positive" responses such as, tingling in the thumb or tongue, light flashes, buzzing sounds, mouthing movements, or finger flexion. Although "negative" errors are much more common during language mapping, both "positive" and "negative" errors may be encountered during language mapping and opportunities for both types of errors must be arranged. Positive signs of language disruption include such things as semantic or phonemic paraphasic substitution errors, circumlocutions, and perseverations. One of the difficulties with language mapping is that disruption is most often evidenced by "negative" signs. These include an inability to perform the task normally, speech arrest during ongoing speech, hesitations of speech. Thus, it is important during mapping to accurately note the consistency of both positive and negative error types.

With regard to the rules of thumb for stimulus intensity, Luders and colleagues (1994) have suggested an absence of signs or symptoms when stimulating at 15 mA for 5 seconds or longer is a fairly reliable sign that the cortex being stimulated is actually "silent." The same assumption cannot be made, however, in cortical sites that were never tested at more than 2–3 mA due to afterdischarges. These authors stress that this is especially true for "language electrodes" in the temporal lobe, which typically do not give positive results at intensities below 5–10 mA.

Intraoperative Stimulation Mapping
Intraoperative language mapping has become less common than extraoperative mapping in recent years for a variety of reasons. Each procedure has its particular advantages and disadvantages. Mapping from chronically implanted subdural grid electrodes allows more time for evaluating language (and other higher cortical functions) using a wider variety of electrical stimulation parameters in a less stressful setting than intraoperative monitoring (Ojemann and Engel, 1987). Extraoperative mapping is also easier on children since they are not required to be awake during surgery. Ojemann and Engel (1987) list the advantages of intraoperative language mapping as having (a) greater flexibility in sampling more cortical areas that would be beyond the boundaries of a grid, (b) greater assurance that the electrodes are actually in contact with the cortex (there can be gaps between grid electrodes and the surface of the cortex), and (c) less risk of morbidity associated with chronic insertion of subdural electrodes and intracranial recording and stimulation. There is also at least one, and possibly two, fewer surgeries necessary if mapping is conducted intraoperatively because the additional surgery to implant the subdural grids is not undertaken.

Stimulation Parameters and Methodology

The equipment and stimulation parameters used during intraoperative mapping are similar to that used with extraoperative stimulation mapping. A constant current, biphasic, square wave at 60 Hz using bipolar stimulation between hand-held carbon tipped electrodes (spaced either 5 or 10 mm apart) is delivered using either a Grass or Radionics stimulator. The stimulation probe contacts the cortex for 5–10 seconds. Stimulation of sensory and motor cortex usually begins at 2 mA and is gradually increased until some movement or sensation is elicited in an awake patient up to ~10 mA or until an AD is seen. Simultaneous electrocorticography using several strip electrodes is necessary to determine the AD threshold and to make sure stimulation is working and not spreading to other cortical areas. Simultaneous electrocorticography (ECoG) refers to recording EEG directly from the surface of the cortex.

The electrical current selected for subsequent language mapping is set at 0.5–1 mA below the AD threshold (Ojemann and colleagues, 1989). Following establishment of the AD threshold, 15–20 small (~5 × 5 mm) sterile numbered tags are placed throughout the perisylvian region that includes the suspected language areas and site of the planned resection. At some centers, stimulation mapping begins before placement of the numbered tags, and then the tags are placed only after a response of interest has been elicited. The language and other cognitive tasks used during intraoperative mapping are the same as that used during extraoperative mapping. Similar baseline and interpretative guidelines are applied during both procedures. Moreover, the role of the neuropsychologist is the same in the two mapping procedures; namely, to select, administer, score, and interpret the results of the cognitive testing.

Confrontational visual naming is by far the most common language task used intraoperatively. Cortical stimulation is started before presentation of the item to be named and is continued for 5–10 seconds, after which the stimulus is withdrawn. Each preselected site is stimulated at least twice, and at some centers three times. Stimulation sites that repeatedly result in some form of positive (e.g., paraphasia, circumlocution, perseveration) or negative (e.g., inability to name, speech arrest, hesitations) error during naming are considered vital language areas. Other language tasks (e.g., repetition counting or rhyme recitation, sentence repetition, aural comprehension) and nonlanguage higher cognitive function tests (e.g., finger localization, memory) may also be designed for use during intraoperative mapping of functional cortex as necessary. The rule of thumb that has been followed at the Medical College of

Georgia is that the resection margin should be 1-2 cm away from vital language areas to avoid permanent impairments. Ojemann and Dodrill (1985) found that resections that came within 2 cm of vital language cortex may cause postoperative speech difficulties.

Case Example

Wada Testing Predicts Memory Decline in a Case with Right Hemisphere Language Dominance

Patient History:
A 27-year-old, mixed-handed (wrote with right hand), white man with 12 years of education presented for possible epilepsy surgery with a history of refractory complex partial seizures since the age of 4 years. Etiology was thought to be a series of febrile seizures suffered during the first year of life. There are no other known etiological factors, such as birth complications, cerebrovascular condition, brain infection, concussion, brain tumor, or brain surgery. Seizures began with whistling and behavioral arrest, followed by versive head and eye deviation toward the left side and making circular motions with the left hand and arm. The patient was unresponsive and had no recollection for events that occurred during the spells. Video-EEG monitoring using scalp electrodes recorded multiple typical seizures arising from both the left and right temporal lobe regions. Brain MRIs showed no evidence of mesial temporal lobe (or hippocampal) sclerosis or any structural abnormality. Although one MRI suggested possible dilatation of the left temporal lobe of the lateral ventricle and mild deep mesial temporal lobe atrophy, two subsequent preoperative MRIs were interpreted as normal.

Because the seizure focus could not be accurately localized using scalp electrodes, the patient underwent Phase II video-EEG monitoring after bilateral stereotaxic implantation of depth electrodes into both mesial temporal lobes. Three seizures were recorded, which appeared to develop independently in the left and right mesial temporal lobes by depth electrode EEG; two seizures began on the right and one seizure began on the left.

Wada Testing:
During intracarotid amobarbital (Wada) evaluation, the patient demonstrated right hemisphere dominance for speech and language. There was a significant asymmetry in object memory performance with superior performance following the left injection and deficient memory performance after the right injection; suggesting relative reliance on right MTL structures for memory. Wada results were as follows:

	Left Injection	Right Injection
Expressive Speech:	Normal	Arrest of counting
Receptive Speech:	Normal	Severe impairment
Visual Naming:	Normal	Mild impairment
Repetition:	Normal	Severe impairment
Reading:	Normal	Severe impairment
Paraphasic Errors:	None	Semantic and phonemic errors
Object Recognition Memory:	8 of 8, 0 False Positives (FPs)	0 of 8, 0 FPs

Neuropsychological Testing Results:
The following interpretive guidelines are offered to assist with interpretation of the scores below:

SS	Standard Score ($M = 100$, $SD = 15$); SS ≤ 75 indicates potential impairment.
ss	scaled score ($M = 10$, $SD = 3$); ss ≤ 5 indicates potential impairment.
z	z score ($M = 0$, $SD = 1$); $z -\leq 1.6$ indicates potential impairment.
T	T score ($M = 50$, $SD = 10$); $T \leq 34$ indicates potential impairment,
percentile	Percentile ranks (M = 50th, -1 SD = 16th, $+1$ SD = 84th, Impaired <5th).
RS	Raw score

General Intellectual Functioning:

	Pre-Surgery	Post-Surgery
Verbal Comprehension Index (SS)	79, 8th percentile	73, 4th percentile
Similarities (ss)	5	3
Comprehension (ss)	8	6
Information (ss)	5	5
Working Memory Index (SS)	86, 18th percentile	90, 25th percentile

Digit Span (ss)	9	10
Arithmetic (ss)	6	6
Perceptual Organization Index (SS)	95, 37th percentile	96, 39th percentile
Picture Completion (ss)	9	9
Block Design (ss)	9	8
Picture Arrangement (ss)	9	11
Processing Speed Index (SS)	96, 39th percentile	94, 34th percentile
Digit Symbol (ss)	11	9
Symbol Search (ss)	8	9

Language:

	Pre-Surgery	Post-Surgery
MAE Visual Naming	50, 27th percentile	26, <1st percentile
MAE Letter Fluency (COWA)	35, 40th percentile	24, 6th percentile
MAE Token Test	41, 33rd percentile	42, 45th percentile

Verbal Learning and Memory:

	Pre-Surgery	Post-Surgery
WMS-R Logical Memory I (RS)	15, 8th percentile	4, 1st percentile
WMS-R Logical Memory II (RS)	3, 1st percentile	0, <1st percentile
Selective Reminding (CLTR, RS)	53, 1st percentile	29, <1st percentile
California Verbal Learning Test Trials 1–5	40, 8th percentile	14, <1st percentile
Short-delay Free Recall	4, 2nd percentile	3, 1st percentile
Serial Digit Learning (supraspan)	17, 27th percentile	1, 1st percentile

Nonverbal Learning and Memory:

	Pre-Surgery	Post-Surgery
WMS-R, Visual Reproduction I (RS)	38, 90th percentile	35, 65th percentile

WMS-R, Visual Reproduction II
(RS) .... 34, 74th percentile 5, 1st percentile
Complex Figure 3' Immediate
Recall (RS) .... 17.5, 7th percentile 4, <1st percentile
Complex Figure 30' Delayed
Recall (RS) .... 15, 1st percentile 4, <1st percentile

Visuoperception and Spatial Thinking:

	Pre-Surgery	Post-Surgery
Facial Recognition Test........	45, 51st percentile	49, 85th percentile
Complex Figure Copy	36, >16th percentile	36, >16th percentile
Judgment of Line Orientation...	22, 22nd percentile	25, 56th percentile

Academic Achievement:

	Pre-Surgery	Post-Surgery
WRAT-3 Reading (SS)	63, 1st percentile	51, <1st percentile
WRAT-3 Spelling (SS)	59, <1st percentile	59, <1st percentile
WRAT-3 Arithmetic (SS)............	76, 5th percentile	82, 12th percentile

Surgery and Outcome:

Because seizures recorded with depth electrodes began independently in the left and right mesial temporal lobes (MTL), a decision regarding surgery was deferred. Although more seizures were recorded from the right MTL region, Wada testing suggested the patient would be at risk for severe memory impairment if the right ATL were resected. Preoperative neuropsychological testing was difficult to interpret since it was unknown whether cognitive functions were normally organized (i.e., lateralized) in the cortex due to left-handedness, right language dominance by Wada testing, and early age of seizure onset (~4 years old).

The patient and his family pursued epilepsy surgery at a different institution, and after an appropriate preoperative evaluation, the patient underwent a right anterior temporal lobectomy without intraoperative language mapping. Surgery consisted of a 4.5 cm resection of the lateral temporal neocortex with subpial aspiration of hippocampus back to the posterior margin of the cerebral peduncle. Immediately following the surgery, the patient demonstrated a severe global amnesia syndrome in conjunction with an expressive aphasia. Unfortunately, in addition, there was no significant reduction in seizure frequency postoperatively. Preoperative average seizure frequency was 5–20 per month and postoperative average seizure frequency was seven per month, which is considered an Engel Class IV ("No worthwhile improvement") outcome.

Clinically and psychometrically, the patient was globally amnesic after surgery. Regular follow-up over a period of 5 years after surgery showed persistent language and memory impairments with memory being much worse. He was consistently unable to orient himself to time or date, and as more time went by, to his current age. Episodic memory was extremely poor for events that occurred in recent weeks or even earlier in the same day. For example, he'd forget having gone to church, shopping, or out to dinner with his family. He would go to a movie with his brother and have no recollection of having done so the next day. The patient enjoyed working on word puzzles in the newspaper, magazines, and special puzzle books. After surgery, he would solve the same puzzle from different copies of the same puzzle book without remembering that he had already solved the identical puzzle at an earlier session (see Loring, Hermann, Meador, et al. [1994] for more details of this case).

Postoperative neuropsychological testing corroborated the severe memory loss. The WMS-R General Memory and Delayed Recall Indexes significantly decreased from the low-average (SS = ~85) to severely defective (SS = <50) range. New verbal learning showed a similar deficit; before surgery CVLT Trials 1–5 learning was 40/80, whereas after surgery it was 14/80. Although perhaps not as evident as these examples, there were consistent declines across other tests of verbal and nonverbal memory (see testing results). The acute and severe postoperative aphasia resolved over time, but the patient was left with mild fluency difficulties and substantial word-finding problems as reflected in decline on MAE Visual Naming from a score of 50/60 preoperatively to 26/60 after surgery.

The outcome in this case is exactly the scenario epilepsy neuropsychologists are hired to protect against. Entering the data available on this case into the more recently developed Stroup (2003) memory risk regression equation (see Chapter 9) suggested there would be a 89% chance of a severe postoperative verbal memory decline if the patient underwent a right anterior temporal lobectomy that included the hippocampus. The important predictive risk factors for postoperative memory decline included surgery on the language-dominant side, absence of hippocampal sclerosis in the to-be-resected temporal lobe, normal Wada memory (8/8 recall) after contralateral injection, and significant memory deficits on preoperative neuropsychological testing. In addition to these considerations, the Wada memory score of 0/8 items following ipsilateral (right) injection indicated the functional adequacy of the contralateral hippocampus would be insufficient to support general memory functions if a right temporal lobectomy was undertaken, and this prediction turned out to be true.

11

Medical Aspects of Epilepsy Surgery

Epilepsy surgery has become increasingly safe and efficacious as advances in electroencephalography (EEG) monitoring, neuroimaging, and surgical techniques have occurred. Candidates for epilepsy surgery have disabling seizures with negative medical and psychosocial consequences. Children who are inadequately treated may never recover from the detrimental impact of seizures during critical periods of development. Many primary care physicians steer their patients away from surgical options due to misconceptions about the safety and efficacy of epilepsy surgery. In randomized clinical trials, 64% of anterior temporal lobectomy (ATL) patients were seizure-free compared with 8% of those who continued on medical therapy (Wiebe et al., 2001). Thus, all epilepsy surgery candidates should be identified early in the course of treatment and referred for diagnostic evaluation.

Criteria for Surgical Evaluation

The primary criteria initiating referral for possible epilepsy surgery is that the epilepsy has proven resistant to adequate trials of several antiepileptic drugs (AEDs). Other considerations include the length of AED treatment, frequency of seizures, likelihood of success of surgery, and the possible impact of surgery on the cognitive, psychosocial, and overall quality of life of the patient.

Number of Drugs Failed

When appropriate treatment with AEDs has failed to control a patient's seizures, epilepsy surgery should be considered. Approximately two-thirds of newly diagnosed epilepsy patients will have a good response to AED

Table 11.1 Seizure control with various AED regimens among patients with newly diagnosed epilepsy (N = 470).

ANTIEPILEPTIC DRUG REGIMEN	NO. OF PATIENTS	% OF PATIENTS
Seizure free on first monotherapy agent	222	47%
Seizure free on second monotherapy agent	61	13%
Seizure free on third monotherapy agent	6	1%
Seizure free on two agents	12	3%
Total Seizure Free	301	64%

Adapted from Kwan P, Brodie MJ. Early identification of refractory epilepsy. *New England Journal of Medicine* 2000;342:314–319.
Adapted from: Kwan & Brodie (2000).

treatment. This leaves about one-third who will have incompletely controlled epilepsy despite appropriate AED therapy. Currently, AED failure is widely defined as the inability to obtain control of seizures after consecutive trials of two or three AEDs which have been well-tolerated (Brodie, 2005). Although a good deal of data has been collected on the decreasing likelihood for seizure control with successive AED failure, one of the most frequently cited studies (Kwan and Brodie, 2000) shown in Table 11.1 illustrates this point.

As may be seen in Table 11.1, the overwhelming majority of patients (74%; 222 of 301) who respond to AED treatment will do so on the first monotherapy agent, and a fair number (13%) who do not respond to initial AED therapy may respond by trying a second drug or on polytherapy with two drugs. Thus, the International League Against Epilepsy (ILAE) subcommittee on neurosurgery recommends that drug-resistant epilepsy should be defined as an inadequate response to a minimum of two AEDs, either as monotherapy or in combination, because after this, trying additional medications has little chance (~1%) of producing seizure freedom (Binnie and Polkey, 2000).

Duration of Antiepileptic Drug Therapy

Another issue that frequently arises in defining medical intractability in order to determine appropriateness for epilepsy surgery is the duration of AED treatment. The ILAE has recommended that adult patients be treated with appropriate AEDs for at least 2 years before surgery is considered, but this length of time may be too long for many children with epilepsy due to the negative consequences of repeated seizures upon development. This is true not only for the detrimental effects of frequent epileptic seizures on the developing brain itself, but also because of the secondary psychological and social consequences of repeated seizures which interfere with the acquisition of interpersonal, educational, and later, vocational skills. If surgery is postponed for too long, many of these developmental skills may never be acquired, and patients may become permanently disabled even though they may become seizure-free after surgery later in life. Thus, in many children with intractable seizures, epilepsy surgery should not necessarily be considered the treatment of last resort.

Seizure Frequency
A final consideration in determining medical intractability is seizure frequency; how many seizures per month or per year are sufficient to determine that epilepsy is refractory to antiepileptic drug (AED) treatment? This is probably best settled on a case by case basis in consultation with the patient and their family since, for some, even one or two seizures a year can significantly impact vocational advancement and quality of life. Because complete seizure freedom is the outcome most consistently associated with improvement in quality of life, Kwan and Brodie (2006) suggest a patient's epilepsy may be considered intractable when one or more seizures per year are occurring.

Diagnostic Evaluation for Epilepsy Surgery
The presurgical evaluation for possible epilepsy surgery is typically conducted at a specialized comprehensive epilepsy surgery center and involves clinical, electrophysiological, neuroimaging, and neuropsychological evaluations. Evaluation usually begins with a thorough review of the patient's history, including previous diagnostic studies and prior unsuccessful treatments, to make certain the epilepsy diagnosis is correct and to plan the subsequent diagnostic work-up. As technical advances have been made in structural and functional neuroimaging, and in greater recognition of ictal EEG patterns predicting good surgical outcome, the requirement for inpatient video-EEG monitoring has been relaxed, although this is somewhat controversial. Mesial temporal lobe epilepsy with hippocampal sclerosis, for example, is now often diagnosed and recommended for epilepsy surgery without the need for extended inpatient monitoring stays or invasive monitoring using intracranial electrodes. However, if the initial work-up leaves any doubt regarding the site

of seizure onset, noninvasive simultaneous video-EEG monitoring in an inpatient facility is undertaken. Despite this trend toward relaxed requirements, some centers still require that all epilepsy surgical candidates be monitored before proceeding to surgery.

Noninvasive Video-Electroencephalography Monitoring
Noninvasive simultaneous video and EEG recording is conducted in an inpatient setting with a dedicated epilepsy monitoring unit (EMU) whose nursing and technical staff are trained in epilepsy monitoring procedures. The purpose of monitoring is to determine whether there is a single area of brain where the seizure begins (the so-called epileptogenic zone) that is amenable to surgical resection. Before admission to the hospital, most patients will have already undergone a thorough neurological exam and history, outpatient EEG, and structural magnetic resonance imaging (MRI). After (or shortly before) admission, most patients' antiepileptic drugs are tapered and gradually withdrawn. Continuous, 24-hour, computer-assisted video-EEG is then carried out when the patient is awake and asleep, and ideally, both the electroencephalographic and clinical signs and symptoms of the patient's habitual seizure will be recorded for later review by the epileptologist.

Cognitive Assessment During Video-Electroencephalography-monitored Seizures
In addition to capturing ictal video-EEGs of the patient's typical seizures, interictal EEG recording is also important because interictal activity is thought to primarily reflect a broader epileptic irritative zone surrounding the more discrete epileptic focus seen at the beginning of a seizure. Other more difficult to define behavioral phenomena also may be noted and recorded. This may include subjective auras the patient may experience, precise description of seizure semiology, and assessment of degree of awareness, speech, language, memory, and voluntary motor control during a seizure and postictally. Cognitive assessment protocols vary across epilepsy centers. At the Medical College of Georgia, there are five common objects hanging on the wall along side of each EMU bed. As soon as a seizure begins, a nurse will take one or two objects down and ask the patient to name them. This is followed by simple commands, such as, "tell me your name," "touch your nose," "squeeze my hand," or "point to the ceiling." Level of awareness is assessed by asking orientation questions, determining memory for events which took place during the seizure, and seeing whether the patient can recall the naming

objects after the seizure is over. The ictal EEG, interictal EEG, and cognitive assessment during seizures may all assist in refining diagnosis of seizure type and determining localization of seizure onset.

Electroencephalography Analysis During Inpatient Monitoring
The EEG itself is recorded from the scalp, often with placement of additional sphenoidal electrodes inserted bilaterally underneath the zygomatic arch of the face, typically using EEG recording with a minimum of 32–64 channels. Sphenoidal electrodes are better able to record seizure onset from the mesial or basal temporal lobe regions. Computerized EEG allows epileptologists to digitally reformat between different electrode montages. For example, reviewers may shift from the traditional bipolar chain montage view (systematically displaying recordings from one electrode to the next nearest electrode around the head, e.g., between T7 and T9) which is best for detecting phase reversals, to a variety of referential montages, which allow for better visualization of the distribution of ictal patterns (van Emde Boas and Parra, 2001). An electrocardiogram (ECG) is also routinely recorded because changes in heart rate and pattern are common in certain seizure types. Another channel is used to signal when a manual event-recorder button positioned in the patient's bed has been pressed by the patient or someone else in the room.

Trending EEG data may be used to compress large EEG data sets into quickly reviewable graphic formats on the computer for long (e.g., 24-hour) periods of time. Events of interest may be further investigated by simply clicking a computer mouse on the corresponding area of the trending EEG graph and the raw EEG tracings from that epoch can be reviewed in detail. This shortens the period of time epileptologists are required to search through pages and pages of raw EEG tracings.

Most patients are monitored in the EMU for 3–5 days, depending upon the number of ictal events recorded. Some patients will not have any seizures during the inpatient stay and will either continue monitoring or be discharged with the recommendation to return for further monitoring at a later date. Even though AEDs have been tapered or discontinued altogether, the relatively stress-free environment of the EMU (e.g., lying in bed watching TV or chatting with family and friends) does not evoke seizures in some individuals. There is no agreement on how many seizures need to be recorded before proceeding to surgery. However in patients with a single consistent seizure type, three recorded habitual seizures are usually thought to be sufficient (van Emde

Boas and Parra, 2001). In cases where the MRI has revealed a structural lesion, fewer seizures may need to be recorded if EEG results are consistent with history and all the other test results.

Structural Neuroimaging in the Presurgical Evaluation
Magnetic Resonance Imaging

Magnetic resonance imaging is the structural imaging method of choice in epilepsy. The MRI is not a single radiological modality, but rather is a collection of techniques that are tailored to identify particular pathological substrates. Different MRI sequences (e.g., fluid-attenuated inversion recovery [FLAIR], spin-echo T2-weighted sequences, magnetization-prepared rapid gradient echo [MPRAGE]), orientations, and slice thicknesses are selected to optimized detection of different types of lesions within particular brain regions (Kuzniecky, 2001). Current clinical MRI imaging is typically performed using a modern 1.5-T or 3-T scanner. Some of these MRI methods have already been described in Chapter 4. The MRI sequences are tailored to the patient depending upon age, as well as the type, location, and duration of epilepsy. Post-contrast studies are typically performed in new onset seizures in adults, but not in children since brain tumors are a rare cause of childhood epilepsy. High resolution T1 and T2 images through the hippocampi are normally conducted in patients thought to have mesial temporal lobe seizures.

Mesial Temporal Lobe Sclerosis. Because the temporal lobe is the most common area of seizure onset in adults, specialized high-resolution, thinly sliced images are commonly taken in coronal section through the anterior temporal lobes to detect mesial temporal sclerosis (MTS). Mesial temporal sclerosis is the most frequent pathological substrate found in adult epilepsy surgery patients, and a variety of MRI techniques may be used to search for it. The MRI sequences commonly used to evaluate MTS and hippocampal atrophy include thin coronal sections (not exceeding 2 mm) using FLAIR, volumetric T1 images, T2 relaxometry, and MPRAGE, among others, depending upon the institution. The FLAIR images are basically T2-weighted images with the high signal of cerebrospinal fluid cancelled out and MTS can usually be identified by increased T2 signal in one of the hippocampi or as being usually bright relative to surrounding neocortex on FLAIR images (Fig. 11.1). Examination of coronal volumetric T1 images will often show an asymmetry in hippocampal volumes as the sclerotic hippocampus is also frequently atrophic.

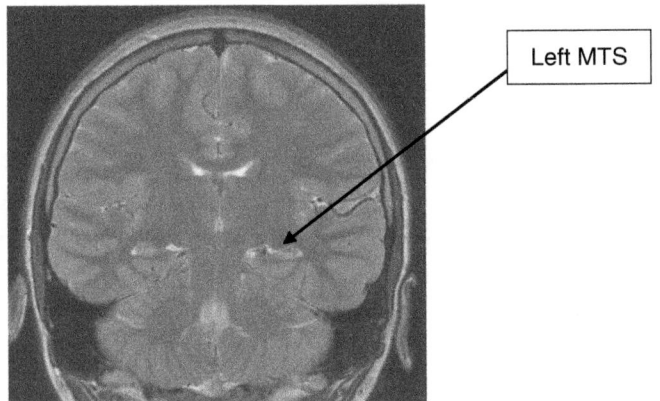

FIGURE 11.1 Magnetic resonance image (MRI) of mesial temporal lobe sclerosis (MTS). T2-weighted MRI showing left hippocampal atrophy with signal change. Post-resection pathology showed MTS. (Courtesy of Y.D. Park, M.D., Medical College of Georgia)

Neoplasms. The MRI methods commonly used to detect brain tumors typically include routine coronal volumetric T1-weighted images through the entire brain and other finer-grained methods (e.g., gradient-echo T2 sequence) within the particular lobe of interest (Knake and Grant, 2006). Contrast dye is typically used in all new-onset adult cases, but since this takes additional time, the high resolution sequences are obtained first and a contrast study may be conducted at a later time.

Cortical Malformations. Large cortical malformations (e.g., schizencephaly, pachygyria) and smooth band heterotopias can usually be diagnosed using routine MRI techniques. Smaller localized developmental lesions, such as focal cortical dysplasia and focal polymicrogyria, often require more high-resolution, thinly sliced images including, coronal T2, axial FLAIR, and volumetric T1 sequences.

Dual Pathology. This pathological substrate occurs when there is evidence of mesial temporal sclerosis in addition to some type of extrahippocampal structural lesion. Extrahippocampal lesions usually include developmental cortical malformations, perinatal vascular injury, vascular malformations (e.g., cavernous angioma), or cysts. Dual pathologies are most often discovered when the MRI procedures, designed to detect MTS through the temporal lobes, are used to image the rest of the brain usually at a higher spatial resolution (not greater than 3mm).

Magnetic Resonance Spectroscopy

^{1}H-magnetic resonance spectroscopy (^{1}H-MRS) may be performed on most 1.5-T scanners and provides information about the biochemical status of specified brain regions of interest. Specifically, ^{1}H-MRS yields information about neuronal function such as N-acetyl aspartate, membrane turnover (choline), total energy stores (creatine), and presence of cerebral lactate (Hetherington et al., 2002; Kantarci et al., 2002). An MRS may improve prediction of seizure-relief outcome and help with lateralization and localization of the epileptic focus. The most common ^{1}H-MRS measures used in epilepsy surgery patients are the ratio of aspartate (NAA) and choline to creatine (Cr). In about 90% of MTS cases, NAA/Cr ratio decreases occur in the ipsilateral temporal lobe (Cendes et al., 1997). Loss of NAA and a decrease in NAA/Cr ratio have been associated with abnormal mitochondrial metabolism and neuronal loss. Because ^{1}H-MRS reflects the biochemical activity in the brain, it can be affected by antiepileptic drug levels, ketogenic diet, or recent seizure activity, and thus should be interpreted carefully with other possible influences in mind.

Diffusion Tensor Imaging

Diffusion tensor MRI is an imaging methodology used to evaluate the integrity of white matter fiber tracts. Diffusion tensor imaging (DTI) may be used to detect abnormalities of axonal organization resulting from damage to the cortex. It is in its investigational stage for use in evaluating epilepsy surgery patients. Early studies in epilepsy have revealed white matter abnormalities in patients with temporal lobe epilepsy and epilepsy caused by cortical dysplasia (Eriksson et al., 2001; Arfanakis et al., 2002). The incremental usefulness of DTI in the evaluation of epilepsy surgery patients has not yet been verified.

Functional Neuroimaging in the Presurgical Evaluation

Positron Emission Tomography

Positron emission tomography (PET) uses a radioactive tracer introduced into the blood stream to measure various aspects of brain function, such as glucose utilization, cerebral blood flow, or neurotransmitter synthesis, depending upon the particular radiotracer used. The most common radiotracer in PET scanning is (^{18}F) fluoro-2-deoxyglucose, or (^{18}F) FDG, which measures glucose utilization. The PET may be used in epilepsy surgery to identify the epileptogenic focus. Interictally, the seizure focus is hypometabolic during PET scanning in the majority of temporal lobe epilepsy patients. The radiotracers used for PET are fast acting, and thus, the seizure onset is difficult to capture and scanning must take place shortly after injection. The PET has a spatial resolution of 2–3 mm. This makes PET less suitable for capturing an ictal event which would show hypermetabolism of the epileptic focus.

Single-photon Emission Computed Tomography
Many centers attempt to obtain an ictal single-photon emission computed tomography (SPECT) during inpatient video-EEG monitoring to assist in seizure onset localization. In the ictal SPECT procedure, a radioisotope drug is injected as soon as possible (e.g., within 5–20 seconds) after a seizure has begun while the patient is in the EMU. The maximum uptake of the radiopharmaceutical within the brain is between 30 and 60 seconds, so that the ictal perfusion pattern is evident shortly after injection. The patient may then be transported to the radiology department for SPECT scanning within the next few hours. The radiopharmaceutical perfusion pattern will remain visible in the brain up to 4 hours after injection since SPECT radiotracers have a longer half-life than PET radiotracers. The most commonly used SPECT radiotracer is ^{99m}Tc-Hexamethyl-propyleneamine oxime (^{99m}Tc-HMPAO).

Similar to PET, SPECT will generally show *hypo*metabolism in the epileptogenic area interictally, and *hyper*metabolism of the epileptic focus during a seizure. Ictal SPECT is a common procedure at most epilepsy surgery centers since it localizes mesial temporal lobe seizures in 86%–97% of patients. Ictal SPECT information may thus be reliably used in conjunction other medical findings to tailor a focal cortical resection or to assist in planning the placement of intracranial electrodes during Phase II evaluation.

Subtraction Ictal SPECT Co-registered with MRI
Although ictal SPECT has been shown to localize mesial temporal lobe seizures with a high degree of accuracy, its ability to localize extratemporal or neocortical seizures is less successful. Subtraction Ictal SPECT Co-registered with MRI, called *SISCOM*, was developed to provide further localizing

information in these more difficult cases (Lewis et al., 2000). To accomplish this, a resting interictal SPECT study, a high-resolution structural MRI, and an ictal SPECT study must first be conducted. Next, both the interictal and ictal SPECT studies are co-registered to the patient's own MRI, and then the interictal SPECT is subtracted from the ictal SPECT scan after normalization.

Injection of the radiotracer for the ictal SPECT must be accomplished as soon as possible after seizure onset. This usually means a nurse or EEG technician must sit at the patient's bedside in the epilepsy monitoring unit waiting for a seizure to begin. Rapid injection (ideally <20 seconds after seizure onset) of the radiotracer is important for accurate seizure localization. Later injections can lead to false localization due to the rapid spread of the seizure to other areas of the brain. Overall, SISCOM improves sensitivity for localization of ictal seizure onset, especially in those cases that are more difficult to localize; such as those with neocortical, extratemporal, or multiple foci. A SISCOM example is given in Figure 11.2. The SISCOM shows an area of hyperperfusion in the left parieto-occipital region that corresponds to the ictal onset zone in this patient.

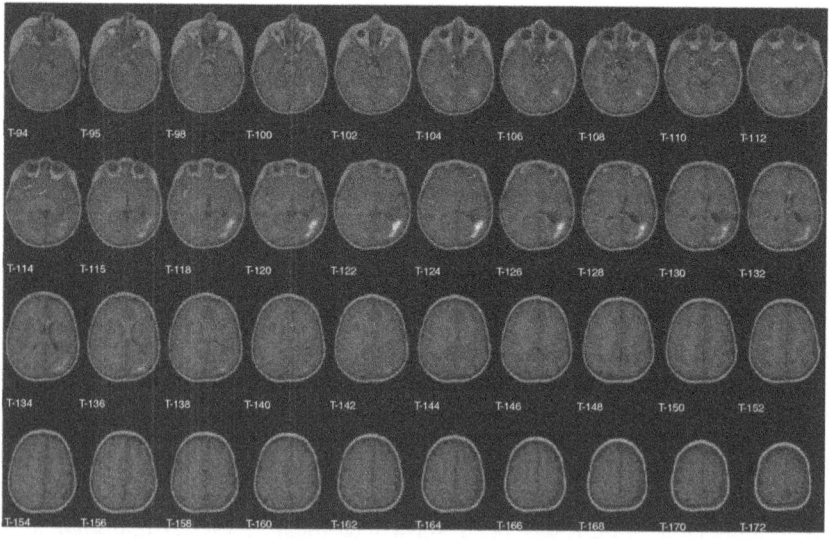

FIGURE 11.2. Axial subtraction ictal SPECT co-registered with MRI (SISCOM) showing the hypermetabolic epileptogenic region in the posterior regions of the left hemisphere. (Courtesy of Y.D. Park, M.D., Medical College of Georgia)

Magnetoencephalography

Magnetoencephalography (MEG) is a complementary physiologic measure similar to EEG, except it records weak magnetic field dipoles generated by neurons instead of averaged excitatory postsynaptic potentials. It is a noninvasive procedure where the patient's head is placed into a concave helmet that has 100–300 channels of magnetic sensors placed within the inner surface of the helmet that convert magnetic fields into electrical signals. Currently, the most common use of MEG in epilepsy surgery is to localize interictal epileptic waveforms similar to those obtained with scalp EEG but with finer-grained spatial resolution. The MEG epileptogenic dipoles are typically transposed onto the patient's own MRI scan; this process is then referred to as *magnetic source imaging* or *MSI*.

A MEG is less often used to record ictal onset of seizures because it is rare for a patient to have a spontaneous seizure during an outpatient recording. Furthermore, it can be a poor localizer of ictal events since MEG sensors are not attached directly to the scalp, and thus, the patient cannot move if an accurate recording is to be obtained. However, the use of MEG for seizure onset localization is particularly valuable with extratemporal seizures and when other tests (e.g., MRI, Wada) are normal or incongruous. Like other neuroradiologic tests, MEG has poor detection capacity around the base of the brain (e.g., orbitofrontal or basal temporal cortices). In addition to its value in seizure onset detection, MEG is also a very sensitive and accurate measure for cortical mapping of sensorimotor and language areas (Tracy and Shah, 2007).

Intracranial Electrodes: Invasive Video-EEG (Phase II) Monitoring with Grid, Strip, or Depth Electrodes

When results of the various noninvasive tests are contradictory, or when surface EEG recordings are unable to define the epileptogenic zone, the patient with intractable seizures may need to undergo invasive EEG monitoring with intracranial depth, or subdural grid or strip, electrodes. Within the epilepsy surgery community, intracranial EEG monitoring is referred to as phase II evaluation. Phase I evaluation includes inpatient video-EEG (scalp) monitoring, MRI, Wada assessment, and preoperative neuropsychological testing. The final phase, phase III, refers to the epilepsy surgery itself.

Phase II invasive video-EEG monitoring is undertaken in patients who have a high likelihood of being able to benefit from epilepsy surgery, but whose noninvasive test data has been insufficient to clearly localize the epileptic focus. Invasive implantation of electrodes is almost always preceded by noninvasive monitoring that has hopefully identified one, or several related, broad epileptogenic zones. Strip and grid electrodes overlie the surface of the cortex, whereas depth electrodes are implanted deep within the substance of the brain. Monitoring with invasive electrodes provides a more finely grained analysis to zero-in on the precise location of the epileptic focus. The increase in fidelity of invasive monitoring over scalp EEG occurs because the electrodes themselves are closer to the epileptogenic focus, the dampening interference of the skull has been removed, and there is less muscle artifact.

Subdural Strip and Grid Electrodes

Grid and strip electrodes are composed of stainless steel or platinum discs encased within polyurethane coverings that are flexible so they may be easily inserted and will conform to the surface of the cortex. Each electrode is between 2 and 4 mm in diameter and are usually spaced 10 mm apart. Strip electrodes consist of a single row of 4–11 electrode contacts covered in flexible material, such as polyurethane. Each strip can be inserted subdurally through a burr hole so they can be placed bilaterally. The risk of infection or hemorrhage from strip electrode insertion is reportedly less than 1% (Wyler, 1991). Common sites for insertion of strip electrodes for EEG recording include the mesial (in the interhemispheric fissure) or orbital region of the frontal lobes and the basal, lateral, or polar regions of the temporal lobes when a neocortical seizure focus is suspected.

Subdural grid electrodes are composed of multiple strips arranged in parallel rows and embedded into a single sheet of polyurethane. The larger grids consist of 8 × 8 (64 contacts), or 4 × 8 (32 contacts), electrode arrays that are used to cover large surface areas. These larger grids usually overlay the lateral frontal-central cortex or parieto-occipital regions. Figure 11.3(a) shows electrode grids placed on the surface of the cortex at the time of surgery and Figure 11.3(b) is a lateral skull x-ray showing the same grids in place over the left lateral convexity after surgery.

Smaller grid arrays (e.g., 4 × 4 or 2 × 6 contacts) may be used for EEG recording from the undersurface of the frontal or temporal lobes. Grids are inserted over the surface of the cortex during surgery involving an open craniotomy, and because of this, the risk for complications increases considerably; up to 26% in one study (Hamer et al., 2002). Most complications are transient, and when present include conditions such as infection, epidural hematomas, and increased intracranial pressure. More serious permanent complications, such as infarction (1.5%) and death (0.5%), are uncommon.

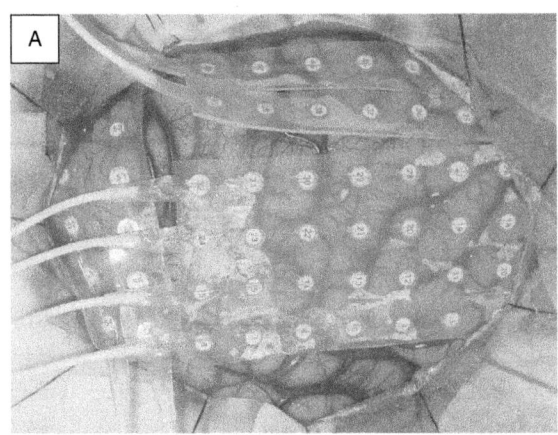

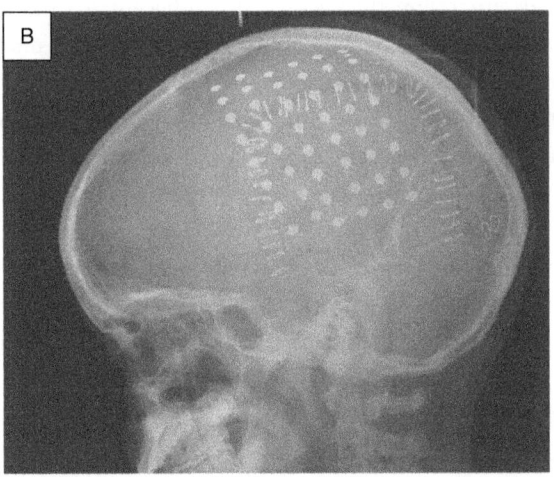

FIGURE 11.3. Multiple electrode grids on cortical surface in operating room (A) and on skull X-ray (B) for intracranial EEG monitoring and electro-cortical stimulation mapping. (Courtesy of Y.D. Park, M.D., Medical College of Georgia)

Chronically implanted strip and grid electrodes stay in place for several days or weeks while EEG recordings are conducted during inpatient video-EEG monitoring.

Depth Electrodes

Depth electrodes are used to detect ictal seizure onset in deep brain structures; most commonly the hippocampus, amygdala, and related mesial temporal lobe structures. The electrodes themselves are long, needle-like thin, flexible wires made of polyurethane, or other similar materials, which typically have eight contacts spaced at 5 or 10 mm intervals along the shaft of the electrode. Depth electrodes are introduced into the substance of the brain under stereotactic guidance through twist-drill holes in the operating room with the patient under general anesthesia (Sperling, 2001). Insertion of depth electrodes does leave a small electrode track in the brain, and there are risks associated with insertion; most notably infection or intracerebral hemorrhage. Complication rates range between 0.5% and 4%. However, because the diameter of the depth electrode is quite small and the electrode so flexible, permanent, clinically significant injury is an extremely rare event.

As with strips and grids, depth electrodes stay in place within the brain for several days or weeks while the patient undergoes simultaneous video-EEG monitoring as an inpatient on the epilepsy monitoring unit. Depth electrodes are introduced either directly into (or just adjacent to) the amygdala and hippocampus via either a lateral approach through the middle temporal gyrus or using a posterior approach through burr holes in the occipital lobe. Ictal EEG onset from mesial temporal structures often precedes scalp EEG onset by as much as 30 seconds. Furthermore, some epileptic EEG abnormalities, such as those associated with auras, may only be seen through depth electrodes (Benbadis et al., 2006). Depth electrodes are sometimes implanted in mesial temporal lobe structures bilaterally. Other brain areas occasionally sampled using depths include orbitofrontal cortex (since this region can be difficult to record from using other methods), near structural lesions, or deep within sulci suspected of harboring the ictal onset zone.

Strip, grid, and depth electrodes are only used when noninvasive methods (e.g., scalp EEG, MEG) produce insufficient localization information to guide surgical resection. Although their primary purpose is to identify the epileptogenic focus and define regions of interictal dysfunction, these chronically implanted electrodes may also be used for electrical stimulation mapping of sensory, motor, language, and other cognitive functions. Since electrical stimulation mapping is a common clinical activity provided by epilepsy neuropsychologists, the details of electrocortical stimulation mapping using strip, grid, and depth electrodes have been discussed in Chapter 8.

Epilepsy Surgery Procedures

After a comprehensive preoperative evaluation for epilepsy surgery has identified the nature, cause, and impact of the patient's epilepsy, the epilepsy surgery team (often composed of the neurosurgeon, neurologist, radiologist, neuropsychologist, nurses, and other professionals involved in the patient's care) meets to review all the preoperative data to decide on a recommended course of action. If the preoperative evaluation has established that the benefits of surgical intervention outweigh the risks and offers hope for improvement in quality of life, the specific surgical therapy is recommended to the patient and family.

The primary goals of epilepsy surgery are to eliminate or reduce seizure frequency, improve quality of life, abolish the negative cognitive and behavioral effects of repeated seizures, and ultimately result in the cessation of AED therapy. Recommended surgical options range from the least invasive, namely implantation of a vagal nerve stimulator (VNS), to probably the most invasive, a hemispherectomy. Most epilepsy surgery involves either removal (resection) of epileptogenic tissue (e.g., lobectomy) or disconnection of pathways which conduct spread of seizures to other brain regions (e.g., corpus callosotomy). This section will briefly review the major forms of epilepsy surgery including vagal nerve stimulation, ATL, frontal lobectomy, parieto-occipital lobectomy, lesionectomy, hemispherectomy, corpus callosotomy, multiple subpial transaction (MST), and implantation of electrical brain stimulators. Special considerations concerning epilepsy surgery in children will also be covered.

Vagus Nerve Stimulation
Vagus nerve stimulation (VNS) is used in patients whose seizures are refractory to treatment AEDs, but who are either unwilling or unable to undergo epilepsy surgery. Use of VNS reduces seizure frequency in many individuals, but very rarely produces complete seizure relief, and thus, VNS is not a replacement for resective epilepsy surgery. The VNS therapy system (Cyberonics, Houston, TX) consists of an electrical pulse generator that is implanted subcutaneously just below the left clavicle and a bipolar electrode wire that runs underneath the skin that is ultimately implanted into the left vagus nerve in the neck (Fig. 11.4).

The pulse generator is powered by a lithium chloride battery which needs to be replaced every 1.5–5 years depending upon the stimulation parameters. Surgery is carried out under general anesthesia and the procedure usually takes about 2 hours. Brain MRIs, if indicated, are usually conducted before VNS implantation because of safety concerns since the MRI could possibly cause the generator or wire to heat up excessively (Fisher and Handforth, 1999). Many centers have, however, found ways to safely obtain brain MRIs in patients with VNS.

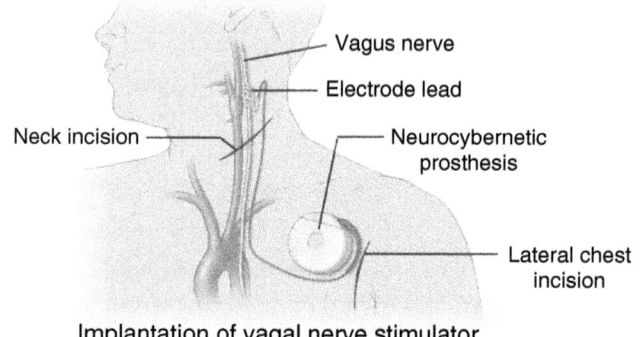

Figure 11.4 Placement of the vagal nerve stimulation (VNS) system showing the electrical pulse generator (neurocybernetic prosthesis) and bipolar electrode implanted into the left vagus nerve. © John Foerster, Department of Neurosurgery, Medical College of Georgia.

Vagus Nerve Stimulation Parameters

The VNS system also comes with a programming wand and software that are attached to a laptop computer. The stimulation parameters are tailored to each patient, but after a ramp-up period of several weeks are typically set to the following: VNS current = between 2.0 and 3.5 mA; frequency = 30 Hz; pulse width = 500 ms; time on = 30–90 seconds; time off = 5–10 minutes (Wheless, 2006). If these parameters are ineffective, the pulse generator can be tried on a rapid cycling pattern consisting of 5- to 7-second stimulation on periods alternating with 14–30 seconds off. Patients are provided with a hand-held magnet that can be externally waved over the generator in an attempt to manually abort a seizure when an aura occurs warning of an impending seizure.

Vagus Nerve Stimulation Mechanism of Action

The precise mechanisms by which the VNS produces its seizure retardant effect is unknown, but based upon the functional neuroanatomy of the vagus nerve, several options exist. The vagal nerve receives visceral afferents in the medulla from the heart, lungs, and gastrointestinal structures which enter the nucleus solitarius. The nucleus of the solitary tract projects back to these organs, as well as to several brainstem nuclei, which in turn, project to thalamus, insula, and cerebral cortex. In addition, nucleus solitarius has direct connections with the reticular formation, hypothalamus, basal forebrain nuclei, amygdala, and hippocampus. Metabolic activation of some combination of these ascending vagal nerve fibers to thalamus, brainstem, and limbic structures is thought to modulate the VNS effect on seizure activity (Fisher and Handforth, 1999).

Vagus Nerve Stimulation Efficacy
The Therapeutics and Technology Assessment Subcommittee of the American Academy of Neurology concluded that sufficient evidence exists to rank VNS for epilepsy as effective and safe based upon several randomized, controlled, double-blind clinical trials (Fisher and Handforth, 1999). Efficacy of VNS is not particularly impressive, however, with an overall mean decline in seizure frequency between 25% and 30% across all clinical trials. The reduction in seizure frequency seems to be evident within the first month or so of treatment, and the effect appears to persist over time. Although several open label studies have suggested VNS may be beneficial in younger children (under the age of 13 years) and for primary generalized seizures, there is insufficient information for these uses at this time. There is positive data for its use in partial seizures among adolescents and adults.

Adverse Effects of Vagus Nerve Stimulation
Contrary to what one might expect, VNS causes no clinically significant effects on heart rate, gastric functions, or respiration. Most patients can sense when the stimulator is on by sensory changes in the throat or neck, and some experience this sensation as throat discomfort. The most common side effect of VNS is hoarseness during the period of stimulation, which occurs in about one-third of patients (Fisher and Handforth, 1999).

Current Status of Vagus Nerve Stimulation
Use of VNS is recommended for adults and children (over the age of 12 years) who have medically refractory seizures but are not candidates for resective epilepsy surgery. Improvement in seizure control is thought to be roughly equivalent to that expected with the newer AEDs, but lower than that expected following temporal lobectomy.

Anterior Temporal Lobectomy
Anterior temporal lobectomy is by far the most common form of epilepsy surgery because mesial temporal lobe epilepsy is the most common seizure type presenting for surgery in adults. Specific surgical approaches differ from surgeon to surgeon, and changes in standard of care have evolved over time.

Some perform a standardized "en bloc" ATL resection on all patients, which is predetermined before surgery, whereas others tailor the lateral and mesial extent of resection for every patient primarily based upon intraoperative electrocorticography (ECoG). The majority of surgeons, however, use a standard baseline resection of lateral and medial temporal lobe which is modified on the basis of intraoperative ECoG, extent of MRI hippocampal atrophy, risk for memory decline as determined by Wada testing, and risk of language impairment based on electrocortical stimulation mapping (Spencer and Ojemann, 1993).

A typical ATL for mesial temporal lobe epilepsy usually begins by resecting the lateral temporal neocortex (see Figure 10.7); between 4 and 5 cm of the middle and inferior temporal gyri in the dominant temporal lobe and 5–6.5 cm in the nondominant temporal lobe (as measured from the temporal pole). The superior temporal gyrus is often either spared or only the most anterior portion is excised. Epilepsy resective surgery is usually carried out using a combination of suction (ultrasonic aspirator) and fine bipolar forceps, and dissectors, which in lay terms essentially "suck," "burn," and "cut" their way through brain tissue (see the abbreviated description of surgery in the case example in Chapter 9 for details).

After the lateral neocortex is removed, the temporal horn of the lateral ventricle is exposed and entered so that the temporal lobe may be retracted laterally to allow visualization of the medial structures. The amygdala, hippocampus, and parahippocampal gyrus are then gradually carefully resected. Extent of hippocampal resection depends upon how much of it is involved in ictal onset (as determined by intraoperative EEG or ECoG), extent of disease, and risk for functional (namely, memory) postoperative impairment, but typically ranges between 3 and 5 cm. Extent of medial resection is most often measured using the anterior aspect of the pes of the hippocampus (Spencer and Ojemann, 1993). The decision of how far back to carry the hippocampal excision is complicated by the knowledge that the greater the extent of hippocampal resection, the better the seizure relief outcome after surgery, but also the greater the risk for memory impairment in many cases.

Outcome of Anterior Temporal Lobectomy

Combining reports of seizure-relief outcome after ATL from multiple sites though the late 1990s, the seizure-free rate was around 68% at 1 year post surgery. An objective epilepsy surgery outcome classification scheme was devised by a special interest group within the American Epilepsy Society in

Table 11.2 Epilepsy surgery outcome criteria of Engel

Class I: Seizure-free[a]

A. Completely seizure-free since surgery
B. Aura only since surgery
C. Some seizures after surgery, but seizure-free for at least 2 years
D. Atypical generalized convulsion with antiepileptic drug withdrawal only

Class II: Rare seizures ("almost seizure-free")

A. Initially seizure-free but has rare seizures now
B. Rare seizures since surgery
C. More than rare seizures after surgery, but rare seizures for at least 2 years
D. Nocturnal seizures only

Class III: Worthwhile improvement

A. Worthwhile seizure reduction
B. Prolonged seizure-free intervals amounting to greater than half the follow-up period, but not less than 2 years

Class IV: No worthwhile improvement

A. Significant seizure reduction
B. No appreciable change
C. Seizures worse

[a] Excludes early postoperative seizures (first few weeks)
From Engel J Jr. Outcome with respect to epileptic seizures. In Engel J Jr. (ed.), *Surgical treatment of the epilepsies.* New York: Raven Press, 1987:553–571.

the 1980s and has been universally referred to as the Engel criteria since being published in his 1987 chapter (see Table 11.2).

The first randomized, controlled study of anterior-mesial temporal lobe epilepsy was published in 2001 (Wiebe et al., 2001). Of 1,988 patients who underwent ATL, 66% were seizure-free, 19% showed significant improvement, and approximately 13% had no improvement.

Wyler and colleagues (1995) found that the extent of hippocampal resection was critical for seizure control following ATL. Wyler randomly assigned his epilepsy surgery patients to either a conservative hippocampal resection (hippocampectomy to the anterior edge of the cerebral peduncle) or a more extensive resection (hippocampectomy to the level of the superior colliculus) and found the radial hippocampectomy group had a 69% seizure-free outcome, whereas only 38% of the limited hippocampectomy patients were seizure-free. This and similar data led to surgeons to design ATL procedures that provided radical excision of the mesial structures, while preserving as much as possible the lateral temporal neocortex.

Most ATL outcome studies for relief of mesial temporal lobe epilepsy only have a short (1 or 2 year) follow-up period, and thus, do not accurately reflect longer-term outcome and risk for relapse. Reported relapse rates range between 15% at 5 years after surgery and 38% across longer time periods of up to 20 years (Yoon et al., 2003). The highest relapse rates among patients with mesial temporal lobe epilepsy have been reported in symptomatic epilepsy patients with tumors and cavernous malformations.

Common Complications of Anterior Temporal Lobectomy
Death caused by ATL is a rare event that reportedly occurs in approximately 1% of cases worldwide; although mortality rates have declined in more recent years (Jensen, 1975). The most common causes of death from ATL include hemorrhage, infarction, and pulmonary complications.

The most common neurological complications seen following ATL include visual field defects (e.g., superior quadrantanopsia), transient aphasia following dominant ATL, and rare, but well documented, cases of hemiparesis and transitory cranial nerve palsies. Partial contralateral superior quadrantanopsia is seen in approximately 50% of patients following ATL, whereas hemianopic or full quadrantic visual field defects have been reported in between 2% and 4% (Pilcher et al., 1993). These partial quadrantic defects do not affect visual acuity, are rarely noticed or complained about by patients, and are not usually a significant disabling condition.

Transient fluent aphasias are quite common in the immediate postoperative period after ATL. These are often mild anomic aphasia or at times more severe comprehension impaired aphasias. Postoperative aphasias usually resolve within a few days or weeks after surgery. Although the aphasia resolves, many patients are left with a permanent mild residual anomia.

Hemiparesis occurs in 2%–5% of patients undergoing temporal lobe resections (Pilcher et al., 1993). Disabling hemiparesis is thought to occur through surgical manipulation of either branches of the middle cerebral artery, or the more medially placed vessels such as the anterior choroidal artery, resulting in vasospasm and infarction. Transient cranial nerve palsies are even more uncommon, occurring in between <1% and 3% of cases and most often involving cranial nerve III (oculomotor) or VII (facial) cranial nerves.

Frontal Lobectomy

After ATL, the next most common site of surgical resection is of the frontal lobe; accounting for approximately 15%–20% of epilepsy surgeries. Unfortunately, the reported outcomes following frontal lobectomy have been considerably worse than after ATL; roughly 30%–35% of all cases reportedly achieve complete seizure control. There are several reasons for the poorer outcomes. The EEG recordings are often less informative. Ictal seizure symptoms recorded during frontal lobe seizures have less reliable localizing value, and the epileptogenic zone is more difficult to identify in idiopathic and cryptogenic cases (Munari et al., 2001).

Outcome of Frontal Lobectomy

Moreover, these overall poor outcomes are misleading since patients with purely frontal resections have been lumped together with those who have had frontal plus extrafrontal (multilobar) resections, and lesional cases have been lumped together with cryptogenic and idiopathic cases. When purely frontal lobe surgeries are considered alone, rates of seizure freedom are more in line with ATL outcomes; between 60% and 76% of patients have achieved seizure relief (Swartz et al., 1998). In contrast when patients with frontal plus extrafrontal epileptogenic zones have been followed, only 10%–29% of patients have reportedly achieved complete seizure control.

This should not automatically be interpreted to mean the surgery was a complete failure, since considerable improvement in seizure frequency was often obtained; 70% of these multilobar patients had a $\geq 75\%$ reduction of their seizures. Factors associated with good outcomes include either focal structural lesions or focal EEG or PET findings. Poor outcomes have been associated with bilateral epileptogenic foci, incomplete surgical resections (often due to avoidance of eloquent cortex), and Rasmussen encephalitis.

Complications of Frontal Lobectomy

One of the reasons seizure-relief outcome is less impressive in frontal lobe relative to temporal lobe surgery is incomplete resections. This primarily revolves around surgeons' reluctance to include eloquent language or motor areas in the resection. In patients with pure frontal lobe resections, transient minor neurological dysfunction is common, whereas permanent deficits are exceedingly rare. In one series, 40% of frontal cases experienced transient monoparesis, incontinence, abulia, or reduced verbal fluency, and none showed any persistent complications (Swartz et al., 1998). In contrast, 72%

of the multilobar ("frontal plus") patients experienced transient impairments that were persistent in 45% of cases. Permanent complications have included partial visual field cuts, monoparesis, hemiparesis, abulia, aphasia, significantly decreased Verbal IQ, and psychosis. Larger extent of resection has been associated with persistent complications. Death has only rarely been reported (<1%).

Parietal Lobectomy

Epilepsy surgery in the parietal lobe is more uncommon than surgery in the temporal and frontal lobes. In the Montreal Neurological Institute's series of patients (the largest in North America), only 7% of cases underwent surgery in the parietal and occipital lobes. In the earlier years, most patients undergoing parietal resections had some type of lesion identified on preoperative scanning (CT or MRI), while in more recent years, cryptogenic etiologies are more commonly encountered. Grid and strip electrodes are implanted for extraoperative monitoring in most parietal lobe cases, and intraoperative ECoG (most often using a 16-contact electrode grid) also is frequently carried out to identify spike activity and map eloquent somatosensory cortex.

In lesional cases, a margin of tissue surrounding the lesion is usually included in the cortical excision. In nonlesional cases, the size of the resection is typically defined by the extent of the epileptogenic tissue across the cortex. The amount of tissue taken may range from a small resection cavity in a single gyrus to several gyri depending upon the size of the epileptogenic zone.

The postcentral gyrus, which subserves primary somatosensory and some motor functions, is usually included in the resection only when there is an absence of hand motion, hemiparesis of the leg, and a loss of joint position sense. In the dominant parietal lobe, the inferior parietal lobule (which is vital for receptive language, reading, writing, and mathematics, among other functions) is generally considered off-limits for surgical resection, while the dominant superior parietal lobule is often removed. Both the superior and inferior parietal lobules are characteristically included in parietal resections in the nondominant hemisphere, although multimodal hemi-neglect phenomenon and spatial deficits may follow.

Outcome of Parietal Lobectomy

Studies of lesional cases of parietal lobe epilepsy (mostly low-grade gliomas) have reported a postoperative seizure-free rate between 64% and 80%. Similar seizure-relief outcomes have been found in nonlesional cases that included the parietal lobe as part of the resection; 64% were seizure-free and 27% were almost seizure-free (Moriarty et al., 1997). In an exclusively pediatric sample of parietal lobe excisions from the Medical College of Georgia, 69% were seizure-free at 1 year after surgery and an additional 9% experienced only rare seizures. When restricting the analysis to only those children with pure parietal resections ($N = 26$), 5 of 8 (62%) children with medial parietal lobe resections were completely seizure-free, and 12 of 18 (67%) with resections restricted to the lateral parietal region were seizure-free (Strickland et al., 2007). As with frontal lobe cases, pure parietal resections had superior seizure control postoperatively relative to patients who had multilobar, parietal plus extraparietal, resections.

Complications of Parietal Lobectomy

As would be anticipated from knowledge of parietal lobe functional neuroanatomy, complications arising from surgery in the parietal area may include sensorimotor, attentional, and higher cognitive impairments. Similar to that seen with temporal and frontal lobectomies, death is a rare event (<1%) with parietal lobectomy. The following neurological impairments have been reported following parietal lobe resections: (1) visual field cuts (both quadrantanopsia and hemianopsia), (2) motor incoordination, (3) mild sensory association deficits (e.g., astereognosis), (4) contralateral spatial neglect, (5) mild aphasia, right-left disorientation, and (6) worsening of pre-existing deficits including acalculia and alexia (Williamson et al., 1992).

Occipital Lobectomy

Similar to parietal lobe excisions, epilepsy surgery restricted to the occipital lobe is a less common procedure than temporal or frontal lobectomies. Occipital lobe seizures are difficult to evaluate due to rapid seizure propagation to the temporal or parietal lobes which may result in misleading localizing symptoms during seizures. The same is true for interictal and ictal EEG recordings with surface electrodes. Since most occipital onset seizures spread rapidly, it is difficult to localize seizure onset using surface EEG, and further Phase II evaluation with intracranial electrodes is undertaken in most cases.

Some type of visual aura, such as visual illusions, field defects, or blindness, is experienced by most patients (estimates range from 62% to 73%) with occipital lobe seizures. Some common illusions include flashes of light, colored balls, or geometric patterns, but fully formed, complex visual hallucinations are not typical auras (Oliver and Awad, 1993).

Cortical resections in the occipital lobe are carried out in a similar fashion to parietal resections described earlier. Fine bipolar forceps are usually used to demarcate a line of resection centrally within a gyrus. This line is cut and suction is then applied to the tissue in an ever-widening area frequently ending in the nearest sulcal margin (Elisevich and Smith, 2001). Some epilepsy surgeons have come to perform more disconnection and fewer resection procedures in occipital lobe cases.

Outcome of Occipital Lobectomy
Similar to focal cortical resections in other lobes, postoperative seizure-free rates (Engel Class I outcome) after occipital lobe resections generally range between 60% and 72%. Postoperative seizure relief is better among patients with symptomatic epilepsy; most often tumors, vascular malformations, or clearly visualized focal cortical dysplasia. Seizure relief outcome after surgery is worse among patients with idiopathic epilepsy and cortical scarring. The presence of a visual aura before surgery is predictive of a positive outcome. In our series of occipital lobe cases at the Medical College of Georgia, for example, of patients with a visual aura, 19 of 21 (or 90%) had a 1-year Engel score of I or II, whereas only 13 of 24 (or 54%) of those without a visual aura had a 1-year Engel score of I or II.

Complications of Occipital Lobectomy
Since the overwhelming majority of occipital cortex is devoted to either central vision, complex visual analysis, gaze fixation, or tracking through smooth pursuit, resection or disconnection of functional tissue will cause deficits in some of these functions. In many cases, a lesion or previous injury will already have produced a visual field defect that surgery may not worsen. New-onset visual field deficits have been reported in recent surgical series where portions of occipital lobe have been included in the resection in 53%–61% of cases. Patients usually learn to compensate for a postoperative hemianopia or quadrantanopia.

Lesionectomy
Surgical resection for the relief of uncontrolled epilepsy in the context of a focal lesion detected on MRI is commonly referred to as a *lesionectomy*. Detection of a lesion by MRI in patients with intractable epilepsy is generally predictive of a good seizure-relief outcome regardless of the region of brain resected since the lesion is usually the cause of the epilepsy. Common lesions

visualized on MRI that can cause epilepsy include tumors, vascular malformations, scars, or areas of focal atrophy.

Although video-EEG monitoring in an inpatient epilepsy unit is not necessary in all cases, it is important to verify ictal EEG onset with lesion location since they are not identical in all cases. For example, closer examination of seizure semiology and ictal EEG may provide evidence of an epileptogenic zone that extends beyond the margins of a lesion. The extended epileptogenic zone would have to be included in the resection to obtain seizure relief. Another common reason for video-EEG monitoring in lesional cases is to verify a diagnosis of intractable epilepsy in patients with suspected psychogenic pseudoseizures so that unnecessary surgery is not performed.

A typical lesionectomy not only resects the tumor or vascular malformation itself, but also 1–2 cm of cortical tissue surrounding the lesion because it is often epileptogenic. This depends upon the brain region involved, however. In most cases, only the lesion itself is resected when it resides in eloquent cortex. When located in noneloquent brain regions, the standard practice is to excise the lesion, as well as the adjacent cortex up to the sulcal boundaries (Asadi-Pooya and Sperling, 2008). Patients with focal epilepsy caused by a circumscribed lesion have among the highest rates of seizure-free outcome. Several recent series of lesionectomy patients reported 80%–86% achieved complete seizure relief.

Hemispherectomy

Functional hemispherectomy is a radical surgical intervention typically performed in children with severe hemispheric encephalopathy and nonfocal intractable seizures. Children undergoing this surgical procedure usually have large malformations of cortical development involving multiple lobes of a single hemisphere, a history of perinatal stroke, or chronic encephalitic epilepsy, such as that seen in Rasmussen syndrome. The seizures seen in these children often begin in infancy and frequently involve multiple seizure types or seizures that are generalized (e.g., generalized tonic, myoclonic, or atonic seizures) and nonlocalizable.

The surgical procedure involves disconnecting an entire cerebral hemisphere while simultaneously leaving as much of the brain substance intact as possible. The procedure has evolved over time but currently commonly consists first of a temporal lobectomy. This is followed by resection of the frontoparietal convexity and insula and a complete corpus callosotomy.

Finally, the residual frontal and parieto-occipital lobes are disconnected from the remaining brain.

In the 1990s, this procedure was revised to emphasize disconnection over excision and should more accurately be called a *hemispherotomy*. Nonetheless, the procedure is still commonly referred to as a hemispherectomy. The EEG recordings immediately following hemispherectomy may continue to show epileptogenic spikes in the isolated hemisphere, but seizures are not clinically evident due to the complete disconnection of the diseased and epileptic hemisphere (Montes et al., 2006).

Outcome of Hemispherectomy

As might be expected given the magnitude of the brain disconnection, seizure-relief outcomes are quite high; in modern samples usually 60%–90% of children are seizure-free and the remaining children show significant reductions in seizure frequency. Improvements in cognitive function and behavior are also commonly reported, and perhaps more importantly, the negative consequences of ongoing seizures on subsequent development are halted.

Delvin, Cross, Harkness, and colleagues (2003) found that seizure outcome was related to etiology. Vascular causes (e.g., congenital middle cerebral artery infarctions) had the best outcomes, progressive disorders (e.g., Rasmussen encephalitis, Sturge-Weber syndrome) had intermediate outcomes, and developmental etiologies (e.g., diffuse hemispheric neuromigrational disorders, hemimegalencephaly) showed the worst outcomes both in terms of seizure relief and quality of life. It has been suggested that the earlier the surgery, the better the psychosocial outcome. Preoperative level of function is an important predictor of overall outcome; that is, cognitive improvement is typically better in children with higher preoperative intellectual abilities.

Complications of Hemispherectomy

For children with incomplete motor deficits, there is usually a contralateral flaccid hemiplegia after surgery that gradually recovers in a gross fashion within a month or so (Montes et al., 2006). In such cases, the children will usually be able to walk and use their arm, but fine motor function of the hand will not return. Contralateral sensory changes, including a homonymous hemianopsia, will also occur. Mortality generally occurs in between 1% and 5% of cases during, or shortly after, surgery which is usually related to hemorrhage.

Corpus Callosotomy

Corpus callosotomy involves transecting the axonal fibers that connect one hemisphere to the other. Severing the callosal fibers is a nonresective procedure used in patients with generalized motor seizures to reduce the severity, but not completely control, the seizures. Disconnection of the corpus callosum will disrupt propagation of the spreading electrical discharges and confine them to a single hemisphere. Callosotomy has been used primarily in refractory cases of atonic or akinetic seizures with drop attacks, Lennox-Gastaut syndrome, and in those who have multiple seizure types. The goal of the procedure is palliative; that is, to reduce the severity of seizures, particularly those associated with falls and subsequent head injury, by confining them to one hemisphere.

Corpus callosotomy may be performed as either a complete transection or as a partial, usually anterior two-thirds, transection. A complete callosotomy produces better seizure control than partial resection, but also causes a more severe and long lasting neuropsychological disconnection syndrome. Most epilepsy surgery centers initially perform an anterior two-thirds to three-fourths transaction, and then if seizure severity reduction is inadequate, go on to complete the resection in a second operation.

The surgical approach is generally from the nondominant, usually right, side. After initial exposure of the interhemispheric fissure, the medial aspect of the right frontal lobe is retracted. An operating microscope is brought in, and the cingulate gyri are then separated down to expose the corpus callosum. A more posterior retractor may also be used to separate the interhemispheric fissure more extensively. The actual transaction of callosal fibers is performed with a microdissection instrument and suction aspiration (Roberts and Seigel, 2001). It is important that the ventral extent of the sectioning does not enter into one of the lateral ventricles below, since this can cause bleeding and subsequent hydrocephalus.

Outcome of Corpus Callosotomy

The goal of corpus callosotomy is not complete seizure control, but rather to reduce or eliminate incapacitating generalized seizures; thus outcome is not measured by degree of complete seizure freedom. Patients with atonic or tonic seizures with drop attacks seem to benefit the most from the procedure. In one small outcome study, 85% of children with drop attacks showed significant improvement; defined as seizure reduction of 70% or more (Phillips and Sakas, 1991). In patients with generalized tonic-clonic seizures, corpus callosotomy reportedly results in significant improvement in seizure frequency in between 50% and 80% of cases (Gates and dePaola, 1996).

Complications of Corpus Callosotomy
In the first few days after corpus callosum sectioning, patients will often show lethargy, apathy, hypokinesia, and mutism; and may display signs of acute disconnection, such as a left-handed callosal apraxia (Black et al., 1992). The hypokinesia and mutism is thought to be secondary to the extended period of retraction on the medial frontal lobe and anterior cingulate gyrus and may be seen in up to 30% of patients.

The neuropsychological disconnection or "split brain" syndrome may occur after complete corpus callosotomy. It is caused by disconnecting the information from the left visual half-field and the somatosensory information from the left side of the body to the language centers in the left hemisphere. The *alien hand* phenomenon, in which intramanual conflict occurs between the left and right hands, typically resolves in the days and weeks following surgery.

Other complications of corpus callosotomy may include bleeding into the brain from injury to the bridging veins, pericallosal arteries, or superior sagittal sinus or postoperative hydrocephalus or ventriculitis after inadvertent entry into the lateral ventricles during the procedure (Smith et al., 2006). The use of corpus callosotomy has declined in recent years due to the emergence of other less invasive treatments, such as vagal nerve stimulation, ketogenic diet, and the newer antiepileptic drugs. Many epilepsy surgery centers will first attempt one or more of these less invasive treatments before recommending corpus callosotomy.

Multiple Subpial Transection
In the past, patients whose epileptogenic zones overlapped with eloquent cortex, such as motor, somatosensory, or language areas, were most often excluded from epilepsy surgery because resection of these important brain regions would result in life-altering neurologic impairment. Morrell (1991) began using MST to treat epilepsy in children with Rasmussen encephalitis whose epileptic foci were in the language centers of the left hemisphere. Since that time, MST has most often been used to treat not only Rasmussen syndrome but also children with Landau-Kleffner syndrome and cases with seizure onset in the sensory-motor areas. Before undergoing MST, patients must have their seizure focus precisely localized (usually with subdural grid

electrodes) and eloquent cortex identified through detailed functional mapping by electrical cortical stimulation.

The technique involves cutting the intracortical (horizontal) fibers at 5mm intervals and leaving the columnar organization of the cortex intact. The idea behind this is that by transecting the horizontal fibers between cortical columns, the spread of epileptic discharges to adjacent columns would be prevented, which in turn should inhibit the development of a clinical seizure. More importantly, the functional status of the eloquent cortex should be largely preserved since the cortical columns and underlying U-fibers remain intact. An MST may be performed as the only surgical intervention to treat the seizure disorder, but more commonly MST is performed in conjunction with resection of noneloquent cortex since the epileptogenic zone often extends beyond the eloquent brain regions. In these cases, resective epilepsy surgery is typically carried out on up to 1.5–2 cm of the margin of identified eloquent cortex, and then MST is performed within the eloquent cortical areas.

Outcome of Multiple Subpial Transection
In general, MST has a much worse seizure-relief outcome than focal cortical resections. When MST is the sole surgical intervention for complex partial epilepsy, only about one-third of patients become seizure-free and another one-third derive no benefit whatsoever. When MST is used in conjunction resective surgery outside the eloquent cortical areas, seizure-relief outcome improves. Of 68 patients treated with simultaneous MST/resective surgery at Rush Medical University, 48% achieved complete seizure freedom, 33% had rare or greater than 90% seizure reduction, and 18% showed no significant improvement at 2-year follow-up (Smith et al., 2006).

Complications of Multiple Subpial Transection
Use of MST has generally been associated with much greater surgical morbidity than temporal or frontal lobectomies. In the acute postoperative period, cerebral edema usually causes transient neurological impairment that may last for several weeks or even a few months. Depending upon the area of cortical invasion and extent of transection, 13%–20% of patients may show some type of permanent neurologic impairment. Some of the more common impairments reported after MST include mild motor deficits, speech impairment, aphasia, anomia, somatosensory defects, and visual deficits. In 14 children who were treated with MST for Landau-Kleffner syndrome, seven (50%)

experienced significant language recovery and seven (50%) showed partial or no language recovery (Grote et al., 1999).

Selective Amygdalohippocampectomy
When the epileptogenic zone causing complex partial seizures is confined to the mesial basal temporal lobe structures, a standard ATL that includes the lateral neocortical portions of the temporal lobe is not required to achieve complete seizure freedom. The transsylvian route for a selective amygdalohippocampectomy was first advocated by Wieser and Yasargil (1984), and this has become the most common surgical approach in use at this time. In this microsurgical procedure, a cortical incision is made in the sulcus insulae circularis pars inferior after the lateral fissure has been opened to reach and remove the amygdala. The temporal horn of the lateral ventricle is then dissected and the hippocampus and parahippocampal gyrus are removed.

The rationale for the selective amygdalohippocampectomy in carefully defined cases of mesial temporal lobe seizure onset is that the seizures may be relieved with minimum tissue resection, thereby sparing neuropsychological functions as much as possible. The procedure is considered to be more difficult surgically than a standard ATL, and thus needs to be carried out by properly trained neurosurgeons skilled in microsurgical techniques.

Outcome of Selective Amygdalohippocampectomy
Seizure-relief outcome following selective amygdalohippocampectomy is equivalent to ATL after equating for the type of pathology causing the seizures. The procedure had been thought to have a more positive neuropsychological outcome than a standard ATL, but results on this have been mixed. It appears fairly clear that the procedure poses a significant risk to recent memory functions. In the largest reported series, approximately 50% of patients experienced a significant decline in verbal learning and memory functions after left, and 27% showed a nonverbal memory loss after right, selective amygdalohippocampectomy 1 year after surgery (Gleissner et al., 2004). In a smaller series of patients, 78% of those whose Wada test suggested a risk to memory showed a postoperative memory decline after left selective amygdalohippocampectomy (Polkey, 1993).

Weiser (1991) has summarized the most important variables predicting surgical outcome following selective amygdalohippocampectomy. The presence of hippocampal sclerosis, a history of febrile seizures, and the presence of unilateral material-specific memory deficits were all predictive of good surgical outcome. Longer duration of refractory seizures, older age at time of surgery, and the presence of severe memory deficits were associated with poor postoperative seizure-relief outcome.

Implanted Electrical Brain Stimulators

Patients with intractable complex partial seizures who have multiple seizure foci or an epileptogenic zone that includes eloquent cortex that cannot be surgically treated may be candidates for implantation of an experimental electrical brain stimulator. One such device that is currently (2009) being studied in randomized clinical trials is the Responsive Neurostimulation System (RNS) manufactured by NeuroPace (Fountas et al., 2005). This device is inserted within the skull and electrodes are placed into the epileptogenic zones which have been defined either noninvasively or with subdural grid or depth electrodes. The RNS constantly monitors the electrical activity of the brain with either subdural strip or depth electrodes. When the device detects abnormal electrical activity that may signal the beginning of a seizure, it sends out a brief electrical pulse a few milliseconds later which interrupts seizure development (see Figure 11.5).

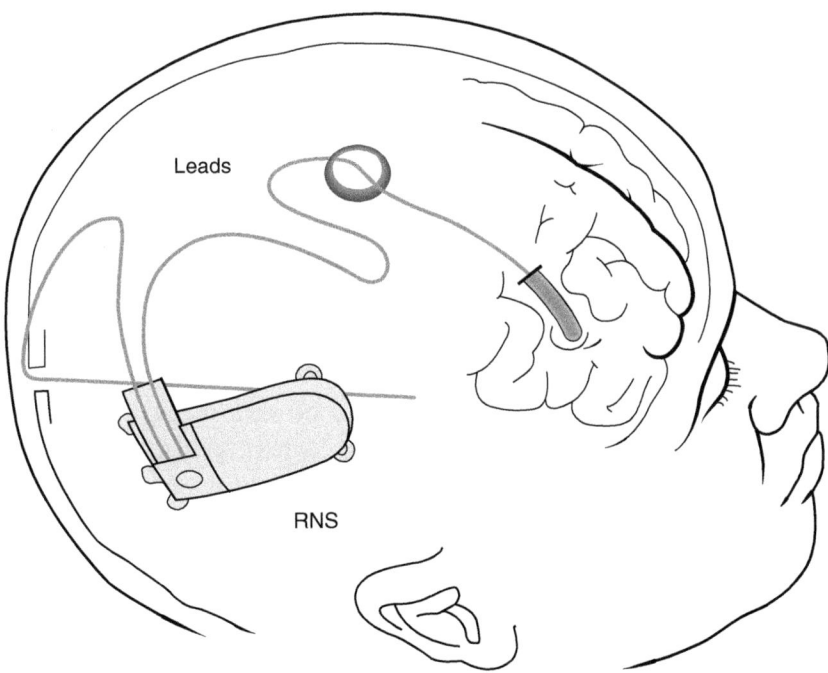

FIGURE 11.5. Closed-loop responsive neurostimulator system (NeuroPace, Inc.). Published with permission of NeuroPace, Inc., Mountain View, CA.

Another type of neurostimulator, which is not a responsive detection system, is also under study. The medical device manufacturer Medtronic is sponsoring clinical trials of its Intercept Epilepsy Control System, which is a deep brain stimulation therapy system for patients with intractable epilepsy. Although these devices have been shown to be feasible from a technical surgical perspective with relatively little risk for mortality and morbidity, their efficacy for controlling refractory seizures is unknown because the clinical trials have not been completed at this time (May 2009).

Case Example

Independent Bilateral Seizure Onset: Candidate for Responsive Neurostimulator Implant

Patient History:
The patient is a 38-year-old, right-handed, white man with intractable complex partial seizures for the past 9 years. Seizures begin with an aura of a "funny, difficult-to-describe feeling" in his head. He then becomes unresponsive, develops a blank stare, and begins picking or grabbing at objects with his hands. There is excessive drooling and bilateral stiffening of the arms, and he has a loss of awareness for events occurring during the seizure. These complex partial spells occur approximately three or four times per month. The partial seizures secondarily generalized about 4 times per year with whole body convulsions, falls, tongue biting, and urinary incontinence.

Etiology of the seizure disorder is presumed to be theophylline (for asthma) toxicity that resulted in a prolonged episode of status epilepticus with anoxic brain injury at age 18 years. The patient was admitted to the intensive care unit where he experienced three cardiac arrest events and days of coma. The patient was seizure-free between the ages of 19 and 29 years but then he developed ongoing complex partial seizures that have been refractory to treatment after adequate trials of eight different antiepileptic drugs.

Medical Test Results:
Brain MRI showed bilateral amygdala and hippocampal atrophy with no evidence of mesial temporal lobe sclerosis. Brain PET scan revealed bilateral decreased uptake in the temporal lobes with maximum reduction in the left mesial temporal lobe. Neurological examinations have been normal in recent

years. EEGs have shown interictal spikes and slow wave activity in both the left and right temporal lobe regions.

Wada Testing:
During intracarotid amobarbital (Wada) evaluation, the patient demonstrated left hemisphere dominance for speech and language. There was no significant asymmetry in object memory performance. There was deficient memory performance after both left and right injections suggesting bilateral mesial temporal lobe dysfunction. Wada results were as follows:

	LEFT INJECTION	RIGHT INJECTION
Expressive Speech:	Arrest of counting	Normal
Receptive Speech:	Severe impairment	Normal
Visual Naming:	Severe impairment	Normal
Repetition:	Severe impairment	Normal
Reading:	Severe impairment	Normal
Paraphasic Errors:	Semantic & phonemic errors	None
Object Recognition Memory:	2 of 8, 0 False Positives (FPs)	1 of 8, 0 FPs

Neuropsychological Testing Results:
The following interpretive guidelines are offered to assist with interpretation of the scores:

SS	Standard Score ($M = 100$, $SD = 15$); SS $\leq$ 75 indicates potential impairment.
ss	scaled score ($M = 10$, $SD = 3$); ss $\leq$ 5 indicates potential impairment.
z	z score ($M = 0$, $SD = 1$); $z \leq 1.6$ indicates potential impairment.
T	T score ($M = 50$, $SD = 10$); $T \leq 34$ indicates potential impairment,
percentile	Percentile ranks ($M = 50^{th}$, -1 SD $= 16^{th}$, $+1$ SD $= 84^{th}$, Impaired $< 5^{th}$).
RS	Raw score

General Intellectual Functioning:

Verbal Comprehension Index (SS)	85, 16th percentile
Similarities (ss)	7
Comprehension (ss)	7
Information (ss)	7
Working Memory Index (SS)	86, 18th percentile
Digit Span (ss)	9
Arithmetic (ss)	6
Perceptual Organization Index (SS)	99, 47th percentile
Picture Completion (ss)	10
Block Design (ss)	10
Picture Arrangement (ss)	9
Processing Speed Index (SS)	96, 39th percentile
Digit Symbol (ss)	11
Symbol Search (ss)	8

Language:

MAE Visual Naming	48, 18th percentile
MAE Letter Fluency (COWA)	17, 1st percentile
MAE Token Test	40, 24th percentile

Verbal Learning and Memory:

WMS-III Logical Memory I (RS)	2, <1st percentile
WMS-III Logical Memory II (RS)	4, 2nd percentile
Selective Reminding (CLTR, RS)	43, <1st percentile
California Verbal Learning Test	
Trials 1–5	46, 16th percentile
List B	4, 7th percentile
Short Delay Free Recall	5, 2nd percentile
Long Delay Free Recall	4, 1st percentile
Long Delay Recognition	13, 2nd percentile
Long Delay False Positives	14, >99th percentile

Nonverbal Learning and Memory:

WMS-R, Visual Reproduction I (ss)	4, 2nd percentile
WMS-R, Visual Reproduction II (ss)	3, 1st percentile
Complex Figure Immediate Recall	16, 7th percentile
Complex Figure 30' Delayed Recall	10, <1st percentile

Visuoperception and Spatial Thinking:

Facial Recognition Test	45, 51st percentile
Complex Figure Copy	34, > 16th percentile
Judgment of Line Orientation	22, 22nd percentile

Executive Functioning:

Wisconsin Card Sorting Test (WCST)	2 categories, 6–10th percentile
% Perseverative Responses	36, 6th percentile
% Perseverative Errors	29, 7th percentile
WISC-III Mazes (RS)	16, 5th percentile

Impressions:
Neuropsychological testing revealed severe deficits of verbal and nonverbal new learning and memory and verbal associative fluency. Frontal-executive functions were borderline. Remainder of the cognitive examination was essentially normal within the context of low-average Verbal IQ and average Performance (perceptual-constructional) IQ.

Video-EEG Monitoring:
The patient was admitted to the Epilepsy Monitoring Unit for continuous video-EEG monitoring and seven habitual seizures were recorded. Ictal EEG showed seizure onset in the left temporal lobe in four of the seizures and onset in the right temporal lobe in three seizures. Interictal EEG showed bilateral independent epileptogenic spikes.

Case Formulation:
All tests and history suggest bilateral mesial temporal lobe (MTL) dysfunction, as well as bilateral MTL seizure onset. Anoxic damage is usually bilateral in nature and the hippocampi are particularly vulnerable to loss of oxygen. MRI revealed bilateral MTL atrophy. The PET showed bilateral MTL hypometabolism interictally. Interictal EEG showed bilateral independent MTL spikes, and simultaneous video-EEG recording revealed bilateral MTL epileptogenic

zones. Neuropsychological testing was consistent with these medical tests in showing verbal and nonverbal memory impairments.

Taken together, these test results suggest this patient would most likely not achieve seizure-freedom if he underwent a unilateral temporal lobectomy (even though there is some evidence the left side is worse). Furthermore, given that Wada testing suggested inadequate contralateral memory support (i.e., deficient *functional reserve*), he may be at risk for a post-temporal lobectomy memory disorder. Therefore, resective epilepsy surgery is not an option for this patient. He does, however, appear to be a good candidate for implantation of a responsive neurostimulator system. He might achieve some degree of seizure control after the responsive neurostimulator system electrodes are placed into the temporal lobes bilaterally.

Appendix I

Traditional Classification of Epileptic Seizures: Description of Seizure Types Not Covered in Body of Text

Simple Partial (or Focal) Seizures

Simple Partial Seizures with Motor Signs

Epileptic foci in the motor areas within the precentral gyrus may result in focal motor seizures characterized by either positive (e.g., excitatory activating symptoms such as clonus) or negative (inhibitory paralytic symptoms such as speech arrest) signs and symptoms (Browne and Holmes, 2004). Motor signs are among the more common manifestations of simple partial seizures. A common motor sign of simple partial seizures restricted to the motor strip (Brodmann's area 4) is clonus. Clonus consists of jerking movements caused by the rhythmic alternation between contraction and relaxation of various muscle groups.

If the electroencephalogram (EEG) discharges remain localized to a particular region of the motor strip, then the symptoms would be restricted to one area of the body; most commonly the contralateral fingers, hand, face, or tongue. Focal motor seizures may not remain strictly focal. They may spread to adjacent cortical regions resulting in sequential involvement of contiguous body parts (e.g., from the muscles of the face to the ipsilateral arm or hand) in an epileptic progression (Kellinghaus et al., 2006). When these seizures spread, they are known as *jacksonian seizures*, and the process is referred to as a *jacksonian march*.

If the epileptic discharges are restricted to the premotor region (Brodmann's area 6) or the premotor regions of the mesial frontal lobe (i.e., supplementary motor area), tonic muscle contractions (i.e., persistent flexion or extension of a restricted group of muscles) are often seen. Sustained tonic muscle contractions may also involve the axial musculature and result in posturing of the limbs or whole body. The supplementary motor area (SMA) is thought to play an important role in simple partial seizures with postural signs. Versive focal motor attacks are characterized by turning of the eyes and head to one side; usually away from the side of the EEG discharges, although ipsilateral head turning may also occur. Versive head and eye movement toward the opposite side from the EEG discharges may be a reliable lateralizing sign when these movements precede secondary generalization of the seizure, but may not provide lateralizing information when the seizure does not secondarily generalize (Chee et al., 1993; Kotagal and Arunkumar, 1998).

Phonatory simple partial seizures may consist of either negative phenomena such as speech arrest (an inability to vocalize) or positive phenomena such as involuntary forced vocalization or perseveration of a syllable or phrase. After a focal motor seizure, there may be a localized paralysis restricted to the area previously involved in the seizure. This is called *Todd paralysis*, and it may persist for several hours after a seizure. When a focal motor seizure is characterized by continuous localized EEG discharges in conjunction with unrelenting focal myoclonus, it is referred to as *epilepsia partialis continua* or partial continuous epilepsy, which is a type of status epilepticus.

Simple Partial Seizures with Somatosensory or Special Sensory Symptoms

Epileptic foci in the postcentral gyrus of the parietal lobe may give rise to focal somatosensory seizures often consisting of tingling, numbness or "pins and needles" sensations in the contralateral mouth, lips, or fingers. Similar to jacksonian seizures, somatosensory seizures may also spread to adjacent cortical regions and continue to evolve into partial complex seizures or even secondary generalized tonic-clonic seizures. Simple partial seizures with somatosensory or special sensory (simple hallucinations or illusory phenomena) symptoms are relatively rare compared to those with motor signs (Oxbury and Duchowny, 2000).

Special sensory seizures may involve the visual or auditory cortical areas, and vary in complexity from elemental to elaborate phenomenon depending upon whether primary or secondary association areas are involved in the seizure. Simple partial seizures confined to the primary visual areas may become evident by either positive or negative phenomena. Common positive visual phenomena include flashes of light, sparks, or zigzag patterns that have color and may move across the visual field or remain stationary. Examples of negative signs of simple partial visual seizures usually consist of visual field defects, such as quadrantanopsias, scotomas, or patches of darkness. When focal seizures are in the secondary visual association areas, fully formed visual hallucinations (e.g., people or complete scenes) or illusions (i.e., distortions of visual experience) may be experienced.

Auditory symptoms also may either be simple or complex depending upon whether the ictal discharges are in the primary or secondary association auditory cortices. Partial seizures arising in the primary auditory cortex most often consist of simple sounds such as roaring, ringing, or humming noises; whereas seizures confined to the secondary auditory association areas may give rise to auditory illusions or hallucinations, such as voices, musical passages, or songs (Oxbury and Duchowny, 2000). Auditory phenomena are rare during simple partial seizures, but when they do occur they most often consist of elemental auditory experiences that may be heard in the contralateral ear or bilaterally (Browne and Holmes, 2004).

Olfactory partial seizures most often give rise to experiences of unpleasant odors often described as a "chemical" or "burning" smell. These seizures are usually confined to the region of the uncus or anterior parahippocampal gyrus. Simple seizures restricted to the primary taste regions in the inferior somatosensory strip and superior insula are also often described as giving rise to unpleasant gustatory sensations such as a bitter, sharp, acid, sickly sweet, or metallic taste. Vertiginous partial seizures may cause a vague dizziness, true vertigo, unsteadiness, or body tilt sensations and are generally thought to initially arise in the vestibular nuclei, nearby brainstem regions, or in the temporal lobe (Oxbury and Duchowny, 2000).

Simple Partial Seizures with Autonomic Symptoms or Signs

Partial seizures may give rise to a variety of autonomic signs and symptoms that may include epigastric distress, unpleasant rising sensations in the abdomen, chest, throat, or head including pain, breathlessness, fullness or pressure, altered heart rate (usually tachycardia), flushing, sweating, shivering, piloerection, pupillary dilatation, anxiety, fear, vomiting, salivation, thirst, urinary urgency, and genital sensations or orgasms. Most of these experiences have not been well-localized to any single cortical region. Epigastric distress, other abdominal and cephalic sensations, and anxiety and fear are common auras in patients with complex partial seizures of temporal lobe origin (Stefan et al., 2002; Oxbury and Duchowny, 2000). Alterations in respiratory rhythms, heart rate and other indicators of sympathetic nervous system activation have been reported in seizures originating in temporolimbic and insular areas of the brain (Isnard et al., 2000).

Simple Partial Seizures with Psychic Signs

The psychic signs referred to in the traditional classification of simple partial seizures include disturbances of higher cognitive functions (i.e., dysphasic, dysmnesic symptoms, dreamy states), affective symptoms, illusions, and structured hallucinations. Focal partial seizures may give rise to many different psychic alterations including changes in speech, language, memory, distortions of time sense, fear, terror, intense sadness, feelings of unworthiness, déjà-vu, sensations of unreality (jamais-vu), depersonalization, illusions, or complex visual or auditory hallucinations (Browne and Holmes, 2004; Oxbury and Duchowny, 2000).

Simple Partial Seizures with Cognitive Signs

Dysphasic disturbances may occur during the seizure itself (ictally) or for a brief period after the seizure has stopped (postictally). Ictal aphasia has been associated with seizures originating in the language dominant (usually left) hemisphere and may consist of most of the language deficits that have been described in the aphasia literature depending upon which cortical region has been affected by the seizure. Thus, more anteriorly located seizure foci may produce muteness, dysarthria, or a nonfluent Broca-like syndrome characterized by effortful output, telegraphic speech, and relatively well-preserved comprehension, whereas a more posteriorly centered epileptic focus may result in a fluent, comprehension-impaired aphasia with semantic paraphasias (Ardila and Lopez, 1988).

Dysmnesic symptoms consist of distortions of memory. The most common dysmnesic seizure symptom is *déjà-vu* (literally "already seen"), which is a strong sense of familiarity or an overwhelming feeling that a novel experience has been experienced before. Although *déjà-vu* may be the sole symptom of a simple partial seizure, it is also a common early manifestation (an aura) of a partial complex seizure of temporal lobe origin (Bancaud et al., 1994). The opposite, less common, experience of feeling that a familiar experience is new or that a repeated experience has never occurred before is called *jamais vu*; it has also been described as the sole symptom of a simple partial seizure or as an aura in a partial complex seizure. Pure ictal amnesic seizures lasting longer than 5 minutes, in which the only clinical manifestation is an inability to retain memory for events occurring during a seizure, have been reported (Palmini et al., 1992), as well as a transient epileptic amnesia that usually lasts longer ($\sim$1 hour) but is present in at least some patients interictally (Butler et al., 2007). These dysmnesic symptoms are most commonly encountered in patients with complex partial seizures of mesial temporal lobe origin.

Simple Partial Seizures with Affective Signs

The so-called cognitive-affective disturbances present in simple partial seizures include dreamy states, distortions of time sense, and sensations of unreality, detachment, or depersonalization. Dreamy states are difficult to describe experiences partly because many patients do not fully recall the experience during their seizure. These states frequently overlap with altered sense of time, experiential hallucinations, forced thinking, or feelings of depersonalization (Dietl et al., 2005). Some patients believe their seizures consist of the same (or very similar) dreamy state each time they experience a seizure, whereas others report it varies from seizure to seizure (Oxbury and Duchowny, 2000). These states are also commonly observed in partial complex seizures where there are alterations of awareness. The EEG investigations have suggested these altered experience states arise from focal electrical disruption of both the medial temporal lobe and lateral temporal neocortex (Gloor, 1990). Since these dreamy states are associated with some alterations in awareness and an inability to fully recall the content of the experience, there is debate as to whether dreamy states should be classified as a complex partial seizure rather than a simple partial seizure.

Fear is the most common affective ictal symptom experienced as part of a simple partial seizure, although as with many of the other simple partial symptoms and signs, it may also be an early warning (an aura) sign of a partial complex seizure. As with most of these simple seizure symptoms, the onset of the fear is sudden, and there is no environmental stimulus that causes it. The intensity of the fear may vary from mild anxiety to abject terror, and it is often accompanied by signs of sympathetic arousal such as pupil dilatation, flushing, piloerection, increased heart rate, sweating, and hypertension (Kellinghaus, et al., 2006). Fearful states have been associated with seizures beginning in the amygdala, medial temporal lobe, and anterior frontal lobe. A variety of other emotions also may be experienced as part of a simple partial seizure including sadness, feelings of worthlessness, irritability, anger, pleasure, or elation. Elation and pleasure are infrequent concomitants of focal seizures. These ictal affective states differ from those seen in psychiatric conditions as they tend to come in attacks that last for only a few minutes. Epileptic affective changes also lack external triggers and tend to resolve quickly after the electrical discharges end.

Simple Partial Seizures with Illusions
Illusions may involve distortions of visual, auditory, or somatosensory perception during a seizure. Ictal involvement of the visual cortex may result in changes in the perception of an object's size (macropsia or micropsia), shape, distance, or color. Visual perseveration (palinopsia) may be experienced as seeing multiple afterimages of an object. Likewise, distortions of sound may be experienced as being either louder (macracusia) or softer (micracusia) during seizures involving the auditory cortices of the temporal lobes. Seizure activity in the region of the somatosensory strip or somatosensory association areas in the parietal lobe may cause phantom sensations or altered perceptions of the size or weight of a limb. Out-of-body illusions have also been described in conjunction with ictal EEG activity in the parietal lobe.

Simple Partial Seizures with Hallucinations
Structured hallucinations during a simple focal seizure are usually visual or auditory perceptions experienced in the absence of an external stimulus. The most frequent are complex visual hallucinations that may consist of fully formed colorful scenes with places and people that may be recollections of past events. Some of these visual hallucinations are "experiential," and patients feel as if they are actually present within the scene and react with various emotions to the unfolding event. Complex auditory hallucinations most often consist of voices, spoken sentences, or music.

Generalized Seizures (Convulsive or Nonconvulsive)

Atypical Absence Seizures
Atypical absence seizures may be differentiated from typical absence attacks by (a) their association with encephalopathic syndromes causing generalized seizures, e.g., Lennox-Gastaut syndrome, Landau-Kleffner syndrome

(see under the Epileptic Syndromes); (b) having multiple other seizure types; (c) mental retardation; (d) a paucity of automatisms; (e) changes in motor tone; and (f) longer seizure durations, as well as (g) differences in characteristic EEG findings (Stefan and Snead, 1997). The seizures themselves have a more gradual onset and slower termination than typical absence attacks. They often begin with a staring spell and variably impaired awareness and responsiveness. The seizure may then progress to motor involvement with changes in tone. Some seizures include clonic components or autonomic changes. Atypical absence attacks are most likely to occur when the patient is drowsy, but are not precipitated by hyperventilation or photic stimulation. The ictal EEG is usually characterized by diffuse, irregular spike-and-wave discharges that may lateralize to one hemisphere. Atypical absence spells are considered among the most difficult to diagnose and treat, and many have poor seizure outcomes. Atypical attacks are most common in the pediatric age group but are rarely occur before 2 years of age. Cognitive impairments, social dysfunction, learning disabilities, and resistance to pharmacotherapy with antiepileptic drugs are common associated features; although the severity of the epilepsy and its neuropsychological consequences depends upon the specific underlying condition causing the seizures. Etiologically, atypical absence seizures are most often caused by symptomatic (identifiable acquired lesion) or cryptogenic (unknown but assumed developmental lesion) conditions as opposed to the idiopathic (genetic) etiology in typical absence seizures.

Myoclonic Seizures
Myoclonic seizures are generalized seizures consisting of sudden, involuntary, shock-like muscle contractions (*myoclonic jerking*) involving the entire body or confined to the face, trunk, or one of more limbs or muscle groups. Myoclonic jerks last for only a fraction of a second. Since myoclonic spasms may occur in conditions other than epilepsy, such as spinal cord disease or upon falling asleep or waking ("nap jerks"), EEG is necessary to confirm myoclonic epilepsy. The EEG activity characteristically seen in epileptic myoclonus consists of generalized epileptiform bursts of multiple, rapid, high-amplitude spikes followed by slow waves (Delgado-Escueta, Serratosa, Medina, 1996). Consciousness is thought to be preserved during the episodes although this is nearly impossible to accurately determine since the seizures are so brief. Although most epilepsies with myoclonic seizures begin in the first 5 years of life, they may occur at any age. Myoclonic

seizures represent a group of disorders that may be caused by a variety of genetic or acquired conditions. For example, myoclonic seizures may be caused by various epileptic encephalopathies including Lennox-Gastaut syndrome or Creutzfeldt-Jacob disease. Thus, the severity and course of myoclonic seizures ranges from relatively benign to catastrophic depending upon etiology of the seizures.

Clonic Seizures

Clonic seizures are similar to generalized tonic-clonic convulsions except there are no tonic muscle contractions. These seizures typically begin with a loss of awareness associated with sudden hypotonia or a brief generalized tonic spasm, followed by one to several minutes of bilateral rhythmic jerks (Browne and Holmes, 2004). The clonic muscle contractions vary greatly in frequency, amplitude, and area of involvement over the course of the seizure. The jerks are often asymmetric and may involve only one limb. In young children, however (especially between the ages of 1 and 3 years), the jerks remain bilateral and synchronous during the attack. After a seizure, recovery may be rapid, or there may be a period of postictal confusion or tiredness. Clonic seizures are rare and occur almost exclusively in neonates and young children. They most often spontaneously remit on their own within a relatively short period of time or evolve into generalized tonic-clonic seizures.

Tonic Seizures

Tonic seizures are brief (~5–20 seconds) generalized epileptic spells characterized by either a sudden or gradual increased tone in the extensor muscles. Electromyography (EMG) is dramatically increased during the episode of rigid muscle extension. The degree of motor involvement varies considerably across patients and even across different seizure episodes in the same individual. Tonic spells vary in severity from a brief upward deviation of the eyes, with or without nystagmus, to a completely generalized symmetric or asymmetric tonic stiffening associated with loss of consciousness, falls, and repeated injury. Tonic seizures often begin with contraction of the neck muscles, widely opened eyes, and jaw clenching or mouth opening. Contraction of the respiratory and abdominal muscles may follow, resulting in a high-pitched scream or period of apnea. The tonic muscle contractions may then spread to the upper extremities causing the shoulders to shrug and the arms to extend outward. Autonomic signs such as increases in blood pressure, heart rate, and respiration may also occur. Postictally, there is typically a period of recovery from confusion, fatigue, and headache that is proportional to the severity of the seizure (Vigevano, Fusco, Kazuichi, et al., 1997).

Tonic spells are usually brief; lasting 10–15 seconds, although they may last up to a minute or so. Generalized tonic seizures are more common in children than adults. These seizures are activated by sleep and typically occur frequently throughout the night during periods of non-rapid eye movement (REM) sleep. Because of their short duration and tendency to occur during sleep, tonic seizures often go undiagnosed, and simultaneous video-EEG inpatient monitoring is often necessary for accurate diagnosis.

Ictal EEG in tonic seizures consists of bilateral, medium- to high-voltage, epileptic spikes with predominance in the frontal regions (Gastaut and Broughton, 1972). Tonic seizures are most often symptomatic (caused by a known disease) and are the most common seizure type in some epileptic encephalopathies. For example, tonic seizures are by far the most common seizure type in children with Lennox-Gastaut syndrome, occurring in 74%–90% of cases (Roger, Dravet, Bureau, 1989). Tonic seizures are frequently refractory to pharmacologic management, and thus, seizure frequency is high and there is a propensity for these children to develop episodes of status epilepticus. Farrell (2001) reported tonic status epilepticus may occur in 54%–97% of patients.

Atonic (Astatic) Seizures

Atonic seizures are a type of generalized epilepsy characterized by a brief and sudden loss of muscle tone. These seizures begin abruptly without warning and typically consist of a loss of postural tone in the flexor and extensor muscles of the trunk, limbs, or neck (Tatum and Farrell, 2006). Atonic seizures may be very brief, lasting only a few seconds (referred to as a *drop attack*) or may last a bit longer, one to several minutes (called an *akinetic seizure*). When a seizure involves all the muscles of the trunk, a drop attack will typically occur, and if standing will result in a hard fall to floor, which may cause injury. *Astatic* seizures involve loss of muscle tone resulting in the inability to stand. Atonic seizures are not always astatic in nature.

Atonic seizures are frequently preceded by one or several myoclonic jerks. Mild atonic seizures last only a second or two and often involve only a head nod or head drop followed by immediate return of normal tone and posture. There is loss of EMG activity in the affected muscles during the attack. The

EEG usually shows rhythmic spike-and-wave discharges soon followed by diffuse, generalized slow waves that are maximal in the vertex and central regions (Gastaut and Broughton, 1972). Consciousness is briefly impaired during the seizure. Postictal confusion is rare in atonic seizures. The period of recovery varies depending upon the severity of the attack, but consciousness typically returns immediately or within a few minutes and the patient can resume a standing position after several seconds. Atonic seizures are more commonly seen in children than adults.

Appendix II

Classification of Epilepsy Syndromes: Description of Seizure Syndromes Not Covered in Body of Text

Idiopathic Localization-related Epilepsies

Benign Childhood Epilepsy with Centrotemporal Spikes

Benign childhood epilepsy with centrotemporal spikes (BECTS), an epilepsy syndrome of childhood, is also called *benign Rolandic epilepsy*. It consists of simple partial seizures with hemifacial motor spasms that often includes somatosensory symptoms. These seizures have a tendency to evolve into generalized tonic-clonic spells and usually begin during sleep. The seizures usually begin around ages 9–10 years (range 3–13 years) in males. A family history of seizures is common. Electroencephalography (EEG) shows high-voltage centrotemporal spikes often followed by slow waves activated by sleep (Commission on Classification and Terminology of the ILAE, 1989). This condition is called benign based upon the ultimate prognosis. Some children with BECTS can demonstrate cognitive deterioration before seizure remission, usually during the teenage years (Berroya, McIntyre, Webster, et al., 2004; Northcott, Connolly, McIntyre, et al., 2006). The likelihood of remission before puberty is high; benign Rolandic epilepsy resolves in the majority of children before the age of 14 years.

Childhood Epilepsy with Occipital Paroxysms
This syndrome is similar to BECTS except that the seizures begin with visual symptoms, such as phosphenes, field defects, or illusions, often followed by unilateral automatisms or clonus on one side of the body. Migrainous headaches are common in the immediate postictal period. The EEG usually shows high-amplitude spike-and-waves or sharp wave activity in the occipital or posterior temporal regions when the eyes are closed. Paroxysmal EEG activity may spread to the central or temporal regions as the seizure progresses. Prognosis is unknown at this time (Commission on Classification and Terminology of the ILAE, 1989).

Reading Epilepsy
Reading epilepsy is a rare form of reflex epilepsy in which seizures are precipitated by reading, either aloud (most common) or silently (Koutroumanidis, Koepp, Richardson, et al., 1998; Radhakrishnan, Silbert, and Klass, 1995). Seizures do not occur spontaneously when the patient is not reading. The most common form of reading epilepsy has been termed *myoclonic reading epilepsy* because these seizures begin with myoclonic jerks involving the jaw, mouth, and throat, which may spread to the arms. If the patient continues to read, the seizures can become secondarily generalized. During reading, the EEG may show a variety of epileptiform discharges, which may be either bilateral synchronous or focal, occurring most often in the left hemisphere.

Although a strong genetic component has been implicated in reading epilepsy by its frequent co-occurrence among first-degree relatives, the mode of inheritance has not been established. Successful antiepileptic drug (AED) treatments for the myoclonic form of reading epilepsy include sodium valproate and clonazepam. The inclusion of reading epilepsy as an idiopathic-genetic epilepsy syndrome in the International Classification of Epilepsies and Epilepsy Syndromes has been disputed since not all cases of reading epilepsy are focal-onset, partial reading seizures. There have been cases of generalized, reading-induced absence seizures, and some reading-induced epilepsies are even symptomatic. The focal onset, nonmyoclonic reading epilepsies have been associated with ictal EEG changes in the posterior language areas in the posterior temporal and inferior parietal lobes. Although there are several forms of reflexive reading epilepsy, the most common form appears to be the genetic myoclonic variety included in the International Classification (Commission on Classification and Terminology of the ILAE, 1989).

As mentioned earlier, the International Classification of the epilepsy syndromes was originally envisioned to be a flexible classification system with an expectation that classifications would change over time as more knowledge was gained. Thus, although only three syndromes are mentioned in the 1989 Classification under the idiopathic, localization-related epilepsy syndromes, two other syndromes probably deserve mention within this subclassification area; namely, hot water epilepsy and autosomal dominant nocturnal frontal lobe epilepsy.

Hot Water Epilepsy
Hot water epilepsy is a type of familial reflex epilepsy found in certain families in southern India (Bebek, Gurses, Gokyigit et al., 2001). The seizures are precipitated by having cups of hot water poured over the head or by bathing in hot water. Some patients with hot water seizures can bring the seizures on voluntarily by recalling earlier bathing experiences. The seizures induced by hot water are usually complex partial seizures but many secondarily generalize into tonic-clonic seizures. Spontaneous seizures that are not precipitated by exposure to hot water may occur later in life in over one-half of the cases. A preponderance of males (3 to 1) have hot water epilepsy, and many express a feeling of pleasure during the seizures. The EEGs show epileptogenic activity over the unilateral temporal regions during hot water–induced seizures.

Autosomal Dominant Nocturnal Frontal Lobe Epilepsy
These seizures usually begin in childhood or adolescence and persist throughout adulthood. This is a genetically transmitted seizure syndrome that is characterized by spells that occur in clusters during sleep, although up to one-quarter of patients experience infrequent attacks when awake (Scheffer, Bhatia, Lopes-Condes, et al., 1995). Most of these seizures occur during light sleep, soon after falling asleep or shortly before waking up. Ictal video-EEG studies have demonstrated the clinical seizures are similar to other frontal lobe–onset epilepsies. The seizures in autosomal dominant nocturnal frontal lobe epilepsy typically begin with an aura (that may consist of somatosensory, special sensory, psychic, or autonomic phenomena). The attacks are usually brief and often begin with groaning, gasping, or some type of vocalization. They typically evolve into prominent motor phenomena such as thrashing, hyperkinetic activity, or tonic stiffening with or without clonic jerking. The majority of these patients report maintenance of awareness during the spells. These seizures are often misdiagnosed as benign nocturnal

parasomnias, sleep paralysis, or psychiatric disorders. The genetic defect has been mapped to chromosome 20q13 and 1q21 with mutations in the genes coding for subunits of the neuronal nicotinic acetylcholine receptor (Hirose, Mitsudome, Okada, et al., 2005). These seizures are reportedly well-controlled with carbamazepine, but apparently do not respond well to sodium valproate.

Symptomatic Localization-related Epilepsy Syndromes

Rasmussen Syndrome (Kojewnikow Syndrome): Chronic Progressive Epilepsia Partialis Continua of Childhood

This is a very rare and severe progressive neocortical disorder of childhood in which the seizures typically begin as focal motor spells that are continuous in many cases (epilepsia partialis continua). The condition then progressively worsens; the seizures become more diffuse, hemiparesis ensues, and there is progressive cognitive deterioration (Commission on Classification and Terminology of the ILAE, 1985). This syndrome is caused by Rasmussen encephalitis, which is a chronic brain infection probably due to either a viral or autoimmune process. No causative virus has been identified, although the condition is often preceded by a bout of otitis media, tonsillitis, or upper respiratory infection 6 months or so before the seizures.

This syndrome is called *Rasmussen syndrome* in English-speaking countries and *Kojewnikow syndrome* in French-speaking countries. It should not be confused with *epilepsia partialis continua,* which is a type of focal motor status epilepticus that may last for months or years and has many different causes. Age of onset of Rasmussen syndrome is between 2 and 10 years old (with a peak onset age of 6 years). The course of the disease usually begins with sporadic focal motor or somatosensory seizures that gradually increase in frequency. Progressive neurologic (usually hemiparesis) and cognitive deterioration usually begins around 3 months after onset of the seizures. Although most of the brain damage usually occurs during the first 8–12 months, progressive worsening may occur over the course of 10 years or so (Bien, Widman, Urbach, et al., 2002). The most effective treatment is surgical hemispherectomy, which seems to stop the progression of the disease (Caplan, Curtiss, Chugani, et al., 1996). Antiepileptic drugs are usually ineffective. High-dose steroids and other immunosuppressive treatments appear to only halt the disease temporarily. After the active stages of the encephalitis have settled down over the course of a few years, the condition usually abates, leaving the child with stable, but permanent neurological deficits.

Reflex Epilepsies
Photosensitive Seizures

The most common type of reflex epilepsy are visually induced seizures that may either be photosensitive (i.e., typically triggered by flickering or flashing lights) or pattern sensitive (e.g., patterns on curtains, wallpaper, or flooring, striped clothing). The most commonly encountered flickering light stimuli include video games, television, computer monitors, discotheque lights, and natural flickering light (e.g., seeing light through the leaves of trees while driving along a tree-lined street, light reflected from moving water).

Photosensitive seizures affect approximately 5% of patients with epileptic seizures (Panayiotopoulos, 2005). Two-thirds are female, and the peak age of onset is 12–13 years old. The exception to this gender difference is in video game–induced seizures, which are more common in boys because more boys than girls play video games. In pure photosensitive epilepsy, the seizures are generalized tonic-clonic in 84% of patients, absence seizures in about 6%, partial motor seizures in 2.5%, and myoclonic in 1.5% of patients (Zifkin and Andermann, 2006).

Television-induced seizures have been more common in Europe and Japan than in the United States. This is thought to be due to differences in the frequency of the alternating current used in the United States (60 Hz) versus Europe and Japan (50 Hz), which result in different alternating flicker frequencies in the United States (30 Hz) and overseas (25 Hz). Intermittent photic stimulation at 25 Hz is more epileptogenic than at 30 Hz. One of the more notable outbreaks of photosensitive seizures occurred in Japan in 1997 when approximately 700 individuals (mostly children) with no history of epilepsy were hospitalized with seizures or seizure-related phenomena after watching a popular cartoon show called *Pokemon* (Hayashi, Ichiyama, Nishikawa, et al., 1998; Ishida, Yamashita, Matsuishi, et al., 1998). Television broadcast standards in the United Kingdom and Japan have been changed since this event to reduce the risk of photosensitive seizures.

From a practical standpoint, photosensitive epilepsy may be treated by environmental manipulation or in more severe cases with AEDs (valproate seems to be the drug of choice). The Epilepsy Foundation of America's (EFA) professional advisory board has issued the following general recommendations for television viewing among individuals who are at risk for photosensitive seizures (Erba, 2006):

- Watch television in a well-lit room to reduce the contrast between the screen light and background light.
- Reduce the brightness of the screen.
- Keep as far back from the screen as possible (minimum 5 feet).
- Use remote controls to ensure proper distance from the television is maintained.
- Use small screens.
- When watching large screens, increase the distance from the screen.

For video game playing, in addition to the above precautions, the EFA professional advisory board recommended the following:

- Players should not play if they are tired, especially if they are sleep deprived.
- Avoid excessive use of alcoholic beverages.
- Take frequent breaks from the game and look away from the screen every once in a while.
- If strange or unusual feelings develop, turn the game off.
- If players start feeling their bodies jerking, they should cover one eye with one hand and immediately look away.

Covering one eye (monocular vision) is a most useful practice because it works in most cases and still allows the subject to see. Simply closing the eyes will not prevent photosensitive seizures because the red-tinted light filtering through the eyelids will be just as stimulating. Another option is to wear special filtering glasses using tinted optical lenses that eliminate the most offensive "deep" red color. These are blue-tinted lenses, and wearers should be cautioned not to use them during activities requiring accurate color vision (e.g., looking at traffic signals while driving a car).

In a recent case study of a 9-year-old girl with a form of photosensitive epilepsy, using a different type of television set eliminated her visuogenic seizures. She was the product of a divorce and lived part-time in two different homes; one with an older model cathode ray tube television set, which had a low-frequency flicker, and the other with newer LCD and plasma screen televisions. Since LCD/plasma screens use a transistor for each pixel, which allows the pixels to maintain a steady state and not manifest any flickering, it was hypothesized that the use of LCD/plasma screens may eliminate television seizures (Sharma and Cameron, 2007).

Musicogenic Epilepsy

Seizures provoked by hearing music are a rare form of reflex epilepsy. The music that provokes a seizure is usually a particular piece of music that is specific for each patient (Zifkin and Andermann, 2006). Musicogenic seizures are most often encountered in patients in response to listening to music, and in some cases thinking about music, and within the context of existing symptomatic localization-related epilepsy. Seizures types are most frequently either simple partial or complex partial seizures. Secondarily generalizing seizures are uncommon. These seizures typically begin after the age of 20 years old. Interictal EEGs usually show sharp and slow wave activity in the temporal lobes, more often on the right side (Browne and Holmes, 2004).

Eating Epilepsy

Seizures induced by eating are another type of reflex epilepsy brought on by the sight or smell of food, by the digestion process after eating food, and sometimes by exposure only to specific foods. These seizures are typically partial (either simple or complex) in nature but secondarily generalize in some cases (Zifkin and Andermann, 2006). Many of these patients have focal symptomatic epilepsy originating in the temporolimbic areas, but some cases remain cryptogenic. Sensory, autonomic, or emotional inputs into these areas during eating appear to trigger these seizures (Browne and Holmes, 2004).

Startle Epilepsy

Startle seizures are considered a symptomatic localization-related epilepsy precipitated by sudden and unexpected stimuli that evoke a startle response; that is, a brief tonic contracture of the axial muscles of the body (Panayiotopoulos, 2005). Sudden noise is the most common triggering stimulus although somatosensory or visual input may provoke seizures in some patients. The majority of patients with startle epilepsy suffer from chronic encephalopathies dating from birth. These seizures are often seen in patients with infantile hemiplegia, Down syndrome, or perinatal anoxia (cerebral palsy); such patients typically have diffuse neurologic and intellectual impairments and/or developmental delays. Neuroimaging often shows unilateral or bilateral medial frontal lobe lesions. Ictal EEGs recorded from depth electrodes frequently show discharges in the supplemental motor area and surrounding medial frontal regions. Interictal EEGs show a variety of diffuse or focal abnormalities that mirror the underlying structural lesions. The startle-induced seizures themselves are usually lateralized and tonic, and less commonly, they may be atonic or myoclonic (Zifkin and Andermann, 2006). Seizures typically occur many times a day and are medically refractory to treatment with AEDs.

Frontal Lobe Epilepsies
Precentral Frontal Lobe Seizures
Precentral frontal lobe seizures begin in primary motor cortex (Brodmann's area 4) and may consist of simple partial (focal motor) seizures or myoclonic jerks with or without secondarily generalizing tonic-clonic convulsions. Precentral seizures often begin with contralateral focal clonic activity in the hand, face, or lower limb and then spread to involve adjacent regions of cortex in a *jacksonian march* fashion. For instance, these seizures may begin with focal jerking of the thumb, followed by clonic movements of the fingers spreading into the hand and then to the forearm. If the seizure spread is confined to the primary motor cortex, it will be a simple focal motor seizure. However, if the seizure focus spreads to other regions, such as the ipsilateral mesial temporal lobe, it will evolve into a complex partial seizure.

Another common occurrence is subsequent seizure spread to motor areas in the contralateral frontal lobe, which causes tonic-clonic movements involving the whole body. In patients whose seizures originate in the lower primary motor strip, speech arrest, forced vocalizations, repetitive swallowing, or tonic-clonic movements of the contralateral face are common signs.

If seizures are confined to the paracentral region (i.e., medial primary motor cortex within the interhemispheric fissure), contralateral movements of the leg may occur. Postictal paresis of the affected body part is commonly observed after these seizures.

Premotor Frontal Lobe Seizures
Premotor frontal lobe seizures originate in the lateral or medial premotor frontal lobe regions including Brodmann's areas 6 and 8. Seizures arising from the premotor cortex are primarily characterized by tonic and postural

motor phenomena, most commonly in the upper extremities and often including upper or whole body turning (adversion).

Supplementary Motor Area Seizures

Supplementary motor area (SMA) seizures begin in the medial premotor area, referred to as the SMA (Brodmann's area 8). Cases of SMA seizures are a relatively rare. Somatosensory auras are common at the start of an SMA seizure. The seizures themselves are brief (usually between 10 and 40 seconds) and typically include postural symptoms, such as abrupt tonic posturing of the arms, turning of the shoulders and pelvis, or adversion of the head and eyes; usually away from the side of seizure onset. During the tonic phase, the patient may cry or moan loudly.

The tonic signs of these seizures are thought to be due to spread of the seizure focus to the lateral premotor areas. If the SMA seizures are confined to the medial premotor region, consciousness will be preserved. Seizure frequency is high; patients may have up to 10 spells every day. In addition, SMA seizures are often refractory to AED treatment, and thus, patients with these seizures are often encountered at epilepsy surgery centers (Browne and Holmes, 2004).

Dorsolateral Prefrontal Lobe Seizures

Symptoms of seizures arising from dorsolateral prefrontal cortex vary greatly depending upon the location of the focal seizure within the prefrontal region. *Dorsolateral frontal lobe seizures* (including frontal polar regions) symptoms consist of forced thinking; loss of contact with an awareness of surroundings; contralateral adversive deviation of the head and eyes; tonic, and less often clonic, movements; and speech arrest.

Forced actions and complex gestural sequences, such as repetitive manual activity that appear to be goal directed (e.g., turning on and lighting a stove, operating factory machinery) or disorganized and nonsense actions involving the whole body (e.g., standing up and then running around a table) may be observed (Browne and Holmes, 2004).

Orbitofrontal Seizures

Seizures that originate from orbitofrontal cortex are usually complex partial seizures that begin with motor or gestural automatisms. These automatisms may include kicking or whole-body thrashing, olfactory hallucinations and illusions, or signs of autonomic over-activity (e.g., facial flushing, rapid breathing). Peri-ictal urination is typical in these seizures. Vocalizations, including either formed or unformed speech such as repetitive swearing, intense fear (or the appearance of fear), and complex motor acts such as getting up and walking around the room also have been described (Williamson, Spencer, Spencer, Novelly, and Mattson, 1985).

Medial Frontal Lobe Seizures

Seizures arising from the *anterior cingulate gyrus* are most often complex partial seizures with complex motor and gestural automatisms observed as early seizure manifestations. Autonomic hyperarousal signs and changes in mood and affect are also common (Browne and Holmes, 2004). Bancaud and Talairach (1992) have described patients with anterior cingulate gyrus seizures who exhibited fear, screaming, complex gestural automatisms, autonomic changes, and loss of responsiveness.

Frontal Opercular Seizures

Seizures that arise in the *frontal opercular region* (i.e., the area of frontal cortex overlying the insula) have been reported to include epigastric aura, fear, salivation, repetitive chewing movements, repetitive swallowing, speech arrest, and autonomic symptoms. Spread of the ictal discharges may be responsible for some of the symptoms of seizures arising in this posterior inferior area of lateral frontal cortex. Frontal opercular seizures are most commonly simple motor partial seizures, especially partial clonic facial seizures, and gustatory hallucinations are often seen with seizures in this area. The focal clonic activity may be ipsilateral as well as contralateral to the seizure focus.

Parietal Lobe Epilepsies

Postcentral Gyrus Seizures

Postcentral gyrus seizures begin in the anterior portion of the superior parietal lobe and are almost always associated with an aura of positive or negative sensory sensations. Positive somatosensory phenomena include (Browne and Holmes, 2004) tingling, feeling of electricity, desire to move a body part, sensation that a body part is being moved, or sensations involving the tongue or face.

Negative sensory auras may include diminished two-point discrimination, numbness, feeling that a body part is missing, or feeling as if a body part is much smaller than normal.

The parts of the body most frequently involved are the hand, arm, and face since these areas have the largest lateral cortical representation on the post-central gyrus. Ictal paresthesias are less common, occurring in less than half of patients (Williamson, et al, 1992).

Pain is a common ictal experience in patients with anterior parietal lobe seizure foci, and contralateral dysesthesias (especially burning sensations) have frequently been reported. The posterior portions of the parietal lobes are largely 'silent' with regard to seizure semiology since many of the symptoms are negative and may not be noticed by the patient. Many observable signs of parietal lobe seizures depend upon where the seizure spreads outside of the parietal lobes. When seizures spread anteriorly into the frontal lobe, tonic motor activity or automatisms are often seen. When seizures extend into the mesial temporal lobe, the seizure will look like a typical mesial temporal lobe seizure (see description above). Patients whose parietal lobe seizures spread posteriorly into the occipital lobe will often experience elementary visual hallucinations, metamorphopsia (visual distortions), or ictal amaurosis (blindness not due to damage to the eye).

Superior Parietal Lobule Seizures
Seizures arising from the superior parietal lobule are also known as *posterior parietal seizures*. These seizures primarily involve Brodmann's areas 5 and 7. Seizures restricted to these areas are characterized by prominent starring and relative immobility (Browne and Holmes, 2004). Visual phenomena, such as hallucinations or metamorphopsias, may also occur.

Inferior Parietal Lobule Seizures
Inferior parietal lobule seizures, also called *inferior parietal seizures*, primarily involve Brodmann's areas 39 (angular gyrus) and 40 (supramarginal gyrus). Seizures arising in the parietal operculum (area 40) may give rise to sexual sensations, severe vertigo, abdominal sensations, gustatory hallucinations, apraxias, disturbances of body image, and spatial disorientation. Patients with dominant inferior parietal lobule seizure onset are more likely to demonstrate receptive language symptoms and associated language symptoms, such as alexia, acalculia, and right–left disorientation. Nondominant parietal lobe epilepsy has been more often associated with metamorphopsia, asomatognosia, and spatial inattention (Browne and Holmes, 2004).

Paracentral Parietal Lobe Seizures
Paracentral seizures originate from the mesial surface of the parietal lobe. Seizure symptoms involving this brain region may include contralateral genital sensations and postural motor activity. Paracentral seizures frequently secondarily generalize since this region has such dense interconnectivity to both ipsilateral and contralateral motor areas.

Idiopathic Generalized Epilepsy Syndromes

Benign Neonatal Familial Convulsions
This is a rare inherited epileptic syndrome in which tonic-clonic seizures are seen during the first few days after birth. Duration of each seizure is generally between 1 and 2 minutes. Seizure frequency may be between 20 and 30 per day. The seizures spontaneously remit in most neonates within 1–2 months, and between 10% and 15% go on to develop other forms of epilepsy (Panayiotopoulos, 2005). There is no increased risk for mental retardation or permanent neurological morbidity.

Benign Neonatal Convulsions (Nonfamilial)
This neonatal epilepsy syndrome is characterized by frequent repetitive seizures beginning around the fifth day of life. Seizures are brief, between 1 and 3 minutes, clonic or apneic seizures that occur one after the other. culminating in clonic status epilepticus. The entire period of repetitive seizures usually lasts from a few hours to a few days, and then never recurs again (Commission on Classification and Terminology of the ILAE, 1989). Etiology is unknown; a genetic cause has not been found. No increased risk for development of seizures later in life has been established. Later neurodevelopment is normal.

Benign Myoclonic Epilepsy of Childhood
The Commission on Classification and Terminology of the ILAE (1989) defines this epilepsy syndrome as consisting of brief generalized myoclonic jerks beginning in the first or second year of life. These children often have a family history of epilepsy. Seizures are usually well-controlled with AEDs. These children may develop generalized tonic-clonic seizures later, typically during adolescence. During childhood, there may be cognitive developmental delay and there is an increased risk for the development of behavior disorders and other psychological conditions.

Juvenile Myoclonic Epilepsy

Juvenile myoclonic epilepsy (JME) is a common genetically-determined syndrome of myoclonic and generalized tonic-clonic seizures. Typical age of onset is between ages 12 and 18 years, although onset may be seen from ages 8–30 years. Juvenile myoclonic epilepsy is thought to be the most common etiology of myoclonic and tonic-clonic seizures in adults. Prevalence is estimated to be around 10% of all the epilepsies, but JME is probably under-diagnosed and may account for up to 30% of all epilepsies (Browne and Holmes, 2004).

Most cases of JME consist of two primary seizure types: myoclonic and tonic-clonic; although some have absence attacks as well. The seizures in JME typically consist of sudden, mild to moderate myoclonic jerks of the shoulders and arms that occur shortly after awakening in the morning. The myoclonic jerks progress to generalized tonic-clonic seizures in the overwhelming majority of JME patients. Absence seizures are also present in approximately one-third of patients with JME.

Each of the three subtypes of JME (i.e., myoclonic, tonic-clonic, absence) may occur in isolation in the same individual at different times. Family history is positive for seizures in about half of the patients, and most evidence suggests genetic inheritance. One form of JME has been linked to abnormalities on chromosome 6. Intelligence is normal. The neurologic exam and imaging studies are normal. The diagnosis is confirmed by the EEG pattern, which consists of bilateral, diffuse 4–6 Hz polyspike and wave discharges. This abnormal EEG pattern is often brought on by photic stimulation.

Although it is a lifelong disorder, JME responds well to medications. Valproic acid is usually an effective monotherapy to control all three subtypes of JME (Browne and Holmes, 2004). Life-long AED treatment must be undertaken since relapse will occur in most patients with drug withdrawal. Several situations are known to precipitate seizures in JME. Patients with JME should thus be warned to avoid these situations, which may include sleep deprivation, early awakening, alcohol intake, fatigue, and in photosensitive patients, flickering lights.

Patients with JME have a high frequency of psychiatric disorders, particularly anxiety, mood, and mild to moderate cluster B personality disorders, such as histrionic, passive-aggressive, and borderline types. There is also a so-called JME psychological syndrome characterized by emotional instability and immaturity, unsteadiness, lack of discipline, hedonism, frequent rapid mood changes, and indifference toward their disease, which has been attributed to structural and neurochemical thalamic-frontal lobe abnormalities seen in juvenile myoclonic epilepsy (Filho, Jackowski, Lin, et al., 2009).

Epilepsy with Tonic-Clonic (Generalized Tonic-Clonic Seizures) Seizures upon Awakening

This epilepsy syndrome most often appears during the adolescent years. As with the other idiopathic epilepsies, a genetic predisposition has been noted. Tonic-clonic seizures most often occur upon awakening, but may occur during the day or in a second-seizure peak in the evening associated with relaxation and drowsiness (Browne and Holmes, 2004). The EEG shows generalized, diffuse spike and wave activity during the seizure. The EEG abnormalities may be precipitated by photic stimulation or sleep deprivation. This seizure syndrome is differentiated from JME by the absence of myoclonic and absence seizure types, which are seen in JME but do not occur in epilepsy with tonic-clonic seizures upon awakening.

Cryptogenic or Symptomatic Generalized Epilepsy Syndromes

West Syndrome (Infantile Spasms)

West syndrome is a severe epilepsy syndrome occurring around 5 months of age (range 4–7 months old) that has been associated with a variety of etiologies. It is characterized by the presence of infantile spasms, arrest of motor development, and hypsarrhythmia (mixture of waveforms) on EEG (Commission on Classification and Terminology of the ILAE, 1989). Infantile spasms (or the more currently preferred term, *epileptic spasms*) are sudden, brief (1–2 seconds), bilateral tonic contractions of the neck, axial, and limb muscles ("jack-knife spasms"). The neck is characteristically flexed downward and the legs are elevated with flexion at the hips and knees. Hypsarrhythmia is an abnormal interictal EEG pattern characterized by a chaotic mixture of high amplitude slow and sharp waves, multifocal spikes, and polyspikes with no recognizable normal rhythms.

Although etiologies may include hypoxic-ischemic birth injury, infections, trauma, cortical malformations, and neurocutaneous diseases, among others, the most common cause of symptomatic West syndrome is a stroke before, during, or after birth; the second most common cause, occurring in about one-third of cases, is congenital brain abnormalities. Prognosis is poor in most cases and depends upon etiology. Across all patients with West syndrome, death occurs in about 5% of cases, two-thirds experience severe cognitive and psychological impairment, and half of the patients have permanent motor deficits (Panayiotopoulos, 2005). Most children with West syndrome go on to develop other seizure types that are resistant to AED treatment, such as Lennox-Gastaut syndrome. In addition to the symptomatic form of West syndrome, there are also idiopathic or cryptogenic forms that generally have a better prognosis than the symptomatic cases.

Epilepsy with Myoclonic-Astatic Seizures

This cryptogenic seizure syndrome is defined by the presence of a myoclonic-atonic type seizure, which is characterized by sudden onset of a symmetrical myoclonic jerks immediately followed by a loss of muscle tone. *Astatic* is an older term used for *atonic*, although the meanings are not precisely identical. These seizures often co-occur with atonic, myoclonic, absence, and generalized tonic-clonic seizures, and status epilepticus is a frequent occurrence. Age of onset is primarily between the ages of 2 and 5 years old, and there is frequently a genetic cause for this syndrome (Commission on Classification and Terminology of the ILAE, 1989). Two-thirds of these children are male. Children are normal prior to the onset of the seizure disorder. The myoclonic-atonic seizures typically spontaneously remit within 1–3 years after onset, while the generalized tonic-clonic seizures tend to continue (Panayiotopoulos, 2005). About one-half of these patients achieve normal development. In those cases where the seizures do not remit, the syndrome probably has a symptomatic cause, and these children have a poor prognosis; severe impairment of cognitive functions, behavior problems, poor motor function, dysarthria, and poor language development may all be seen.

Epilepsy with Myoclonic Absences

This is an exceedingly rare syndrome seen in less than 1% of epilepsy patients. The seizures are absence attacks accompanied by severe bilateral clonic jerks. The EEG shows the bilateral synchronous 3 Hz spike and wave complexes typical for absence seizures (Commission on Classification and Terminology of the ILAE, 1989). Seizure frequency is usually multiple times per day. Median onset age is 7 years old, although onset may occur throughout childhood, and over two-thirds are male. The seizures are resistant to treatment with AEDs. The myoclonic absence attacks remit in about one-third of patients after 5 years or so. In those patients whose seizures continue into adulthood, the seizures usually evolve into other types of epilepsy, such as Lennox-Gastaut syndrome, and there is cognitive deterioration and development of psychological problems in most of these cases.

Appendix III

Wada Assessment Procedures and Rating Criteria at the Medical College of Georgia

This appendix describes the Wada protocol currently in use at the Medical College of Georgia (MCG). (Loring, Meador, Lee, King, 1992). Baseline Wada assessment is usually conducted the day before the actual Wada procedure, so that the patient knows what to expect and the neuropsychologist has a non-drug cognitive test baseline against which change can be measured. On the day of Wada testing, the neuropsychologist approaches the patient in the angiography suite after angiography has been performed on the side of planned amobarbital injection. The patient is lying supine, and the examiner stands behind the patient's head. The procedure is videotaped, so that patient performance may be reviewed later in detail if necessary.

The patient is told to raise both arms into the air and rotate the hands rostrally (backward, facing the examiner). The patient then begins to count from 1 to 20, loudly, slowly, and steadily, and a bolus of 100 mg (in the typical adult) of amobarbital is injected into the internal carotid artery by the radiologist over a 4–5 second interval. After drug administration, degree and type of disruption to repetitive counting is noted, and the neurological effects of the amobarbital are evaluated and recorded (refer to the MCG *Wada Evaluation* scoring form at the end of this appendix). Next, the patient is asked to perform simple commands such as, "open your mouth," "stick out your tongue," "touch your nose," or "tell me your name."

After presenting these early commands, the memory target items are presented, and the patient's level of alertness is evaluated using a modified Glasgow Coma Scale (see MCG *Wada Evaluation* scoring form below). Various language domains are then formally assessed, and motor functions are regularly monitored for return to neurologic baseline. After neurologic functions and language (if affected) have returned to baseline, recognition for the memory items is tested. This must be a minimum of 10 minutes after amobarbital injection. If all language and sensorimotor functions have returned to baseline before 10 minutes has gone by, the examiner will wait until the 10-minture mark before testing memory. Evaluation of one hemisphere usually takes approximately 12 minutes. The same procedure is then conducted in the opposite hemisphere after waiting at least 30 minutes between injections, so that the drug may completely clear. Detailed descriptions of MCG's language and memory evaluations follow.

Procedure for Measuring Wada Language Functions at the Medical College of Georgia

Fluency of Speech

Oral fluency is measured by what happens to initial expressive speech after amobarbital injection (see "Injection 1" on the MCG *Wada Evaluation* scoring form at the end of the appendix). Initial expressive speech is assessed by degree of disruption of repetitive counting (1 through 20) using a 5-point scale: $0 =$ speech arrest (>20 seconds), $1 =$ single number of word perseveration, $2 =$ sequencing errors, $3 =$ counting perseveration with normal sequencing, and $4 =$ normal or slowed or brief pause in counting.

Aural Comprehension

Comprehension is evaluated using a modified (vertically oriented) Token Test card with four different colored shapes (i.e., red circle, blue square, red circle, blue circle). The size of the Token card is 12 × 4.5 inches. Aural commands using the Token card range from the simplest commands ("point to a circle," or "point to a red one") to the most complex two-stage commands using inverted syntax ("point to the red circle after you point to the blue square"). If patients are unable to comply with the simplest Token card commands, they are asked to perform more elemental midline commands such as, "stick out your tongue," or "touch your nose." Scoring is based on a 4-point scale as follows: 0 = severe impairment, 1 = moderate impairment, 2 = mild impairment, and 3 = normal (refer to the *Wada Evaluation* scoring form below). A score of 0 is awarded if the patient cannot perform any Token card commands. A score of 1 is given when the patient can only perform single-stage commands (e.g., "point to the blue square"). A score of 2 occurs when the patient is able to perform two-stage commands (e.g., "point to the red square and then the blue circle"), but not two-stage commands using inverted syntax (e.g., "point to the blue square after you point to the red circle"). Finally, a score of 3 is awarded when the patient can perform all commands successfully relative to baseline. In the case of children or adults who cannot perform the more difficult commands at baseline, these criteria are geared downward as appropriate.

Visual Naming
The primary naming stimuli currently consist of two black and white line drawings presented on a 6.75 × 5.75 inch card; one side shows a jacket and the other a wrist watch. Patients are asked to name the entire object and then various parts of the objects. Scoring is based on a 4-point scale as follows: 0 = severe impairment, 1 = moderate impairment, 2 = mild impairment, and 3 = normal (refer to the *Wada Evaluation* scoring form below). A score of 0 is given is the patient is unable to name any of the items correctly, and a score of 3 is awarded when all items are normally named consistent with baseline. A score of 2 (mild impairment) is applied when patients only make one or two errors, and a 3 (moderate impairment) is given when the patient makes more than a few naming errors (regardless of the type of error). In the case of children or adults with limited naming ability, a series of seven black-and-white line drawings are used with pictures of simple objects, such as a shoe, table, horse, clock, and corn. The pictures are tailored to the individual patient at baseline to ensure there are a sufficient number (at least seven) of stimuli to measure naming ability during the Wada.

Repetition
Repetition ability is evaluated during the Wada using phrases and sentences that range from simple nursery rhymes to more complex sentences of increasing length and more difficult articulatory agility. Examples of the simple rhymes include, "Jack and Jill went up the hill," or "Mary had a little lamb." The more

difficult items have been taken from several aphasia test batteries such as, 'No ifs ands or buts," or 'The phantom soared across the foggy heath.' The repetition items to be used during Wada testing are selected at baseline and will typically include five or six sentences ranging across difficulty levels. As with comprehension and naming, scoring consists of a 4-point scale as follows: 0 = severe impairment, 1 = moderate impairment, 2 = mild impairment, and 3 = normal (refer to the *Wada Evaluation* scoring form below). A score of 0 is given is the patient is unable to accurately repeat any of the items correctly, and a score of 3 is awarded when all items are normally (consistent with baseline) repeated. A score of 2 (mild impairment) is applied when patients only make one or two errors, and a 3 (moderate impairment) is given when the patient makes more than a few repetition errors (regardless of the type of error).

Reading

The current reading items employed at the Medical College of Georgia consist of two sentences: "The car backed over the curb" and "The rabbit hopped down the lane." For children or illiterate patients, reading may not always be administered, depending upon skill level. Scoring is rated along the same 4-point scale as comprehension, naming, and repetition (from 0 = severe impairment to 3 = normal relative to baseline). Scores for reading are assigned using the same criteria as that given for repetition.

Paraphasias

Positive signs of aphasia, such as phonemic and semantic paraphasic errors or circumlocutions, are frequently encountered during Wada language testing and are qualitatively factored into the 4-point qualitative scoring system. In general, clinicians can be much more confident that the language system has been affected by the drug when positive signs of aphasia, such as paraphasic errors, are present.

Procedure for Measuring Wada Memory at the Medical College of Georgia

After injecting 100 mg of sodium amobarbital for adults (for small children and some frail adults, we may begin with 75 mg) into the internal carotid artery and demonstration of hemiparesis, we evaluate eye-gaze deviation and ability to carry out simple midline commands (e.g., "open your mouth," "touch your nose"). The memory stimuli are then presented, which usually occurs 30–45 seconds after injection.

Wada Memory Stimuli

The memory stimuli consist of real objects that are a combination of ordinary household items (e.g., fork, measuring cup, mousetrap), small toys (e.g., troll doll, plastic animals), and plastic food items (e.g., piece of pizza, potato chip). A complete listing of the Wada memory items used at the Medical College of Georgia, Yale University, and the Minnesota Epilepsy Group has been published elsewhere (Lee, Park, Westerveld, et al., 2002). The common, everyday objects are presented in the central visual field and in the visual field ipsilateral to the side of amobarbital injection. Objects are not shown in the contralateral visual field to avoid the confounding effects of possible visual neglect, which is frequently encountered especially after right hemisphere injection. Eight memory items are presented for 4–8 seconds each, and the names of the objects are repeated twice. If inattention, nonresponsiveness, confusion, or obtundation occurs, the patient's eyes may be held open to show the memory stimuli. No verbal response is required of patients during memory item presentation, and the objects are shown during the period of aphasia after dominant hemisphere injection. Language is then evaluated in detail after presentation of the memory items.

Recognition memory for the real objects is assessed after the effects of amobarbital have worn off, at a minimum of 10 minutes after initial injection. Cessation of drug effect is demonstrated by a return to baseline language performance across all five domains assessed and return of 5/5 strength with an absence of pronator drift, asterixis, and bradykinesia. Although we sometimes ask patients to freely recall the memory items, formal scoring and interpretation is based upon recognition memory performance. At the 10-minute mark, and after return to baseline, recognition memory is tested for the eight target objects, which are interspersed with 16 foils. During item recognition, 24 objects (8 targets plus 16 foils) are presented in a pseudorandomized sequence (the examiner pulls objects out of a bag without looking in a different order each time), and patients indicate whether or not each object had been presented earlier.

Scoring of Wada Memory Items

In scoring, a correction of one-half the number of false positive responses is subtracted from the total number of objects accurately recognized to correct for guessing. Although there are no hard and fast rules for deciding Wada memory test failure in individual patients, a single hemisphere score of 2/8 or less (or 25%) has worked reasonably well as a rough guide to define memory failure. Wada asymmetry scores to assess lateralized Wada memory differences between the hemispheres may also be computed. Asymmetry scores are calculated by subtracting the left injection corrected score from the right injection corrected score. Positive scores suggest left temporal lobe dysfunction, and negative scores may reflect right temporal lobe dysfunction. Similarly, there are no universal rules to determine *the* memory score that accurately reflects a true memory asymmetry or significant difference between left and right injections. A difference score between the hemispheres of two to three objects has often proved useful in determining lateralized impairment, where the hemisphere with the lower score represents the dysfunctional one. Greater predictive confidence may be placed on larger memory asymmetries. Although probably obvious, it should be reiterated that the absence of a Wada memory asymmetry (left minus right), reflected in a score of 0, may either mean that both hippocampi are healthy (left injection = 8; right injection = 8) or both hippocampi are dysfunctional (left injection = 0; right injection = 0).

Summary of Behavioral Assessment During Wada Testing at the Medical College of Georgia

- Amobarbital administered
- Neurologic drug effects evaluated
- Initial expressive language (counting speech arrest?)
- Simple (early) commands (Receptive Language)
- Presentation (encoding) of memory items (Object Presentation)
- Level of arousal (Modified Glasgow Coma Scale) assessed
- Aural comprehension (modified Token test)
- Visual confrontational naming
- Sentence repetition
- Sentence reading
- Repeat abnormal language tests until normal (or until return to baseline)
- Monitor for return to neurologic baseline (Motor Strength, Asterixis)
- Recognition (recall) of memory items (at least 10 minutes post-injection)

MEDICAL COLLEGE OF GEORGIA
WADA EVALUATION

Name: _____ Hospital Number: _____ Date: _____

Age: _____ Sex: _____ Handedness: _____ Family Handedness: _____

Date of last seizure:

BASELINE

MEMORY PRESENTATION: Time: _____

Set Administered: **A B C**

PRESENTATION ORDER (1st 4 objects)

_____

_____

BASELINE LANGUAGE: Scoring Scale: 0 = severe, 1 = moderate, 2 = mild, 3 = normal

COMPREHENSION: (Token Card) 0 1 2 3

NAMING: watch _____, watchband _____, buckle _____, hands _____, winder _____;

coat/jacket _____, sleeve _____, collar _____, buttons _____ 0 1 2 3

REPETITION: No ifs ands or buts / Methodist Episcopal / Mary had a little lamb /
Pry the tin lid off / The spy fled to Greece / The phantom soared across the foggy heath /
The President lives in Washington / Jack and Jill went up the hill

(3=normal, 2=mild, 1=moderate, 0=severe) 0 1 2 3

_____

_____

_____

DYSARTHRIA: (3=normal, 2=mild, 1=moderate, 0=severe) 0 1 2 3

READING: The car backed over the curb. 0 1 2 3

Response: _____

The rabbit hopped down the lane.

Response: _____

MOTOR STRENGTH:

Left _____ 0 1 2 3

Right _____ 0 1 2 3

Pronator drift?   Y   N   Side   L   R

Asterixis?   Y   N   Side   L   R

© 1993–2010 Section of Behavioral Neurology, Medical College of Georgia

Name: _____ Date: _____

BASELINE

OBJECT/WORD FREE RECALL: Time: _____

_____
_____
_____
                _____

BASELINE OBJECT RECOGNITION: Circle Set: A B C   Time: _____

_____ Y N	_____ Y N	_____ Y N		
_____ Y N	_____ Y N	_____ Y N		
_____ Y N	_____ Y N	_____ Y N		
_____ Y N	_____ Y N	_____ Y N		
_____ Y N	_____ Y N	_____ Y N	1st 4 =	/4
_____ Y N	_____ Y N	_____ Y N	2nd 4 =	/4
_____ Y N	_____ Y N	_____ Y N	Total =	/8
_____ Y N	_____ Y N	_____ Y N	False Positives =	

Baseline Object Corrected Score = _____

Comments:

Modified Glasgow Coma Scale = 14? Y N   If not, give score: _____

© 1993–2010 Section of Behavioral Neurology, Medical College of Georgia

Name: _____   Date: _____

INJECTION 1

Side: Right   Left   Dose (mg): _____   Time Test Begun: _____

Flaccid Hemiparesis Y N   If not, describe: _____

INITIAL EXPRESSIVE LANGUAGE (Counting) .... 0 1 2 3 4   Running Time: 0 sec

(4 = normal, slowed or brief pause; 3 = counting perseveration with normal sequencing; 2 = sequencing errors;
1 = single number or word perseveration; 0 = arrest)

Scoring Scale: 0 = severe, 1 = moderate, 2 = mild, 3 = normal

LATERAL GAZE PALSY   NONE   R   L
  Spontaneous   ____ ____ ____ 0 1 2 3
  Command   ____ ____ ____ .... 0 1 2 3 N/A

RECEPTIVE LANGUAGE (Initial-Simple): .... 0 1 2 3   Time _____

1. ⌐Are both your arms in the air?¬   Y   N   N/A
2. For Right ISAs: ⌐Point with your right hand to your left hand.¬ Record where the patient initially points, including direction and height (e.g., where hand is now on bed; where hand was in air at beginning of injection; denote other location; or no response).

3. Is either one of your arms weak?   Y   N   N/A   Which?   L   R   Both
4. Put patient=s paralyzed hand in good space and ask ⌐What is this?¬

SERIES I OBJECT PRESENTATION (Begin at 30-60 seconds)

   Set Administered: A B C   Time for 1st 4 Objects (~1 min post) _____

1st 4 objects: _____

2nd 4 objects: _____

                                                         Time for 2nd 4 Objects _____

MODIFIED GLASGOW COMA SCALE: If Ptosis, note severity: .... 0 1 2 3 (3 = Normal)
(scored during presentation of 1st 4 objects)   If akinesia, note severity: .... 0 1 2 3 (3 = Normal)

Motor:	Ipsi Arm	Contra Arm	Eye Opening:		Eye Movements:	
spontaneous	6	6	spontaneous	4	follows objects well	4
localizes	5	5	to loud noise	3	follows objects variably	3
withdraws	4	4	to pain	2	orients to loud noise and light	2
abnormal flexion	3	3	nil	1		
extensor response	2	2			nil	1
nil	1	1				

                                                         Modified Glasgow Coma Score =   /14

LANGUAGE I: Scoring Scale: 0 = severe, 1 = moderate, 2 = mild, 3 = normal   Time _____

COMPREHENSION: (Token Card) .... 0 1 2 3 (3 = Normal)

NAMING: watch _____, watchband _____, buckle _____, hands _____, winder _____;

           coat/jacket _____, sleeve _____, collar _____, buttons _____ 0 1 2 3

© 1993–2010 Section of Behavioral Neurology, Medical College of Georgia

Wada Assessment

Name: _____ Date: _____

INJECTION 1   Side: R   L

REPETITION: "No ifs ands or buts" **or** ᴀJack and Jill went up the hill☺ .... 0 1 2 3 N/A (3 = Normal)

Responses:_____

"Mary had a little lamb" .... 0 1 2 3 N/A (3 = Normal)

Responses:_____

"Pry the tin lid off" .... 0 1 2 3 N/A (3 = Normal)

Response:_____

"The spy fled to Greece" .... 0 1 2 3 N/A (3 = Normal)

Response:_____

"Methodist Episcopal" .... 0 1 2 3 N/A (3 = Normal)

Responses:_____

"The phantom soared across the foggy heath" .... 0 1 2 3 N/A (3 = Normal)

Response:_____

ᴀThe President lives in Washington☺ .... 0 1 2 3 N/A (3 = Normal)

Response:_____

DYSARTHRIA: .... 0 1 2 3 N/A (3 = Normal)

READING: The car backed over the curb. .... 0 1 2 3 N/A (3 = Normal)

Response:_____

The rabbit hopped down the lane.   0 1 2 3 N/A (3 = Normal)

Response:_____

MOTOR STRENGTH I: Time _____

Contralateral Strength .... 0 1 2 3 (3 = Normal)

Pronator drift? Y   N   Asterixis: Contralateral?   Y   N   Ipsilateral?   Y   N

1. ᴀAre both your arms in the air?☺   Y   N   N/A
2. For Right ISAs: ᴀPoint with your right hand to your left hand.☺ Record where the patient initially points, including direction and height (e. g., where hand is now on bed; where hand was in air at beginning of injection; denote other location; or no response).

3. Is either one of your arms weak?   Y   N   N/A   Which?   L   R   Both
4. Put patient=s paralyzed hand in good space and ask ᴀWhat is this?☺

LANGUAGE II: Scoring Scale: 0 = severe, 1 = moderate, 2 = mild, 3 = normal   Time _____

COMPREHENSION: (Token Card) .... 0 1 2 3 N/A (3 = Normal)

NAMING: watch _____, watchband _____, buckle _____, hands _____, winder _____,

coat/jacket _____, sleeve _____, collar _____, buttons _____ .... 0 1 2 3 N/A

© 1993–2010 Section of Behavioral Neurology, Medical College of Georgia

Name: _____ Date: _____

INJECTION 1   Side: R L

REPETITION:

"Mary had a little lamb" .... 0 1 2 3 N/A (3 = Normal)

Responses: _____

"No ifs ands or buts" **or** ⌐Jack and Jill went up the hill⌐ .... 0 1 2 3 N/A (3 = Normal)

Responses: _____

"Pry the tin lid off" .... 0 1 2 3 N/A (3 = Normal)

Response: _____

"The spy fled to Greece" .... 0 1 2 3 N/A (3 = Normal)

Response: _____

"Methodist Episcopal" .... 0 1 2 3 N/A (3 = Normal)

Responses: _____

"The phantom soared across the foggy heath" .... 0 1 2 3 N/A (3 = Normal)

Response: _____

⌐The President lives in Washington⌐ .... 0 1 2 3 N/A (3 = Normal)

Response: _____

DYSARTHRIA: .... 0 1 2 3 N/A (3 = Normal)

READING:

The car backed over the curb. .... 0 1 2 3 N/A (3 = Normal)

Response: _____

The rabbit hopped down the lane. .... 0 1 2 3 N/A (3 = Normal)

Response: _____

MOTOR STRENGTH II: Time _____

Contralateral Strength .... 0 1 2 3 (3 = Normal)

Pronator drift? Y   N   Asterixis: Contralateral? Y   N   Ipsilateral? Y   N

1. ⌐Are both your arms in the air?⌐   Y   N   N/A
2. For Right ISAs: ⌐Point with your right hand to your left hand.⌐ Record where the patient initially points, including direction and height (e.g., where hand is now on bed, where hand was at beginning, denote other location, or no response).

3. Is either one of your arms weak?   Y   N   N/A   Which?   L   R   Both
4. Put patient=s paralyzed hand in good space and ask ⌐What is this?⌐

If Language and Motor <u>not</u> normal on above testing, then record time that they return to normal:

Language:	Time:	**Motor Strength:**	Time:
Comprehension:	_____	5/5	_____
Naming:	_____	Drift resolved	_____
Repetition:	_____	Asterixis resolved	_____
Reading:	_____		

If patient became aphasic, note time of first intelligible vocalization: _____

© 1993–2010 Section of Behavioral Neurology, Medical College of Georgia

Wada Assessment

Name: _____ Date: _____

MEMORY PERFORMANCE - INJECTION 1: R   L

OBJECT/WORD FREE RECALL: Time: _____

_____

EARLY OBJECT RECOGNITION: Circle Set: A B C   Time: _____

_____ Y N   _____ Y N   _____ Y N
_____ Y N   _____ Y N   _____ Y N
_____ Y N   _____ Y N   _____ Y N
_____ Y N   _____ Y N   _____ Y N
_____ Y N   _____ Y N   _____ Y N   1st 4 =   /4
_____ Y N   _____ Y N   _____ Y N   2nd 4 =   /4
_____ Y N   _____ Y N   _____ Y N   Total =   /8
_____ Y N   _____ Y N   _____ Y N   False Positives =

                                                                         Object Corrected Score =

_____

Comments on Denial, Anosognosia and Recall of Deficits:

Talking: _____

Weakness/Numbness: _____   Time First Study Completed: _____

Understanding: _____

_____

_____ Affect Changes: _____

_____ Response Perseverations: _____

_____ Paraphasic Errors: _____

_____ Attentional Deficits: _____

_____ Crossflow Information: _____

© 1993–2010 Section of Behavioral Neurology, Medical College of Georgia

Name: _____ Date: _____

INJECTION 2

Side: Right   Left   Dose (mg): _____   Time Test Begun: _____

Flaccid Hemiparesis Y N   If not, describe: _____

INITIAL EXPRESSIVE LANGUAGE (Counting) .... 0 1 2 3 4   Running Time:
                                                                                                                                        0 sec

(4 = normal, slowed or brief pause; 3 = counting perseveration with normal sequencing; 2 = sequencing errors;
1 = single number or word perseveration; 0 = arrest)

Scoring Scale: 0 = severe, 1 = moderate, 2 = mild, 3 = normal

LATERAL GAZE PALSY   NONE   R   L
    Spontaneous   ____ ____   ____ 0 1 2 3
    Command   ____ ____   ____ .... 0 1 2 3 N/A

RECEPTIVE LANGUAGE (Initial-Simple): 0 1 2 3
                                                                                                                    Time _____
1. AAre both your arms in the air?   Y   N   N/A
2. For Right ISAs: APoint with your right hand to your left hand. Record where the patient initially points, including direction and height (e.g., where hand is now on bed; where hand was in air at beginning or injection; denote other location; or no response).

3. Is either one of your arms weak?   Y   N   N/A   Which?   L   R   Both
4. Put patient=s paralyzed hand in good space and ask AWhat is this?

SERIES I OBJECT PRESENTATION (Begin at 30-60 seconds)

    Set Administered: A B C   Time for 1st 4 Objects (~1 min post) _____

1st 4 objects: _____

2nd 4 objects: _____

                                                            Time for 2nd 4 Objects _____

MODIFIED GLASGOW COMA SCALE: If Ptosis, note severity: .... 0 1 2 3 (3 = Normal)
(scored during presentation of 1st 4 objects)   If akinesia, note severity: .... 0 1 2 3 (3 = Normal)

	Ipsi Arm	Contra Arm	Eye Opening:		Eye Movements:	
Motor:						
spontaneous	6	6	spontaneous	4	follows objects well	4
localizes	5	5	to loud noise	3	follows objects variably	3
withdraws	4	4	to pain	2	orients to loud noise	
abnormal flexion	3	3	nil	1	and light	2
extensor response	2	2			nil	1
nil	1	1				

                                                            Modified Glasgow Coma Score =   /14

LANGUAGE I: Scoring Scale: 0 = severe, 1 = moderate, 2 = mild, 3 = normal   Time _____

COMPREHENSION: (Token Card) .... 0 1 2 3 (3 = Normal)

NAMING: watch _____, watchband _____, buckle _____, hands _____, winder _____,

           coat/jacket _____, sleeve _____, collar _____, buttons _____ 0 1 2 3

© 1993–2010 Section of Behavioral Neurology, Medical College of Georgia

Wada Assessment

Name: _____ Date: _____

INJECTION 2   Side: R   L

REPETITION: "No ifs ands or buts" **or** ʌJack and Jill went up the hillꜞ .... 0 1 2 3 N/A (3 = Normal)

Responses:_____

"Mary had a little lamb" .... 0 1 2 3 N/A (3 = Normal)

Responses:_____

"Pry the tin lid off" .... 0 1 2 3 N/A (3 = Normal)

Response:_____

"The spy fled to Greece" .... 0 1 2 3 N/A (3 = Normal)

Response:_____

"Methodist Episcopal" .... 0 1 2 3 N/A (3 = Normal)

Responses:_____

"The phantom soared across the foggy heath" .... 0 1 2 3 N/A (3 = Normal)

Response:_____

ʌThe President lives in Washingtonꜞ .... 0 1 2 3 N/A (3 = Normal)

Response:_____

DYSARTHRIA: .... 0 1 2 3 N/A (3 = Normal)

READING: The car backed over the curb. .... 0 1 2 3 N/A (3 = Normal)

Response:_____

The rabbit hopped down the lane.   0 1 2 3 N/A (3 = Normal)

Response:_____

MOTOR STRENGTH I: Time _____

Contralateral Strength .... 0 1 2 3 (3 = Normal)

Pronator drift?   Y   N   Asterixis: Contralateral?   Y   N   Ipsilateral?   Y   N

1. ʌAre both your arms in the air?ꜞ   Y   N   N/A
2. For Right ISAs: ʌPoint with your right hand to your left hand.ꜞ   Record where the patient initially points, including direction and height (e.g., where hand is now on bed, where hand was in air at beginning of injection; denote other location; or no response).

_____

3. Is either one of your arms weak?   Y   N   N/A   Which?   L   R   Both
4. Put patient=s paralyzed hand in good space and ask ʌWhat is this?ꜞ

_____

_____

LANGUAGE II: Scoring Scale: 0 = severe, 1 = moderate, 2 = mild, 3 = normal   Time: _____

COMPREHENSION: (Token Card) .... 0 1 2 3 N/A (3 = Normal)

NAMING: watch _____, watchband _____, buckle _____, hands _____, winder _____,

coat/jacket _____, sleeve _____, collar _____, buttons _____ .... 0 1 2 3 N/A

© 1993–2010 Section of Behavioral Neurology, Medical College of Georgia

Name: _____ Date: _____

INJECTION 2   Side: R   L

REPETITION:
"Mary had a little lamb"... 0 1 2 3 N/A (3 = Normal)

Responses:_____

"No ifs ands or buts" **or** ΑJack and Jill went up the hill..................................... 0 1 2 3 N/A (3 = Normal)

Responses:_____

"Pry the tin lid off".. 0 1 2 3 N/A (3 = Normal)

Response: _____

"The spy fled to Greece".. 0 1 2 3 N/A (3 = Normal)

Response: _____

"Methodist Episcopal"... 0 1 2 3 N/A (3 = Normal)

Responses:_____

"The phantom soared across the foggy heath".. 0 1 2 3 N/A (3 = Normal)

Response: _____

ΑThe President lives in Washington.. 0 1 2 3 N/A (3 = Normal)

Response: _____

DYSARTHRIA: .... 0 1 2 3 N/A (3 = Normal)

READING: The car backed over the curb... 0 1 2 3 N/A (3 = Normal)

Response:_____

The rabbit hopped down the lane... 0 1 2 3 N/A (3 = Normal)

Response:_____

MOTOR STRENGTH II: Time _____

Contralateral Strength .... 0 1 2 3 N/A (3 = Normal)

Pronator drift? Y   N   Asterixis: Contralateral?   Y   N   Ipsilateral?   Y   N

1. ΑAre both your arms in the air?   Y   N   N/A
2. For Right ISAs: ΑPoint with your right hand to your left hand. Record where the patient initially points, including direction and height (e.g., where hand is now on bed; where hand was in air at beginning of injection; denote other location; or no response).

3. Is either one of your arms weak?   Y   N   N/A   Which?   L   R   Both
4. Put patient=s paralyzed hand in good space and ask ΑWhat is this?

If Language and Motor <u>not</u> normal on above testing, then record time that they return to normal:

Language:	Time:	**Motor Strength:**	Time:
Comprehension:	_____	5/5	_____
Naming:	_____	Drift resolved	_____
Repetition:	_____	Asterixis resolved	_____
Reading:	_____		

If patient became aphasic, note time of first intelligible vocalization: _____

© 1993–2010 Section of Behavioral Neurology, Medical College of Georgia

Wada Assessment

Name: _____ Date: _____

MEMORY PERFORMANCE - INJECTION 2: R L

OBJECT/PICTURE FREE RECALL: Time: _____

_____

EARLY OBJECT RECOGNITION: Circle Set: A B C   Time: _____

_____ Y N	_____ Y N	_____ Y N		
_____ Y N	_____ Y N	_____ Y N		
_____ Y N	_____ Y N	_____ Y N		
_____ Y N	_____ Y N	_____ Y N		
_____ Y N	_____ Y N	_____ Y N	1st 4 =	/4
_____ Y N	_____ Y N	_____ Y N	2nd 4 =	/4
_____ Y N	_____ Y N	_____ Y N	Total =	/8
_____ Y N	_____ Y N	_____ Y N	False Positives =	
			Object Corrected Score =	

_____

Comments on Denial, Anosognosia and Recall of Deficits:

Talking: _____ Time Second Study Completed: _____

Weakness/Numbness: _____

Understanding: _____

_____ Affect Changes: _____

_____ Response Perseverations: _____

_____ Paraphasic Errors: _____

_____ Attentional Deficits: _____

_____ Crossflow Information: _____

Impressions: _____

© 1993–2010 Section of Behavioral Neurology, Medical College of Georgia

WADA TEST MEMORY ITEMS

List A:

Targets: frog, nut cracker, comb, toy gun, slipper, baby, duck, clothes pin.

Foils; oven mitt, clip, bowl, airplane, jump rope, potato chip, dinosaur, hose nozzle, sun glasses, shoe laces, turtle, bug, padlock, cassette tape, top, key chain.

Target Cards: pen, broom, pear, lamp.

Foil Cards: telephone, sword, cow, bird, piano, bed, bridge, corn.

Recognition Order: telephone, sword, *pear*, cow, *pen*, bird, piano, *broom*, bed, *lamp*, bridge, corn.

List B:

Targets: rolling pin, cup, pizza, paint brush, shark, fork, ash-tray, grandpa.

Foils: hamburger, whistle, rattle, banana, yo-yo, bicycle horn, switch plate, mouse trap, kangaroo, bottle opener, pencil, beret, motorcycle, spool of thread, doorstop, paddle.

Target Cards: candle, book, pie, hat.

Foil Cards: apple, bug, cake, wheel, mirror, horn, bird cage, ladder.

Recognition Order: apple, bug, *pie*, cake, wheel, *book*, mirror, *candle*, horn, bird cage, *hat*, ladder.

List C:

Targets: hotdog, drain, chair, lobster, troll, scrub pad, eyebolt, turkey baster.

Foils: golf ball, spatula, football, ruler, curler, flip flop, ice tray, baseball cards, baby block, snake, tape, glove, mini frisbee, measuring cup, orange, electrical plug.

Target Cards: cat, matches, clock, house.

Foil Cards: fork, stool, box, table, window, bread, foot, bell.

Recognition Order: boot, stool, *matches*, box, *cat*, table, window, bread, *house*, foot, *clock*, bell.

List D:

Targets: basket, playing cards, tongs, razor, car, eraser, marker, stapler.

Foil Objects: fish, penny, mirror, guitar, tractor, scarecrow, wrench, roller skate, salt shaker, raisin man, tank, train, egg beater, pacifier, tonka man, tiger.

Target Cards: light bulb, fence, sweater, key.

Foil Cards: stairs, nose, pail, cigarette, bottle, hand, mountain, pants.

Recognition Order: stairs, nose, *key*, pail, cigarette, bottle, *fence*, hand, *light bulb*, mountain, *sweater*, pants.

© 1993–2010 Section of Behavioral Neurology, Medical College of Georgia

Glossary

AED – an acronym for antiepileptic drug; an agent that prevents seizures.
Absence seizure – a "typical" absence attack is a generalized seizure most common in children; consists of a lapse of consciousness with a blank stare lasting only a few seconds; often accompanied by rapid eye blinking.
AED – abbreviation for antiepileptic drug.
Aicardi syndrome – an epileptic syndrome seen only in females, usually appearing in the first year of life, typically consisting of infantile (flexor) spasms; associated with agenesis of the corpus callosum.
Anticonvulsant – older term for medications that prevent or arrest convulsions.
Alpha rhythm – an EEG frequency within the 8–13 Hz bandwidth.
Astatic seizure – older term for a drop attack; see *Atonic seizure*.
Atonic seizure – a drop attack; generalized seizure characterized by complete loss of muscle tone.
Audiogenic seizures – reflex seizures produced by auditory stimuli.
Aura – a sensation that precedes or signals the beginning of a seizure. It is a simple partial seizure that may involve epigastric distress, uneasiness, sensory illusions, dizziness, or déjà-vu.
Automatism – involuntary, nondirected movements during complex partial and atypical absence seizures usually involving the hands ("picking motions"), mouth ("lip smacking") or voice (repetitive vocalization).
Autonomic seizure – autonomic signs are often a component of partial seizures; often an aura in complex partial seizures; signs may include epigastric distress, flushing, sweating, pupillary dilatation, piloerection, salivation, bradycardia, tachycardia, palpitations, and changes in respiration and blood pressure.
Benign familial neonatal convulsions – an epilepsy syndrome inherited by autosomal dominant transmission beginning in the first week of life and spontaneously resolving within a few weeks or months in most cases; approximately ≤20% of patients continue to experience seizures.
Benign Rolandic epilepsy – a common idiopathic, localization-related epilepsy syndrome of childhood; typically resolves naturally by age 14 years. Seizure semiology varies but most often consists of focal

seizures with sensorimotor symptoms involving the face, mouth, tongue, or pharynx.

β activity – EEG activity in the frequency range above 12 Hz; considered fast electrical activity and a normal component of the waking (and some parts of the sleeping) EEG.

BOLD – acronym for "blood oxygenation level-dependent"; serves as the basis for generating images during functional magnetic resonance imaging (fMRI).

Carbamazepine – trade names include Tegretol or Carbatrol; an antiepileptic drug related to tricyclic antidepressants, useful in the treatment of partial and secondary generalized seizures. Also used to treat bipolar disorder and neuropathic pain.

Catamenial epilepsy – occurs in women when there is a tendency for seizures to occur at the time of menstruation.

Centrencephalic seizures – an older term referring to generalized seizures hypothesized to originate in the "central encephalon," which includes the nonspecific nuclei of the thalamus, diencephalon and mesencephalon.

Clonazepam – trade name is Klonopin; a benzodiazepine used as an antiepileptic drug in myoclonic epilepsies, absence seizures, and photosensitive seizures; may be administered acutely or chronically.

Clonic seizures – seizures characterized by either focal or generalized rhythmic clonic jerks that varying frequency and amplitude; typically occur in young children with symptomatic epilepsies and with febrile seizures.

Complex partial seizure – a seizure localized to one part of the brain that usually begins with a blank stare and sometimes automatisms; patients are unresponsive for a brief period of time and have no recollection for events that occurred during the seizure.

Convulsion – refers to the generalized jerking movements seen in generalized tonic-clonic seizures.

Corpus callosotomy – sectioning of the largest interhemispheric commissure aimed at reducing the frequency and severity of debilitating generalized (bilateral) seizures; most often used as a palliative treatment for drop attacks with tonic or atonic seizures. Also called a commissurotomy.

Cortical dysplasia – abnormal formation of the cellular layers of cortex during development; a condition that can cause epilepsy.

Cryptogenic – of unknown cause, but presumably symptomatic; cryptogenic epilepsy syndromes are most often caused by malformations of cortical development.

Cursive seizure – seizures with ambulatory automatisms involving gait; patients may walk or run into obstacles; thought to be due to limbic epileptic foci.

Dacrystic seizure – a rare, nonconvulsive simple partial seizure whose main symptom is either a brief episode of crying or an uncontrollable urge to cry without any external stimulus.

δ activity – an EEG frequency bandwidth of less than 4 Hz; abnormal in waking nonelderly. Diffuse δ activity may be associated with diffuse toxic, metabolic, or infectious encephalopathies.

Depth electrodes – thin wire electrodes inserted deep into the brain, used to detect seizure activity that cannot be recorded from the surface of the brain.

Diazepam – trade name Valium; the first benzodiazepine antiepileptic drug used to treat status epilepticus. It has been replaced by lorazepam as the drug of first choice for status epilepticus.

Diffusion tensor imaging (DTI) – an imaging technique using MRI that reflects white matter (myelin) fiber tract integrity in vivo.

Dilantin – trade name for phenytoin.

Drop attacks – seizures that cause falls; most commonly atonic spells although falling may also occur secondary to tonic seizures.

Dual pathology – occurs when two independent pathological processes are present, such as mesial temporal lobe sclerosis in conjunction with cortical dysplasia; may be associated with poorer outcomes following epilepsy surgery.

ECoG – abbreviation for electrocorticography

Electrocortical stimulation (ECS) mapping – the procedure where by electrical stimulation of sensorimotor or language areas of cortex are used to define the cortical representation of these functions; maybe conducted either extraoperatively using subdural electrodes or intraoperatively using a bipolar electrical stimulus probe.

Electrocorticography – recording of electroencephalographic (EEG) potentials directly from the brain; usually in the operating room on surgically exposed brain.

Electroencephalography (EEG) – amplified recording of electrical signals produced by cerebral neuronal activity; most important diagnostic tool in epilepsy.

EMU – epilepsy monitoring unit

Encephalopathy – diffuse cerebral dysfunction secondary to the systemic effects of a variety of etiologies including toxic-metabolic, anoxic, hypoxic, degenerative, infectious, inflammatory, or neoplastic causes.

Epilepsy – a chronic neurological condition characterized by recurrent seizures.

Epileptic focus – localized area of the brain from which partial seizures arise.

Epileptic syndrome – a cluster of signs and symptoms customarily occurring together including such things as seizure type, etiology, anatomy, precipitating factors, onset age, severity, chronicity, diurnal cycling, and sometimes prognosis.

Epilepsia partialis continua – a type of focal motor status epilepticus with preserved consciousness that may last for months or years.

Epilepsy – two or more unprovoked seizures.

Epilepsy with continuous spike wave during slow sleep (CSWS) – an epilepsy syndrome characterized by nearly continuous spike-and-wave epileptiform discharges on EEG occurring during slow wave sleep; electrical status epilepticus during slow sleep.

Epileptogenic zone – region from which seizure discharges arise on EEG.

Ethosuximide – trade name Zarontin; a member of the succinimide class of antiepileptic drugs; used as a first-line drug for many years to treat typical absence seizures in young children.

Febrile seizure – a seizure occurring during a high fever in infants and young children.

Focus – the epileptic focus is the area of brain from which epileptic EEG discharges arise; same as "epileptogenic zone."

Frontal lobe epilepsy – site of onset for localization-related partial seizures; second most common site for partial seizures after the temporal lobes comprising about 20% of focal onset seizures.

Functional adequacy model – measures the memory capacity of the hippocampus ipsilateral to the proposed side of temporal lobectomy.

Functional brain mapping – a term used to describe electro-cortical stimulation mapping where electrical stimulation of sensorimotor or language areas of cortex are used to define the cortical representation of these functions.

Functional reserve model – measures the memory capacity of the hippocampus contralateral to the proposed side of temporal lobectomy; is an estimate of risk for postsurgical global amnesia.

γ-aminobutyric acid (GABA) – an important inhibitory CNS neurotransmitter implicated in the genesis of epilepsy; selective inhibition of GABAergic transmission may produce convulsions and GABA agonist drugs protect against the development of seizures.

Gabapentin – trade name is Neurontin; an antiepileptic drug used as an adjunctive agent to treat partial and secondarily generalized seizures in patients over the age of 12 years.

Gelastic seizure – a simple partial seizure where the primary symptom is an uncontrollable urge to laugh with no external provocation.

Generalized seizure – an epileptic seizure that involves most or all of the brain bilaterally.

Gliosis – an excess of neuroglial cells in damaged areas of the central nervous system.

Grand mal seizure – an older term for a generalized tonic-clonic seizure.

Gyratory seizures – rare type of epilepsy consisting of turning of the entire body around its vertical axis 180 degrees or more in what has been described as a "ballet-like" movement; seizures are thought to emanate from the contralateral frontal lobe.

Hamartomas – developmental brain abnormalities in which histologically normal tissue resides in an abnormal location; a form of developmental neuropathology seen in resected brain tissue after epilepsy surgery. Hypothalamic hamartomas are uncommon lesions associated with gelastic epilepsy.

Hemispherectomy – a resective form of epilepsy surgery that removes a major portion of a single hemisphere for relief of severe debilitating seizures.

Heterotopia – abnormal cellular migration during in utero development resulting in displacement of gray matter into white matter; a common cause of focal-onset seizures.

Hippocampal sclerosis – also called *mesial temporal sclerosis*; most common underlying neuropathology causing intractable complex partial seizures of temporal lobe origin; febrile seizures and other early insults may cause neuronal atrophy and sclerosis of the hippocampus involving the CA1 and CA2 subfields and the dentate gyrus.

Hirsutism – abnormal hairiness; particularly an adult male pattern of hair distribution in a female; a side effect of prolonged use of some antiepileptic drugs, for example, of phenytoin.

Hypsarrhythmia – abnormal interictal EEG pattern characterized by a chaotic mixture of high amplitude slow and sharp waves, multifocal spikes and polyspikes with no recognizable normal rhythms; typically seen in West syndrome.

Ictus – a sudden neurological event, such as a stroke or an epileptic seizure.

Ictal – during a seizure, or events due to a seizure; also refers to any paroxysmal event such as a migraine.

Ictal-onset zone – cortical region from which seizures arise as demonstrated by EEG or SPECT recordings; also called the epileptogenic zone.

Idiopathic – etiology unknown; idiopathic epilepsy syndromes are most often due to inherited (genetic) factors.

Inhibitory seizures – seizure that causes a loss of function or "negative symptoms," such as aphasia, deafness, blindness, amnesia, neglect, or numbness.

Irritative zone – a region surrounding the epileptogenic zone from which occasional epileptic discharges may be recorded on EEG.

Intracranial EEG – EEG recorded directly from the surface (e.g., grid electrodes) or deep within (e.g., depth electrodes) the brain; used in planning epilepsy surgery or for functional cortical mapping.

Jacksonian seizures – partial seizures characterized by a progressive spread of clinical phenomenon from one part of the body to another ("jacksonian march") without alterations of consciousness.

Juvenile myoclonic epilepsy (JME) – a common idiopathic generalized epilepsy syndrome with a polygenetic cause characterized by myoclonic jerks, generalized tonic-clonic seizures, and sometimes absence seizures.

Ketogenic diet – a high-fat, low-carbohydrate diet that mimics the metabolism of prolonged fasting, used to treat childhood epilepsies that are refractory to AED treatment; most commonly used in young children to treat intractable absence, atonic seizures, infantile spasms, myoclonic seizures, and mixed seizure types associated with Lennox-Gastaut syndrome.

Kindling – occurs in animals when repeated, low-intensity (subconvulsive) electrical stimulation later causes spontaneous epileptogenic discharges; the kindling effect persists long after repetitive stimulation ceases. Secondary epileptogenesis from kindling has not yet been shown to exist in humans.

Lamotrigine – trade name Lamictal; chemically a phenyltriazine that is thought to have an antiepileptogenic effect through its inhibitory effect on glutamate release; approved for use as adjunctive therapy in children 2 years and older for partial and generalized tonic-clonic seizures and for monotherapy in patients 16 years and older with partial seizures.

Landau-Kleffner syndrome – an acquired epileptic encephalopathy characterized by progressive aphasia, epileptiform EEG discharges, and seizures in most (but not all) cases. Normal development occurs with a loss of language functions beginning after two years of age. Typical age of onset is between 3 and 8 years; more common in males.

Lennox-Gastaut syndrome – is a severe epileptic encephalopathy of childhood (onset most common between ages 2 and 6 years) characterized by (1) mixed generalized seizure types including tonic seizures, atonic seizures, and atypical absences, (2) interictal EEG abnormalities with slow spike-and-wave discharges during wakefulness, abnormal background rhythm, and paroxysmal rapid fast activity during non-REM sleep, and (3) cognitive dysfunction. Prognosis is poor with worsening seizures and progressive deterioration of cognitive development and behavior.

Levetiracetam – trade name Keppra; currently approved as an adjunctive therapy for partial seizures in adults and children over the age of 4 years, but has also been shown to be an effective monotherapy for multiple seizure types (i.e., partial and generalized onset seizures including myoclonic seizures).

Lissencephaly – a common neuronal migrational disorder characterized by a diffusely absent or reduced gyrational pattern with poorly organized cortex and reduced total surface area of cortex. Profound mental retardation and intractable seizures are seen in the majority of patients.

Lorazepam – trade name Ativan; a benzodiazepine used as a first-line drug to treat generalized convulsive status epilepticus.

Magnetoencephalography (MEG) – noninvasive measure of the magnetic fields of brain neurons; similar to EEG, MEG is used to detect interictal epileptogenic activity, localize the seizure focus, and may also be used for functional cortical mapping of sensorimotor and language functions. MEG is thought to be superior to EEG with regard to spike-detection sensitivity and extratemporal seizure localization.

Mesial – towards the midline.

Mesial Temporal Sclerosis (MTS) – see *Hippocampal sclerosis*.

Multiple subpial transection (MST) – nonresective epilepsy surgery used for seizure control when epileptogenic focus lies within eloquent cortex; technique selectively interrupts the horizontal propagation of seizures by disconnecting the horizontally oriented intracortical fibers while preserving the columnar organization of cortex. When used alone, seizure outcome is typically worse with MST than resective surgeries.

Musicogenic epilepsy – a rare form of complex partial seizures (usually of temporal lobe origin) precipitated by specific melodies, certain types of music, or repetition of music.

Myoclonic epilepsy – a family of epileptic syndromes in which myoclonus occurs during seizures. Course and prognosis differ considerably depending upon etiology of the specific syndrome and range from benign myoclonic epilepsy in infancy (with good outcome) to severe progressive myoclonic epilepsy (with severe physical, cognitive, and behavioral deficits).

Myoclonic seizure – these generalized seizures consist of brief, violent muscle contractions that are usually bilateral without any apparent impairment of consciousness.

Neonatal seizures – occur during the first 4 weeks of life and usually consist of tonic, myoclonic, or clonic seizures which may be secondary to any insult to brain.

Nonepileptic seizures (NES) – behavioral spells without simultaneous EEG evidence of an epileptic brain event; two etiological types: psychogenic or physiological nonepileptic seizures.

Oxcarbazepine – trade name Trileptal; an analog of carbamazepine used for monotherapy or adjunctive therapy in both adults and children over the age of 4 years to control partial seizures, with or without secondary generalization, and for generalized tonic-clonic seizures in both adults and children.

Panayiotopoulos syndrome – a common idiopathic focal onset epilepsy syndrome of childhood arising from the occipital lobes. These are for the most part normal children with infrequent nocturnal seizures associated with vomiting, other autonomic signs, and eye deviation; prognosis is very good.

Paroxysm – a sudden attack, spasm, or fit; such as a convulsion.
Partial seizure – a focal seizure involving only one part of the brain; classified as either simple partial or complex partial.
Peri-ictal – an event occurring either immediately before or after a seizure.
Petit mal – older term for an absence seizure.
Phenobarbital – among the oldest anticonvulsant drugs; indicated for all forms of epilepsy except typical absence seizures; use is limited when other AEDs are available due to its strong sedative properties.
Phenytoin – trade name Dilantin; the most widely used AED in the U.S.; effective for partial onset and generalized tonic-clonic seizures; ineffective in absence and myoclonic seizures; has significant acute and long-term side effects.
Photosensitive seizures – seizures induced by intermittent light sources, such as flickering lights, television, video games; a type of reflex epilepsy. Photosensitivity may be observed in patients with partial or generalized seizure disorders; usually develops in late childhood/early adolescence and disappears by the third decade of life.
PLEDs – an acronym for periodic lateralized epileptiform discharges.
Polypharmacy – the use of two or more antiepileptic drugs simultaneously in the same individual; up to one-third of seizures are uncontrollable using a single (monotherapy) AED. Polypharmacy increases likelihood of cognitive side effects of AEDs.
Positron emission tomography (PET) – a functional neuroimaging technique that visualizes the energy metabolism of cerebral neurons using a variety of radioisotopes; in epilepsy, FDG (2–deoxy-2[18]fluoro-D-glucose) PET studies show hypometabolism of the partial seizure focus during the interictal period and hypermetabolism during the ictal period.
Postictal – an event occurring after a seizure has ended.
Physiological nonepileptic seizures – behavioral spells without simultaneous EEG evidence of an epileptic brain event; common causes include cardiac, syncopal, metabolic conditions and sleep disorders.
Pregabalin – trade name Lyrica; similar to gabapentin; has high affinity for certain voltage-gated calcium channels of CNS neurons; indicated as an adjunctive therapy for partial-onset seizures; also useful in nonepileptic conditions that involve neuropathic pain associated with diabetic neuropathy and postherpetic neuralgia.
Primidone – trade name Mysoline; similar to phenobarbital; used in patients who do not respond to first-line or second-line antiepileptic drugs; useful for adjunctive therapy in partial-onset, generalized tonic-clonic, and myoclonic seizures; has significant sedative and other short- and long-term, dose related side effects.
Psychogenic nonepileptic seizure (PNES) – episodes of seizure phenomena, such as unresponsiveness or other behavioral abnormalities, in the absence

of EEG changes; also called "functional" seizures, "dissociative" seizures, or "stress-induced" seizures.

Pyknolepsy – a term used primarily in Europe as a synonym for childhood absence epilepsy.

Ramsey Hunt syndrome – a rare inherited progressive disorder composed of two subtypes; Type 1, called dyssynergia cerebellaris myoclonia, is a severe condition characterized by epilepsy, tremor, action myoclonus, incoordination, progressive gait disturbance and cognitive impairment. Type 2 is a less severe condition with infrequent seizures, ataxia, and mild cognitive changes.

Rasmussen syndrome – also called Rasmussen encephalitis; a rare disorder of childhood characterized by progressive inflammatory changes and hemispheric atrophy associated with epilepsy, cognitive dysfunction, and hemiparesis. Seizures are usually focal motor seizures with secondarily generalized tonic-clonic seizures; one of the most common indications for hemispherectomy.

Reading epilepsy – a type of reflex epilepsy triggered by the act of reading (often only out loud); seizures are localization-related with focal partial seizures that may secondarily generalize into tonic-clonic seizures; seizures arise only when reading. Etiologies are either cryptogenic or symptomatic.

Reflex epilepsy – is a condition in which seizures are consistently induced by a specific environmental (e.g., music) or intrinsic (e.g., thinking) trigger; seizure type is most commonly partial although generalized seizures may occur; causes are often either genetic or symptomatic. Photosensitive seizures are the most common type of reflex epilepsy.

Refractory seizures – seizures that are not controlled with adequate trials of antiepileptic drugs; also called intractable or pharmacoresistant.

Rolandic epilepsy – see *Benign Rolandic epilepsy*; also called benign childhood epilepsy with centrotemporal spikes.

Schizencephaly – is a thick cleft in the cortex that extends to the lateral ventricle and lies near the Sylvian fissure. It is a neurodevelopmental brain malformation that may cause epilepsy.

Sclerosis – hardening due to hyperplasia and glial overgrowth most often caused by inflammation; often used interchangeably with the term *gliosis*.

Seizure – sudden onset of abnormal, uncontrolled electrical discharges of neurons in the brain.

Semiology – seizure semiology refers to the signs and symptoms evidenced during a seizure.

Simple partial seizure – seizure activity localized to one part of the brain causing elemental motor or sensory symptoms without alternations of awareness or memory loss; usually involves jerking in the arm, leg, or face or simple sensations or illusions.

Single photon emission computed tomography (SPECT) – a functional neuroimaging technique to measure regional cerebral blood flow (rCBF); uses systemically injected radio-labeled tracer, typically HMPAO (hexamethylpropylene amine oxime) with ^{99m}Tc (technetium), to produce images of rCBF; similar to PET scanning, but less expensive.

Status epilepticus – continuous, severe seizures that are potentially life-threatening.

Stevens-Johnson syndrome – is a rare and serious abnormal systemic immune response, usually to certain antiepileptic drugs in epilepsy patients, characterized by severe rash over the face, nose, mouth, eye, and urethral, vaginal, gastrointestinal, and lower respiratory tracts; may cause permanent morbidity or death.

Sturge-Weber syndrome – a congenital neurocutaneous syndrome with excess calcium deposits in blood vessels and perivascular tissue; causes hypoxic-induced calcified cortex. Seizures commonly begin in first year life; mental retardation and cognitive impairments are typical.

Subtraction ictal SPECT coregistered with MRI (SISCOM) – in this technique the change from resting (interictal) SPECT to SPECT during a seizure (ictal SPECT) is measured then projected onto the patients own structural MRI scan; used in focal onset seizures to localize the seizure focus.

Symptomatic epilepsy – an acquired form of epilepsy with an identifiable, known cause; symptomatic epilepsy syndromes have some detectable structural lesion.

Symptomatic zone – region(s) of the brain that is associated with the behavioral symptoms of a seizure; propagation from the ictal-onset zone may either be nearby (through depolarization of neighboring neurons) or distant (propagated through intracortical association pathways) to the site of seizure onset.

Syndromes – epileptic syndromes are a constellation of signs and symptoms that customarily occur together. Cluster of signs and symptoms may include seizure type, etiology, precipitating factors, age of onset, severity, chronicity, and prognosis.

Theta activity – the EEG frequency range between 4 and 8 Hz is referred to as the EEG theta band; abnormal in healthy, awake adults where it may indicate some diffuse nonspecific cerebral dysfunction.

Tonic seizures – characterized by sustained muscular contraction; most often seen in children with symptomatic generalized epilepsies.

Tonic-clonic seizure – a generalized seizure most known for bilateral rhythmic convulsions of the limbs and face; evolves across three stages: tonic phase, tonic-clonic phase, and postictal phase.

Topiramate – trade name Topamax; a unique antiepileptic drug that has been approved for adjunctive therapy for partial and generalized seizures in adults and children ages 2 years and older; also used in children with Lennox-Gastaut syndrome and infantile spasms.

Tuberous sclerosis – an inherited multisystem neurocutaneous syndrome that causes benign tumors (tubers) to grow in the brain among other organs which frequently presents with severe epilepsy and mental retardation; often as infantile spasms in the first year of life.

Uncinate fits – an outdated term used to describe partial-onset seizures that begin with an olfactory aura.

Unresponsiveness – an inability to respond to external stimuli; usually assessed during a seizure by the inability to follow simple commands.

Vagal nerve stimulator – a device consisting of a surgically implanted pacemaker with a bipolar electrode attached to the left vagus nerve in the neck; pacemaker is programmed to deliver regular electrical stimulation of the vagal nerve which reduces the onset of complex partial seizures in some patients.

Valproate – sodium valproate's trade name is Depakote; is a first-line drug in generalized or unclassified epilepsy and may be used as monotherapy or adjunctively in absence or complex partial seizures or in patients with multiple seizure types; also used to treat bipolar disorder and migraine headache. Not recommended for use in pregnant women due to teratogenic effects on the fetus.

Wada testing – also called the intracarotid amobarbital procedures (IAP); a preoperative procedure in which a barbiturate drug is injected into the internal carotid artery to evaluate hemisphere lateralization of language and memory functions; named for the physician who devised the procedure, Juhn Wada.

Zonisamide – trade name Zonegran; an antiepileptic drug with multiple cellular mechanisms of action used to treat patients with refractory partial epilepsies.

References

Chapter 1

Annegers JF. Epidemiology of epilepsy. In: E. Wyllie, ed. *The treatment of epilepsy: Principles and practice* (2nd ed.). Philadelphia: Lippincott Williams & Wilkins, 1996;165–172.

Barr WB. Epilepsy and neuropsychology: Past, present, and future. *Neuropsychol Rev* 2007;17(4):381–383.

Browne TR, Holmes GL. *Handbook of epilepsy* (3rd ed.). Philadelphia: Lippincott Williams & Wilkins, 2004.

Epilepsy Foundation of America. http://www.epilepsyfoundation.org/about/statistics.cfm, 2008.

Gastaut H. *Dictionary of epilepsy*. Geneva, Switzerland: World Health Organization, 1973.

Hauser AW, Annegers JF, Kurland LT. Incidence of epilepsy and unprovoked seizures in Rochester, Minnesota 1935–1984. *Epilepsia* 1993;34:453–468.

Hauser AW, Hesdorffer DC. *Epilepsy: Frequency, causes and consequences*. Landover, MD: Epilepsy Foundation of America Publications, 1990.

ILAE. International League Against Epilepsy Commission report: The epidemiology of the epilepsies: Future directions. *Epilepsia* 1997;38(5):614–618.

Lassonde M, Sauerwein HC, Gallagher A, Theriault M, Lepore F. Neuropsychology: Traditional and new methods of investigation. *Epilepsia* 2006;47(Suppl. 2):9–13.

Loring DW, Meador KJ, Lee GP, King DW. *Amobarbital effects and lateralized brain function: The Wada test*. New York: Springer-Verlag, 1992.

Reynders HJ, Baker GA. A review of neuropsychological services in the United Kingdom for patients being considered for epilepsy surgery. *Seizure* 2002;11:217–223.

Walczak TS, Boop FA, Cascino GD, Labiner DM, Olson DM. Report of the National Association of Epilepsy Centers: Guidelines for essential services, personnel, and facilities in specialized epilepsy centers in the United States. *Epilepsia* 2001;42(6):804–814.

Chapter 2

Annegers JF. The epidemiology of epilepsy. In: E Wyllie, ed. *The treatment of epilepsy: Principles and practice* (2nd ed.). Philadelphia: Lippincott Williams & Wilkins, 1996:165–172.

Annegers JF, Hauser WA, Beghi E, Nicolosi A, Kurland LT. The risk of unprovoked seizures after encephalitis and meningitis. *Neurology* 1988;38:1407–1410.

Annegers JF, Grabow JD, Groover RV, Laws ER, Elveback LR, Kurland LT. Seizures after head trauma: a population study. *Neurology* 1980;30:683–689.

Ardila A, Lopez MV. Paroxysmal aphasias. *Epilepsia* 1988;29:630–634.

Bancaud J, Brunet-Bourgin F, Chauvel P, Halgren E. Anatomical origin of *déjà-vu* and vivid 'memories' in human temporal lobe epilepsy. *Brain* 1994;117:71–90.

Bell GS, Sander JW. The epidemiology of epilepsy: the size of the problem. *Seizure* 2001;10:306–314.

Bleasel A, Kotagal P, Kankirawatana P, Rybicki L. Lateralizing value and semiology of ictal limb posturing and version in temporal and extratemporal epilepsy. *Epilepsia* 1997;38:168–174.

Browne TR, Holmes GL. *Handbook of epilepsy* (3rd ed.). Philadelphia: Lippincott Williams & Wilkins, 2004.

Butler CR, Graham KS, Hodges JR, Kapur N, Wardlaw JM, Zeman AZJ. The syndrome of transient epileptic amnesia. *Annals Neurology* 2007;61:587–598.

Chee MW, Kotagal P, Van Ness PC, Gragg L, Murphy D, Luders HO. Lateralizing signs in intractable partial epilepsy: blinded multiple-observer analysis. *Neurology* 1993;43:2519–2525.

Christensen J, Pedersen MG, Pedersen CB, Sidenius P, Olsen J, Vestergaard M. Long-term risk of epilepsy after traumatic brain injury in children and young adults: a population-based cohort study. *Lancet* 2009;373(9669):1105–1110.

Chow SY, His MS, Tang LM. Epilepsy and intracranial meningiomas. *Chinese Medical Journal* 1995;55:151–155.

Commission on Classification and Terminology, International League Against Epilepsy. Proposal for revised clinical and electrographic classification of epileptic seizures. *Epilepsia* 1981;22:489–501.

Commission on Classification and Terminology, International League Against Epilepsy. Proposal for revised classification of epilepsies and epileptic syndromes. *Epilepsia* 1989;30:389–399.

Commission on Epidemiology and Prognosis, International League Against Epilepsy. Guidelines for epidemiologic studies on epilepsy. *Epilepsia* 1993;34:592–596.

Delgado-Escueta AV, Serratosa JM, Medina MT. Myoclonic seizures and progressive myoclonus epilepsy syndromes. In: Wylllie E, ed. *The treatment of epilepsy: principles and practice* (2nd ed.). Philadelphia: Lippincott Williams & Wilkins, 1996:467–483.

Dietl T, Bien C, Urbach H, Elger C, Kurthen M. Episodic depersonalization in focal epilepsy. *Epilepsy & Behavior* 2005;7:311–315.

Engel J Jr., Kuhl De, Phelps ME. Patterns of human local cerebral glucose metabolism during epileptic seizures. *Science* 1982;218:64–66.

Engel J Jr., Pedley TA, Acardi J, Dichter MA, Moshé S. *Epilepsy: a comprehensive textbook* (2nd ed.). Philadelphia: Lippincott Williams & Wilkins, 2007.

Farrell K. Tonic and atonic seizures. In: Wylllie E, ed. *The treatment of epilepsy: principles and practice* (3rd ed.). Philadelphia: Lippincott Williams & Wilkins, 2001:405–413.

Fisch BJ, Olejniczak PW. Generalized tonic-clonic seizures. In: Wyllie E, Gupta A, Lachhwani DK, eds. *The treatment of epilepsy: Principles and practice* (4th ed.). Philadelphia: Lippincott Williams & Wilkins, 2006:279–304.

Fisch BJ, Pedley TA. Generalized tonic-clonic epilepsies. In: Luders IJ, Lesser RP, eds. *Epilepsy: electroclinical syndromes*. New York: Springer-Verlag, 1987:151–185.

Gastaut H, Broughton R. *Epileptic seizures: clinical and electrographic features, diagnosis, and treatment*. Springfield, IL: Charles C. Thomas, 1972.

Gloor P. Experiential phenomena of temporal lobe epilepsy. *Brain* 1990;120:183–192.

Haerer AF, Anderson DW, Schoenber BS. Prevalence and clinical features of epilepsy in a biracial United States population. *Epilepsia* 1986;27:66–75.

Hauser WA, Annegers JF, Kurkland LT. Incidence of epilepsy and unprovoked seizures in Rochester, MN: 1935–1984. *Epilepsia* 1994;34:453–468.

Helmsteadter C, Elger CE, Lendt M. Postictal courses of cognitive deficits in focal epilepsies. *Epilepsia* 1994;35:1073–1078.

Isnard J, Guenot M, Ostrowsky K, Sindou M, Mauguiere F. The role of the insular cortex in temporal lobe epilepsy. *Annals of Neurology* 2000;48:614–623.

Jenssen S, Gracely EJ, Sperling MR. How long do most seizures last? A systematic comparison of seizures recorded in the epilepsy monitoring unit. *Epilepsia* 2006;47:1499–1503.

Kellinghaus C, Luders HO, Wyllie E. Classification of seizures. In: Wyllie E, Gupta A, Lachhwani DK, eds. *The treatment of epilepsy: Principles and practice* (4th ed.). Philadelphia: Lippincott Williams & Wilkins, 2006:217–228.

Kinnunen E, Wikstrom J. Prevalence and prognosis of epilepsy in patients with multiple sclerosis. *Epilepsia* 1986;27:729–733.

Kotagal P, Arunkumar GS. Lateral frontal lobe seizures. *Epilepsia* 1998;39 (Suppl 4):S62–S68.

Kotagal P, Loddenkemper T. Focal seizures with impaired consciousness. In: Wyllie E, Gupta A, Lachhwani DK, eds. *The treatment of epilepsy: Principles and practice* (4th ed.). Philadelphia: Lippincott Williams & Wilkins, 2006:241–255.

Lechtenberg R. *The diagnosis and treatment of epilepsy*. New York: Macmillan, 1985.

Lennox WG, Cobb S. Aura in epilepsy: a statistical review of 1,359 cases. *Arch Neurol Psychiatry* 1933;30:374–387.

Manford M, Hart YM, Sander JW, Shorvon SD. National General Practice Study of Epilepsy (NGPSE): partial seizure patterns in a general population. *Neurology* 1992;42:1911–1917.

McLachlan RS. The significance of head and eye turning in seizures. *Neurology* 1987;37:1617–1619.

McVicker RW, Shanks OEP, McClelland RJ. Prevalence and associated features of epilepsy in adults with Down's syndrome. *British Journal of Psychiatry* 1994;164:528–532.

Oxbury JM, Duchowny M. Diagnosis and classification, In: Oxbury JM, Polkey CE, Duchowny M, eds. *Intractable focal epilepsy*. London: WB Saunders, 2000:11–23.

Palmini AL, Gloor P. The localizing value of auras in partial seizures: a prospective and retrospective study. *Neurology* 1992;42:801–808.

Palmini AL, Gloor P, Jones-Gotman M. Pure amnestic seizures in temporal lobe epilepsy. *Brain* 1992;115(Pt. 3):749–769.

Penfield W, Erikson TC, Taylov I. Relation of intracranial tumors and symptomatic epilepsy. *Archives of Neurology & Psychiatry* 1940;44:300–315.

Penry JK, Porter RJ, Dreifuss FE. Simultaneous recording of absence seizures with videotape and electroencephalography. A study of 374 seizures in 48 patients. *Brain* 1975;98:427–440.

Placencia M, Shorvon SD, Paredes V, Epileptic seizures in an Andean region of Ecuador: incidence and prevalence and regional variation. *Brain* 1992;115:771–782.

Rasmussen T. Localization aspects of epileptic seizure phenomenon. In: Thompson RA, Green JR, eds. *New perspectives in cerebral localization*. New York: Raven Press, 1982:177–203.

Roger J, Dravet C, Bureau M. The Lennox-Gastaut syndrome. *Cleveland Clinic Journal of Medicine* 1989;56(Suppl 2):S172–S180.

Romanelli MF, Morris JC, Ashkin K, Coben LA. Advanced Alzheimer's disease is a risk factor for late onset seizures. *Archives of Neurology* 1990;47:847–850.

Rwiza HT, Kilonzo GP, Haule J, Matuja WPB, Mteza I, Mbena P, Kilma PM, et al. Prevalence and incidence of epilepsy in Ulanga, a rural Tanzanian district: a community-based study. *Epilepsia* 1992;33:1051–1056.

Salazar AM, Jabbari B, Vance SC, Grafmann J, Amin D, Dillon JD. Epilepsy after penetrating head injury, I: clinical correlates. *Neurology* 1985;35:1406–1414.

Sander JW, Hart YM, Shorvon SD, Johnson AL. National General Practice Study of Epilepsy: newly diagnosed epileptic seizures in the general population. *Lancet* 1990;336:1267–1271.

Sander JW. The epidemiology of epilepsy revisited. *Current Opinion in Epilepsy* 2003;16:165–170.

Sell SH. Long term sequelae of bacterial meningitis in children. *Pediatric Infectious Disease* 1983;2:90–93.

Sirven JI, Sperling MR, French JA, O'Conner M. Significance of simple partial seizures in temporal lobe epilepsy. *Epilepsia* 1996;37:450–454.

Sperling MR, O'Connor MJ. Auras and subclinical seizures: characteristics and prognostic significance. *Ann Neurology* 1990;28:320–329.

Stefan H, Pauli E, Kerling F, Schwarz A, Koebnick C. Autonomic auras: left hemispheric predominance of epileptic generators of cold shivers and goose bumps? *Epilepsia* 2002;43:41–45.

Stefan H, Snead III OC. Absence seizures. In: Engel J Jr., Pedley TA, eds. *Epilepsy: a comprehensive textbook*. Philadelphia: Lippincott-Raven, 1997:579–590.

Sveinbjörnsdottir S, Duncan JS. Parietal and occipital lobe epilepsy. *Epilepsia* 1993;34:493–521.

Tatum WO, Farrell K. Atypical absence, myoclonic, tonic, and atonic seizures. In: Wyllie E, Gupta A, Lachhwani DK, eds. *The treatment of epilepsy: Principles and practice* (4th ed.). Philadelphia: Lippincott Williams & Wilkins, 2006:317–331.

Vigevano F, Fusco L, Kazuichi Y. Tonic seizures. In: Engel J Jr., Pedley TA, eds. *Epilepsy: a comprehensive textbook*. Philadelphia: Lippincott-Raven, 1997:617–625.

Chapter 3

American Academy of Pediatrics. Practice parameter: The neurodiagnostic evaluation of the child with a first simple febrile seizure. American Academy of Pediatrics. Committee on Quality Improvement, Subcommittee on Febrile Seizures. *Pediatrics* 1996;97:769–772.

American Academy of Pediatrics. Practice parameter: Long-term treatment of the child with simple febrile seizures. American Academy of Pediatrics. Committee on Quality Improvement, Subcommittee on Febrile Seizures. *Pediatrics* 1999;103:1307–1309.

Bancaud J, Talairach J. (1992). Clinical semiology of frontal lobe seizures. In Chauvel P, Delgdo-Escueta AV, Halgren E, Bancaud J. (eds.), *Frontal lobe seizures and epilepsies*. New York: Raven Press, 1992:3–58.

Bebek N, Gurses C, Gokyigit A, Baykan B, Ozkara C, Dervent A. Hot water epilepsy: clinical and electrophysiologic findings based on 21 cases. *Epilepsia* 2001;42:1180–1184.

Berroya AG, McIntyre J, Webster R, Lah, S, Lawson J, Bleasel AF, Bye AME. Speech and language deterioration in benign rolandic epilepsy. *Journal of Child Neurology* 2004;19:53–58.

Bien CG, Widman G, Urbach H, Sassen R, Kuczaty S, Wiestler OD, Schramm J, Elger CE. The natural history of Rasmussen's encephalitis. *Brain* 2002;125:1751–1759.

Browne TR, Holmes GL. *Handbook of epilepsy* (3rd ed.). Philadelphia: Lippincott Williams & Wilkins, 2004.

Callenbach PM, Geerts AT, Arts WF, Donselaar CA, Peters ACB, Stroink H. Brouwer OF. Family occurrence of epilepsy in children with newly diagnosed multiple seizures: Dutch Study of Epilepsy in Childhood. *Epilepsia* 1998;39:331–336.

Caplan R, Curtiss S, Chugani HT, Vinters HV. Pediatric Rasmussen encephalitis: social communication, language, PET and pathology before and after hemispherectomy. *Brain and Cognition* 1996;32:45–66.

Cendes F. Febrile seizures and mesial temporal sclerosis. *Current Opinion in Neurology* 2004;17:161–164.

Commission on Classification and Terminology of the International League Against Epilepsy. Proposal for revised classification of epilepsy and epileptic syndromes. *Epilepsia* 1985;26:268–278.

Commission on Classification and Terminology of the International League Against Epilepsy. Proposal for revised classification of epilepsies and epileptic syndromes. *Epilepsia* 1989;30:389–399.

DeLorenzo RJ, Towne AR, Pellock JM, Ko D. Status epilepticus in children, adults, and the elderly. *Epilepsia* 1992;33 (Suppl 4):15–25.

DeLorenzo RJ, Pellock JM, Towne AR, Boggs JG. Epidemiology of status epilepticus. *Journal of Clinical Neurophysiology* 1995;12:316–325.

Duchowny M. Febrile seizures. In Wyllie E, Gupta A, Lachhwani DK (eds.), *The treatment of epilepsy: Principles and practice* (4th ed.). Philadelphia: Lippincott Williams & Wilkins, 2006:511–520.

Engel J Jr. A proposed diagnostic scheme for people with epileptic seizures and with epilepsy: A report of the International League Against Epilepsy Task Force on Classification and Terminology. *Epilepsia* 2001;42:796–803.

Elger CE. Semiology of temporal lobe seizures. In Oxbury JM, Polkey CE, Duchowny M. (eds.), *Intractable focal epilepsy*. London: W.B. Saunders, 2000:63–68.

Erba G. Shedding light on photosensitivity, one of epilepsy's most complex conditions. http://www.epilepsyfoundation.org/epilepsyusa/photosensitivity 20060306.cfm, 2006.

Ficker DM, So EL, Shen WK, Annegers JF, O'Brien PC, Cascino GD, Belau PG. Population-based study of incidence of sudden unexplained death in epilepsy. *Neurology* 1998;51:1270–1274.

Gil-Nagel A, Risinger MW. Ictal semiology in hippocampal versus extrahippocampal temporal lobe epilepsy. *Brain* 1997;120:183–192.

Hayashi T, Ichiyama T, Nishikawa M, Isumi H, Furukawa S. Pocket Monsters, a popular television cartoon, attacks Japanese children. *Annals of Neurology* 1998;44:427–428.

Hirose S, Mitsudome A, Okada M., Kaneko S. Epilepsy Study Group, Japan. Genetics of idiopathic epilepsies. *Epilepsia* 2005;46(Suppl. 1):38–43.

Ishida S, Yamashita Y, Matsuishi T, Ohshima M, Ohshima H, Kato H, Maeda H. Photosensitive seizures provoked while viewing "Pocket Monsters," a made-for-television animation program in Japan. *Epilepsia* 1998;39:1340–1344.

Jallon P, Loiseau P, Loiseau J. Newly diagnosed unprovoked epileptic seizures: presentation at diagnosis in CAROLE study. *Epilepsia* 2001;42:464–475.

Kossoff EH, Boatman D, Freeman JM. Landau-Kleffner syndrome responsive to levetiracetam. *Epilepsy & Behavior* 2003;4:571–575.

Koutroumanidis M, Koepp MJ, Richardson MP, Camfield C, Agathonikou A, Ried S, Papadimitriou A, Plant GT, Duncan JS, Panayiotopoulos CP. The variants of reading epilepsy: A clinical and video EEG study of 17 patients with reading-induced seizures. *Brain* 1998;121:1409–1427.

Manford M, Hart YM, Sander J, Shorvon SD. National General Practice Study of Epilepsy (NGPSE): partial seizure patterns in a general population. *Neurology* 1992;42:1911–1917.

Neuspiel DR, Kuller LH. Sudden and unexpected natural death in childhood and adolescence. *Journal of the American Medical Association* 1985;254:1321–1325.

Neville B, Cross JH. Continuous spike and wave of slow sleep and Landau-Kleffner syndrome. In Wyllie E, Gupta A, Lachhwani DK (eds.), *The treatment of epilepsy: Principles and practice* (4th ed.). Philadelphia: Lippincott Williams & Wilkins, 2006:455–462.

Panayiotopoulos CF. *The epilepsies: seizures, syndromes and management.* Chipping Norton, Oxfordshire, UK: Bladon Medical Publishing, 2005.

Radhakrishnan K, Silbert PL, Klass DW. Reading epilepsy. An appraisal of 20 patients diagnosed at the Mayo Clinic, Rochester, Minnesota, between 1949 and 1989, and delineation of the epileptic syndrome. *Brain,* 1995;118(Pt 1): 75–89.

Rasmussen T. Surgery for epilepsy arising in regions other than the temporal and frontal lobes. In Purpura DP, Penry JK, Walter RD (eds.), *Neurosurgical management of the epilepsies.* New York: Raven Press, 1975:207–226.

Salanova V, Andermann F, Rasmussen T, Olivier A, Quesney LF. Parietal lobe epilepsy: clinical manifestations and outcome in 82 patients treated surgically between 1929 and 1988. *Brain* 1995;118:607–627.

Scheffer IE, Bhatia KP, Lopes-Condes I, Fish DR, Marsden CD, Andermann E, Andermann F, et al. Autosomal dominant nocturnal frontal lobe epilepsy: A distinctive clinical disorder. *Brain* 1995;118 (Pt. 1): 61–73.

Sharma A, Cameron D. Reasons to consider a plasma screen television – photosensitive epilepsy. *Epilepsia* 2007;48(10):1528.

So EL, Bainbridge J, Buchhalter JR, Donalty J, Donner EJ, Finucane A, Graves NM, Hirsch LJ, Montouris GD, Temkin NR, Wiebe S, Sierzant TL. Report of the American Epilepsy Society and the Epilepsy Foundation joint task force on sudden unexplained death in epilepsy. *Epilepsia* 2009;50:917–922.

Verity CM, Butler NR, Golding J. Febrile convulsions in a national cohort followed up from birth. I. Prevalence and recurrence in the first five years of life. *British Medical Journal* 1985;290:1307–1310.

Verity CM, Greenwood R, Golding J. Long-term intellectual and behavioral outcomes of children with febrile convulsions. *New England Journal of Medicine* 1998;338:1723–1728.

Wallace SJ, Farrell K. (eds.), *Epilepsy in children* (2nd edition). New York: Oxford University Press, 2004.

Williamson PD, Spencer D, Spencer S, Novelly R, Mattson R. Complex partial seizures of extra temporal origin. *Annals of Neurology* 1985;18: 497–504.

Williamson PD, Thadani VM, Darcey TM, Spencer DD, Spencer SS, Mattson RH. Occipital lobe epilepsy: clinical characteristics, seizure spread patterns, and results of surgery. *Annals of Neurology* 1992;31:3–13.

Williamson PD, Engel J, Munari C. Anatomic classification of localization related epilepsies. In Engel J Jr., Pedley TA (eds.), *Epilepsy: a comprehensive textbook*. New York: Lippincott-Raven, 1997:2405–2416.

Working Group on Status Epilepticus, Epilepsy Foundation of America. Treatment of convulsive status epilepticus. *Epilepsia* 1993;270:854–859.

Wyllie E, Gupta A, Lachhwani DK (eds.), *The treatment of epilepsy: Principles and practice* (4th ed.). Philadelphia: Lippincott Williams, & Wilkins, 2006.

Zifkin B, Andermann F, Rowan AJ (eds.). *Reflex epilepsies and reflex seizures*. New York: Lipppincot-Raven Press, 1998.

Zifkin B, Andermann F. Epilepsy with reflex seizures. In: Wyllie E, Gupta A, Lachhwani DK, eds. *The treatment of epilepsy: Principles and practice* (4th ed.). Philadelphia: Lippincott Williams & Wilkins, 2006:463–475.

Chapter 4

American Clinical Neurophysiology Society. Guidelines in electroencephalography, evoked potentials and polysomnography. *Journal of Clinical Neurophysiology* 1994;11:1–147.

Anslow P, Oxbury J. Diagnostic radiology. In Oxbury JM, Polkey CE, Duchowny M (eds.), *Intractable focal epilepsy*. London: WB Saunders, 2000:297–309.

Binnie CD, Prior PF. Electroencephalography. *Journal of Neurology, Neurosurgery & Psychiatry* 1994;57:1308–1319.

Blume WT, Kaibara M. *Atlas of adult electroencephalography*. New York: Raven Press, 1995.

Blume WT, Kaibara M. *Atlas of pediatric electroencephalography* (2nd ed.). Philadelphia: Lippincott-Raven, 1999.

Chabolla DR, Cascino GD. Application of electroencephalography in the diagnosis of epilepsy. In Wyllie E, Gupta A, Lachhwani DK (eds.), *The treatment of epilepsy: Principles and practice* (4th ed.). Philadelphia: Lippincott Williams & Wilkins, 2006:169–182.

Commission on Neuroimaging of the International League Against Epilepsy. Recommendations for neuroimaging of patients with epilepsy. *Epilepsia* 1997;38:1255–1256.

Cook M, Stevens JM. Imaging in epilepsy. In: A Hopkins, S Shorvon, G Cascino (eds.), *Epilepsy*. London: Chapman & Hall, 1995:143–169.

Homan RW, Herman J, Purdy P. Cerebral location of international 10–20 system electrode placement. *Electroencephalography & Clinical Neurophysiology* 1987;66:376–382.

Kennett RP. Neurophysiologic investigation of adults. In Oxbury JM, Polkey CE, Duchowny M (eds.), *Intractable focal epilepsy*. London: WB Saunders, 2000:333–362.

Luders H, Noachtar S. *Atlas and classification of electroencephalography*. Philadelphia: WB Saunders, 2000.

Niedermeyer E, Lopez da Silva FH (eds.). *Electroencephalography: basic principles, clinical applications and related fields* (3rd ed.). Baltimore, MD: Urban & Schwarzenberg, 1993.

Pillai J, Sperling MR. Interictal EEG and the diagnosis of epilepsy. *Epilepsia* 2006;47(Suppl 1):14–22.

Salinsky M, Kanter R, Dasheiff RM. Effectiveness of multiple EEGs in supporting the diagnosis of epilepsy: An operational curve. *Epilepsia* 1987;28:331–334.

Chapter 5

Aikia M, Jutila L, Salmenpera T, Mervaala E, Kålviåinen R. Long-term effects of tiagabine monotherapy on cognition and mood in adult patients with chronic partial epilepsy. *Epilepsy & Behavior* 2006;8:750–755.

Berroya AG, McIntyre J, Webster R, Lah S, Lawson J, Bleasel AF, Bye AME. Speech and language deterioration in benign rolandic epilepsy. *Journal of Child Neurology* 2004;19:53–58.

Beyreuther BK, Freitag J, Heers C, Krebsfanger N, Scharfenecker U, Stohr T. Lacosamide: a review of preclinical properties. *CNS Drug Reviews* 2007;13:21–42.

Camfield CS, Camfield PR. Initiating drug therapy. In Wyllie E (ed.), *The treatment of epilepsy: Principles and practice* (2nd ed.). Philadelphia: Lippincott Williams & Wilkins, 1997:763–770.

Fastenau PS, Johnson CS, Perkins SM, Byars AW, deGrauw TJ, Austin JK, Dunn DW. Neuropsychological status at seizure onset in children: risk factors for early cognitive deficits. *Neurology* 2009;73:526–534.

Faught E, Duh MS, Weiner JR, Guerin A, Cunnington MC. Nonadherence to antiepileptic drugs and increased mortality: findings from the RANSOM study. *Neurology* 2008;70:1572–1578.

French JA, Kanner AM, Bautista J, Abou-Khalil B, Browne T, Harden CL, Theodore WH et al. Efficacy and tolerability of the new antiepileptic drugs I: treatment of new onset epilepsy: report of the Therapeutics and Technology Assessment Subcommittee and Quality Standards Subcommittee of the American Academy of Neurology and American Epilepsy Society. *Neurology* 2004;62:1252–1260.

Gates JR. Nonepileptic seizures: classification, coexistence with epilepsy, diagnosis, therapeutic approaches, and consensus. *Epilepsy & Behavior* 2002;3:28–33.

Ghaemi SN, Hsu DJ, Thase ME, Wisniewski SR, Nierenberg AA, Miyahara S, Sachs G. Pharmacological treatment patterns at study entry for the first 500 STEP-BD participants. *Psychiatric Services* 2006;57:660–665.

Harden CL, Pennell PB, Koppel BS, Hovinga CA, Gidal B, Meador KJ, Hopp J et al. Practice Parameter update: Management issues for women with epilepsy – focus on pregnancy (and evidenced-based review): Vitamin K, folic acid, blood levels, and breastfeeding. Report of the Quality Standards Subcommittee and Therapeutics and Technology Assessment Subcommittee of the American Academy of Neurology and American Epilepsy Society. *Epilepsia* 2009;50:1247–1255.

Hauser WA, Anderson VE, Loewenson RB, McRobert SM. Seizure recurrence after a first unprovoked seizure. *New England Journal of Medicine* 1982;307:522–528.

Hindmarch I, Trick L, Ridout F. A double-blind, placebo- and positive-internal-controlled investigation of the cognitive and psychomotor profile of pregabalin in healthy volunteers. *Psychopharmacology* 2005;183:133–143.

Holmes LB, Wyszynski DF, Lieberman E. The AED (Antiepileptic Drug) Pregnancy Registry: a 6-year experience. *Archives of Neurology* 2004;61:673–678.

Kaaja E, Kaaja R, Hiilesmaa V. Major malformations in offspring of women with epilepsy. *Neurology* 2003;60:575–579.

Kaddurah A, Moorjani BI. Benign childhood epilepsy. *eMedicine* 2009 March. http://www.emedicine.com/neuro/topic641.htm. Accessed May 29, 2009.

Kossoff EH, Laux LC, Blackford R, Morrison PF, Pyzik PL, Hamdy RM, Turner Z, Nordli DR Jr. When do seizures usually improve with the ketogenic diet? *Epilepsia* 2008;49:329–333.

Loring DW, Marino S, Meador KJ. Neuropsychological and behavioral effects of antiepileptic drugs. *Neuropsychological Review* 2007;17(4):413–425.

Mattson RH, Cramer JA, Collins JF, Smith DB, Delgado-Escueta AV, Browne TR, Williamson PD et al. Comparison of carbamazepine, phenobarbtial, phenytoin, and primidone in partial and secondarily generalized tonic-clonic seizures. *New England Journal of Medicine* 1985;313:145–151.

Mattson RH, Cramer JA. The choice of antiepileptic drugs in focal epilepsy. In Wyllie E (ed.), *The treatment of epilepsy: Principles and practice* (2nd ed.). Philadelphia: Lippincott Williams & Wilkins, 1997:771–778.

Mattson RH, Cramer JA, Collins JF. A comparison of valproate with carbamazepine for the treatment of complex partial seizures and secondarily generalized tonic-clonic seizures in adults. The Department of Veterans Affairs Epilepsy Cooperative Study No. 264 Group. *New England Journal of Medicine* 1992;327:765–771.

Meador KJ, Baker GA, Browning N, Clayton-Smith J, Combs-Cantrell DT, Cohen M, Kalayjian LA et al. Cognitive function at 3 years of age after fetal exposure to antiepileptic drugs. *New England Journal of Medicine* 2009;360(16):1597–1605.

Meador KJ, Browning N, Cohen MJ. Cognitive outcomes at 2 years old in children of women with epilepsy differ as a function of in utero antiepileptic drug. *Neurology* 2007;68(Suppl 1):A337.

Meader KJ, Gevins A, Loring DW, McEvoy LK, Ray PG, Smith ME, Motamedi GK, Evans BS, Baum C. Neuropsychological and neurophysiologic effects of carbamazepine and levetiracetam. *Neurology* 2007;69:2076–2084.

Meador KJ, Loring DW, Ray PG, Murro AM, King DW, Nichols ME, Deer EM, Goff WT. Differential cognitive effects of carbamazepine and gabapentin. *Epilepsia* 1999;40:1279–1285.

Melvin CL, Carey TS, Goodman F, Oldham JM, Williams Jr. JW, Ranney LM. Effectiveness of antiepileptic drugs for the treatment of bipolar disorder: findings from a systematic review. *Journal of Psychiatric Practice* 2008;14(Suppl 1):9–14.

Miyamoto T, Kohsaka M, Koyama T. Psychotic episodes during zonisamide treatment. *Seizure* 2000;9:65–70.

Morrell M. Hormones, catamenial epilepsy, and reproductive and bone health in epilepsy. In Wyllie E, Gupta A, Lachhwani DK (eds.), *The treatment of epilepsy: Principles and practice* (4th ed.). Philadelphia: Lippincott Williams & Wilkins, 2006:695–703.

Morrow JI, Russell AJC, Irwin B. The safety of antiepileptic drugs in pregnancy: results of the U.K. epilepsy and pregnancy register. *Epilepsia* 2004;45(Suppl 3):57.

Mula M, Trimble MR, Thompson P, Sander JWAS. Topiramate and word-finding difficulties in patients with epilepsy. *Neurology* 2003;60:1104–1107.

Nadkarni S, LaJoie J, Devinsky O. Current treatments of epilepsy. *Neurology* 2005;64(Suppl 3):S2–S11.

Nordli DR Jr., DeVIvo DC. The ketogenic diet. In Wyllie E, Gupta A, Lachhwani DK (eds.), *The treatment of epilepsy: Principles and practice* (4th ed.). Philadelphia: Lippincott Williams & Wilkins, 2006:961–967.

Panayiotopoulos CF. *The epilepsies: seizures, syndromes and management.* Chipping Norton, Oxfordshire, UK: Bladon Medical Publishing, 2005.

Patsalos PN, Berry DJ, Bourgeois BFD, Cloyd JC, Glauser TA, Johannessen SI, Leppik IE, Tomson T, Perucca E. Antiepileptic drugs-best practice guidelines for therapeutic drug monitoring: A position paper by the subcommission on therapeutic drug monitoring, ILAE Commission on Therapeutic Strategies. *Epilepsia* 2008;49:1239–1276.

Pennell P. Hormones, seizures and lamotrigine: oh, my! *Epilepsy Currents* 2008;8(1):8–10.

Pressler RM, Binnie CD, Coleshill SG, Chorley GA, Robinson RO. Effect of lamotrigine on cognition in children with epilepsy. *Neurology* 2006;66:1495–1499.

Rosenfeld WE, Bramley TJ, Meyer KL. Patient compliance with topiramate vs. other antiepileptic drugs: a claims database analysis. *Epilepsia* 2004;45:238.

Rowan AJ, Ramsay RE, Collins JF, Pryor F, Boardman KD, Uthman BM, Spitz M et al. New onset geriatric epilepsy: A randomized study of gabapentin, lamotrigine, and carbamazepine. *Neurology* 2005;64:1868–1873.

Sankar R. Initial treatment of epilepsy with antiepileptic drugs: pediatric issues. *Neurology* 2004;63(Suppl 4):S30–S39.

Thiele EA. Assessing the efficacy of antiepileptic treatments: the ketogenic diet. *Epilepsia* 2003;44(Suppl 7):26–29.

Thompson PJ, Baxendale SA, Duncan JS, Sander JWAS. Effects of topiramate on cognitive function. *Journal of Neurology, Neurosurgery & Psychiatry* 2000;69:636–641.

U.S. Food and Drug Administration. http://www.fda.gov/cder/drug/InfoSheets/HCP/antiepilepticsHCP.htm. 2008.

Viinikainen K, Heinonen S, Eriksson K, Kälviäinen R. [NB7]Fertility in women with active epilepsy. *Neurology* 2007; 69:2107–2108.

Wheless JW, Clarke DF, Carpenter D. Treatment of pediatric epilepsy: expert opinion. *Journal of Child Neurology* 2005;20(Suppl 1):S1–S56.

Chapter 6

Aarts JH, Binnie CD, Smith AM, Wilkins AJ. Selective cognitive impairment during focal and generalized epileptiform EEG activity. *Brain* 1984;107:293–308.

Aldenkamp AP, Arends J. Effects of epileptiform EEG discharges on cognitive function: Is the concept of "transient cognitive impairment" still valid? *Epilepsy & Behavior* 2004;5:S25–S34.

American Academy of Neurology, American Epilepsy Society, Epilepsy Foundation of America. Consensus statements, sample statutory provisions, and model regulations regarding driver licensing and epilepsy. *Epilepsia* 1994;35:696–705.

Baker GA, Taylor J. Neuropsychologic effects of seizures. In Schachter SC, Holmes GL, Kasteleijn-Nolst Trenité, DGA (eds.), *Behavioral aspects of epilepsy: principles & practice*. New York: Demos, 2008:93–98.

Baron IS. *Neuropsychological evaluation of the child*. New York: Oxford University Press, 2004.

Bernhardt BC, Worsley KJ, Kim H, Evans AC, Bernasconi A, Bernasconi N. Longitudinal and cross-sectional analysis of atrophy in pharmacoresistant temporal lobe epilepsy. *Neurology* 2009;72:1747–1754.

Buck D, Smith M, Appleton R, Baker GA, Jacoby A. The development and validation of the Epilepsy and Learning Disabilities Quality of Life (ELDQOL) scale. *Epilepsy & Behavior* 2007;10:38–43.

Buelow JM, Ferrans CE. Quality of life in epilepsy. In Ettinger AB, Kanner AM (eds.), *Psychiatric issues in epilepsy: a practical guide to diagnosis and treatment*. Philadelphia: Lippincott Williams & Wilkins, 2001:307–318.

Camfield C, Breau L, Camfield P. Impact of pediatric epilepsy on the family: a new scale for clinical and research use. *Epilepsia* 2001;42:104–112.

Cascino GD. Temporal lobe epilepsy is a progressive neurological disorder: time means neurons! *Neurology* 2009;72:1718–1719.

Chaplin JE, Yepez R, Shorovom S, Floyd M. A quantitative approach to measuring social effect of epilepsy. *Neuroepidemiology* 1990;9:151–158.

Cramer JA, Westbrook L, Devinski O, Perrine K, Glassman MB, Camfield C. Development of the Quality of Life in Epilepsy Inventory for Adolescents: the QOLIE-AD-48. *Epilepsia* 1999;40:1114–1121.

Devinski O, Gershengorn J, Brown E, Perrine K, Vazquez B, Luciano, D. Frontal functions in juvenile myoclonic epilepsy. *Neuropsychiatry, Neuropsychology & Behavioral Neurology* 1997;10:243–246.

Devinski O, Vickrey BG, Cramer J, Perrine K, Hermann B, Meador KJ, Hays RD. Development of the quality of life in epilepsy inventory. *Epilepsia* 1995;36:1089–1104.

Devinski O, Westbrook L, Cramer J, Glassman M, Perrine K, Camfield C. Risk factors for poor health-related quality of life for adolescents with epilepsy. *Epilepsia* 1999;40:1715–1720.

Dodrill CB. Correlates of generalized tonic-clonic seizures with intellectual, neuropsychological, emotional, and social function in patients with epilepsy. *Epilepsia* 1986;27:399–411.

Dodrill CB. Neuropsychological effects of seizures. *Epilepsy & Behavior* 2004;5:S21–S24.

Dodrill CB, Batzel LW, Queisser HR, Temkin NR. An objective method for the assessment of psychological and social problems among epileptics. *Epilepsia* 1980;21:123–135.

Dunn DW, Austin JK, Harezlak J, Ambrosius WT. ADHD and epilepsy in childhood. *Developmental Medicine & Child Neurology* 2003;45:50–54.

Fastenau PS, Johnson CS, Perkins SM, Byars AW, deGrauw TJ, Austin JK, Dunn DW. Neuropsychological status at seizure onset in children: risk factors for early cognitive deficits. *Neurology* 2009;73:526–534.

Fastenau PS, Shen J, Dunn DW, Austin JK. Academic underachievement among children with epilepsy: proportion exceeding psychometric criteria for learning disabilities and associated risk factors. *Journal of Learning Disabilities* 2008;41:195–207.

Hansotia P, Broste SK. The effect of epilepsy or diabetes on the risk of automobile accidents. *New England Journal of Medicine* 1991;324:22–26.

Hecaen H, Albert ML. *Human neuropsychology*. New York: J Wiley, 1978.

Hermann BP, Jones J, Dabbs K, Allen CA, Sheth R, Fine J, McMillan A, Seidenberg M. The frequency, complications, and aetiology of ADHD in new onset paediatric epilepsy. *Brain* 2007;130(Pt 12):3135–3148.

Hermann BP, Wyler AR, Richey ET. Wisconsin Card Sorting Test performance in patients with complex partial seizures of temporal-lobe origin. *Journal of Clinical & Experimental Neuropsychology* 1988;10:467–476.

Jokeit H, Ebner A. Long term effects of refractory temporal lobe epilepsy on cognitive abilities: a cross-sectional study. *Journal of Neurology, Neurosurgery & Psychiatry* 1999;67:44–50.

Jokeit H, Schachter M. Neuropsychological aspects of type of epilepsy and etiological factors in adults. *Epilepsy & Behavior* 2004;5:S14–S20.

Kavros PM, Clarke T, Strug LJ, Halperin JM, Dorta NJ, Pal DK. Attention impairment in rolandic epilepsy: systematic review. *Epilepsia* 2008;49:1570–1580.

Kockelmann E, Eldger CE, Helmsteadter C. Cognitive profile of topiramate as compared with lamotrigine in epilepsy patients on antiepileptic drug (AED) polytherapy: Relationships to blood serum levels and comedication. *Epilepsy & Behavior* 2004;5:716–721.

Krumholz A. Driving issues in epilepsy: past, present, and future. *Epilepsy Currents: The Journal of the American Epilepsy Society* 2009;9:31–35.

Lee GP, Loring DW, Thompson JL. Construct validity of material-specific memory measures following unilateral temporal lobectomy. *Psychological Assessment: A Journal of Consulting & Clinical Psychology* 1989;1:192–197.

Loring DW, Meador KJ, Lee GP. Effects of temporal lobectomy on generative fluency and other language functions. *Archives of Clinical Neuropsychology* 1994;9:229–238.

Mandelbaum DE, Burack GD. The effect of seizure type and medication on cognitive and behavioral functioning in children with idiopathic epilepsy. *Developmental Medicine & Child Neurology* 1997;39:731–735.

Meador KJ. Cognitive outcomes and predictive factors in epilepsy. *Neurology* 2002;58(Suppl 5):S21–S26.

Mesulam M (ed.). *Principles of behavioral and cognitive neurology*. New York: Oxford University Press, 2000.

Ojemann GA, Ojemann JG, Lettich E, Berger M. Cortical language localization in left dominant hemisphere: an electrical stimulation mapping investigation in 117 patients. *Journal of Neurosurgery* 1989;71:316–326.

Perrine K. A new quality-of-life inventory for epilepsy patients: interim results. *Epilepsia* 1993;34(Suppl. 4):S28–S33.

Perrine K, Gershengorn J, Brown ER. Interictal neuropsychological function in epilepsy. In Devinsky O, Theodore WH (eds.), *Epilepsy and behavior*. New York: Wiley-Liss, 1991:181–193.

Piazzini A, Turner K, Chifari R, Morabito A, Canger R, Canevini MP. Attention and psychomotor speed decline in patients with temporal lobe epilepsy: a longitudinal study. *Epilepsy Research* 2006;72:89–96.

Sabaz M, Lawson JA, Cairns DR, Duchowny MS, Resnick TJ, Dean PM, Bye AME. Validation of the Quality of Life in Childhood Epilepsy questionnaire in American epilepsy patients. *Epilepsy & Behavior* 2003;4:680–691.

Sabaz M, Lawson JA, Cairns DR, Duchowny MS, Resnick TJ, Dean PM, Bleasel AF., Bye A. The impact of epilepsy surgery on quality of life in children. *Neurology* 2006;66:557–561.

Sanchez-Carpintero R, Neville BGR. Attentional ability in children with epilepsy. *Epilepsia* 2003;44:1340–1349.

Sankar R. Initial treatment of epilepsy with antiepileptic drugs: pediatric issues. *Neurology* 2004;63(Suppl 4):S30–S39.

Stores G, Hart J. Reading skills of children with generalized or focal epilepsy attending ordinary school. *Developmental Medicine & Child Neurology* 1976;18:705–716.

Vickrey BG. A procedure for developing a quality-of-life measure for epilepsy surgery patients. *Epilepsia* 1993;34(Suppl. 4):S22–S27.

Vingerhoets G. Cognition. In Schachter SC, Holmes GL, Kasteleijn-Nolst Trenité, DGA (eds.), *Behavioral aspects of epilepsy: principles & practice*. New York: Demos, 2008:155–163.

Zacks J, Pascale M, Vettel J, Ojemann J. Functional reorganization of spatial transformations after a parietal lesion. *Neurology* 2004;63:287–292.

Zelnik N, Sa'adi L, Silman-Stolar Z, Goikhman I. Seizure control and educational outcome in childhood-onset epilepsy. *Journal of Child Neurology* 2001;16:820–824.

Chapter 7

Ballenger JC, Burrows GD, DuPont RL, Lesser IM, Noyes R Jr., Pecknold JC, Rifkin A, Swinson RP. Alprazolam in panic disorder and agoraphobia: results from a multicenter trial. *Archives of General Psychiatry* 1988;45:413–422.

Barry JJ, Lembke A, Huynh N. Affective disorders in epilepsy. In Ettinger AB & Kanner AM (eds.), *Psychiatric issues in epilepsy: a practical guide to diagnosis and treatment*. Philadelphia: Lippincott Williams & Wilkins, 2001:45–71.

Bear DM, Fedio P. Quantitative analysis of interictal behavior in temporal lobe epilepsy. *Archives of Neurology* 1977;34:454–467.

Ghaemi SN, Hsu DJ, Thase ME, Wisniewski SR, Nierenberg AA, Miyahara S, Sachs G. Pharmacological treatment patterns at study entry for the first 500 STEP-BD participants. *Psychiatric Services* 2006;57:660–665.

Harden CL, Goldstein MA. Mood disorders in patients with epilepsy: epidemiology and management. *CNS Drugs* 2002;16:291–302.

Harris EC, Barraclough B. Suicide as an outcome for mental disorders: a meta-analysis. *British Journal of Psychiatry* 1997;170:205–228.

Hermann BP, Whitman S. Neurobiological, psychosocial, and pharmacological factors underlying interictal psychopathology in epilepsy. In Smith D, Treiman D & Trimble M (eds.), *Advances in Neurology, Volume 55: Neurobehavioral problems in epilepsy*. New York: Raven Press, 1991:439–452.

Kanner AM, Kozak AM, Frey M. The use of sertraline in patients with epilepsy: is it safe? *Epilepsy & Behavior* 2000;1:100–105.

Krishnamoorthy ES, Trimble MR, Blumer D. The classification of neuropsychiatric disorders in epilepsy: a proposal by the ILAE Commission on Psychobiology of Epilepsy. *Epilepsy & Behavior* 2007;10:349–353.

Lopez-Rodriquez F, Altshuler L, Kay J, Delarhim S, Mendez M, Engel J. Personality disorders among medically refractory epileptic patients. *Journal of Neuropsychiatry & Clinical Neuroscience* 1999;11:464–469.

Mendez MF, Grau R, Doss RC, Taylor JL. Schizophrenia in epilepsy: seizure and psychosis variables. *Neurology* 1993;43:1073–1077.

Monaco F, Cavanna A, Magli E, Barbagli D, Collimedaglia L, Cantello R, Mula M. Obsessionality, obsessive-compulsive disorder, and temporal lobe epilepsy. *Epilepsy & Behavior* 2005;7:491–496.

Mungas D. Interictal behavior abnormality in temporal lobe epilepsy: a specific syndrome or nonspecific psychopathology. *Archives of General Psychiatry* 1982;39:108–111.

Schachter SC. Psychiatric comorbidity of epilepsy. In Wyllie E, Gupta A, Lachhwani DK (eds.), *The treatment of epilepsy: Principles and practice* (4th ed.). Philadelphia: Lippincott Williams & Wilkins, 2006:1197–1200.

Schmitz B, Trimble M. Psychosis and forced normalization. In Schachter SC, Holmes GL, Kasteleijn-Nolst Trenité, DGA (eds.), *Behavioral aspects of epilepsy: principles & practice*. New York: Demos, 2008:235–243.

Schondienst M & Reuber M. Epilepsy and anxiety. In Schachter SC, Holmes GL, Kasteleijn-Nolst Trenité, DGA (eds.), *Behavioral aspects of epilepsy: principles & practice*. New York: Demos, 2008:219–226.

Scicutella A. Anxiety disorders in epilepsy. In Ettinger AB & Kanner AM (eds.), *Psychiatric issues in epilepsy: a practical guide to diagnosis and treatment*. Philadelphia: Lippincott Williams & Wilkins, 2001:95–109.

Slater E, Beard AW, Glithero E. The schizophrenia-like psychosis of epilepsy. *British Journal of Psychiatry* 1963;109:95–150.

Swinkels WAM, Duijsens IJ, Spinhoven Ph. Personality disorder traits in patients with epilepsy. *Seizure* 2003;12:587–594.

Swinkels WAM, Kuyk J, van Dyck R, Spinhoven Ph. Psychiatric comorbidity in epilepsy. *Epilepsy & Behavior* 2005;7:37–50.

Torta R, Keller R. Behavioral, psychotic, and anxiety disorders in epilepsy: etiology, clinical features, and therapeutic implications. *Epilepsia* 1999;40 (Suppl 10):S2–S20.

Trimble MR. Interictal psychosis of epilepsy. In Smith D, Treiman D, Trimble M (eds.), *Advances in neurology: neurobehavioral problems in epilepsy*. New York: Raven Press, 1991:143–152.

Tucker GJ, McDavid J. Neuropsychiatric aspects of seizure disorder. In Ydofsky SC, Hales RE (eds.), *Textbook of neuropsychiatry* (3rd ed.). Washington, DC: American Psychiatric Press, 1997:561–582.

Waxman SG, Geschwind N. The interictal behavioral syndrome in temporal lobe epilepsy. *Archives of General Psychiatry* 1975;32:1580–1586.

Wells KB, Golding JM, Burnam MA. Psychiatric disorder in a sample of the general population with and without chronic medical conditions. *American Journal of Psychiatry* 1988;145:976–981.

Chapter 8

Bell WL, Park YD, Thompson EA, Radtke RA. Ictal cognitive assessment of partial seizures and pseudoseizures. *Archives of Neurology* 1998;55:1456–1459.

Binder LM, Kindermann SS, Heaton RK, Salinsky MC. Neuropsychological impairment in patients with nonepileptic seizures. *Archives of Clinical Neuropsychology* 1998;13:513–522.

Binder LM, Salinsky MC. Psychogenic nonepileptic seizures. *Neuropsychology Review* 2007;17:405–412.

Bortz JJ, Prigatano GP, Blum D, Fisher RS. Differential response characteristics in nonepileptic and epileptic seizure patients on a test of verbal learning and memory. *Neurology* 1995;45:2029–2034.

Cragar DE, Berry DTR, Fakhoury TA, Cibula JE, et al. A review of diagnostic techniques in the differential diagnosis of epileptic and nonepileptic seizures. *Neuropsychology Review* 2002;12:31–64.

Cragar DE, Berry DT, Schmitt FA, Fakhoury TA. Cluster analysis of normal personality traits in patients with psychogenic nonepileptic seizures. *Epilepsy & Behavior* 2005;6:593–600.

Dodrill CB, Wilkus RJ, Batzel LW. The MMPI as a diagnostic tool in non-epileptic seizures. In Rowan AJ, Gates JR (eds.), *Non-epileptic seizures*. Boston: Butterworth-Heinemann, 1993:211–219.

Drane DL, Williamson DJ, Stroup ES, Holmes MD, Jung M, Koerner E, Chaytor N, Wilensky AJ, Miller JW. Cognitive impairment is not equal in patients with epileptic and psychogenic nonepileptic seizures. *Epilepsia* 2006;47:1879–1886.

Goldstein LH, Deale AC, Mitchell-O'Malley SJ, Toone BK, Mellers JDC. An evaluation of cognitive behavioral therapy as a treatment for dissociative seizures. *Cognitive & Behavioral Neurology* 2004;17:41–49.

LaFrance WC Jr., Devinski O. The treatment of nonepileptic seizures: historical perspectives and future directions. *Epilepsia* 2004;45(Suppl 2), 15–21.

Locke, DEC, Berry, DTR, Fakhoury, TA, Schmitt, FA. Relationship of indicators of neuropathology, psychopathology, and effort to neuropsychological results in patients with epilepsy or psychogenic non-epileptic seizures. *Journal of Clinical and Experimental Neuropsychology* 2006;28:325–340.

Reuber M, Howlett S, Kemp S. Psychologic treatment for patients with psychogenic nonepileptic seizures. *Expert Reviews in Neurotherapy* 2005;5:737–752.

Reuber M, Elger CE. Psychogenic nonepileptic seizures: an overview. In Schachter SC, Holmes GL, Kasteleijn-Nolst Trenité, DGA (eds.), *Behavioral aspects of epilepsy: principles & practice*. New York: Demos, 2008:411–419.

Syed TU, Azozullah AM, Loparo KL, Jamasebi R, Suciu GP, Griffin C, Mani R, Syed I, Loddenkemper T, Alexopoulos AV. A self-administered screening instrument for psychogenic nonepileptic seizures. *Neurology* 2009;72:1646–1652.

Westbrook LE, Devinski O, Geocadin R. Nonepileptic seizures after head injury. *Epilepsia* 1998;39:978–982.

Chapter 9

Busch RM, Frazier TW, Haggerty KA, Kubu CS. Utility of the Boston Naming Test in predicting ultimate side of surgery in patients with medically intractable temporal lobe epilepsy. *Epilepsia* 2005;46:1773–1779.

Busch RM, Frazier TW, Lampietro MC, Chapin JS, Kubu CS. Clinical utility of the Boston Naming Test in predicting ultimate side of surgery in patients with medically intractable temporal lobe epilepsy: a double cross-validation study. *Epilepsia* 2009;50:1270–1273.

Chelune GJ. The role of neuropsychological assessment in the presurgical evaluation of the epilepsy surgery candidate. In Wyler AR, Hermann BP (eds.), *The surgical management of epilepsy*. Boston: Butterworth-Heinemann, 1994:78–89.

Fargo JD, Schefft BK, Szaflarski JP, Howe SR, Hwa-Shain Y, Privitera MD. Accuracy of clinical neuropsychological versus statistical prediction in the classification of seizure types. *The Clinical Neuropsychologist* 2008;22:181–194.

Helmstaedter C. Neuropsychological aspects of epilepsy surgery. *Epilepsy & Behavior* 2004;5:S45–S55.

Helmstaedter C, Kurthen M, Lux S, Johansen K, Quiske A, Schramm J, Elger CE. Chronic epilepsy and cognition: a longitudinal study in temporal lobe epilepsy. *Annals of Neurology* 2003;54:425–432.

Keary TA, Frazier TW, Busch RM, Kubu CS, Iampietro M. Multivariate neuropsychological prediction of seizure lateralization in temporal lobe epilepsy surgery cases. *Epilepsia* 2007;48:1438–1446.

Kim H, Yi S, Son EI, Kim J. Lateralization of epileptic foci by neuropsychological testing in mesial temporal lobe epilepsy. *Neuropsychology* 2004;18:141–151.

Kneebone AC. Presurgical neuropsychological evaluation for localization and lateralization of the epileptogenic zone. In Luders HO, Comair YG (eds.), *Epilepsy surgery* (2nd ed.). Philadelphia: Lippincott Willliams & Wilkins, 2001:487–496.

Kneebone AC, Chelune GJ, Luders HO. Individual patient prediction of seizure lateralization in temporal lobe epilepsy: a comparison between neuropsychological memory measures and the intracarotid amobarbital procedure. *Journal of the International Neuropsychological Society* 1997;3:159–168.

Loring DW, Chelune GJ. Neuropsychological evaluation in epilepsy surgery. In Luders HO, Comair YG (eds.), *Epilepsy surgery* (2nd ed.). Philadelphia: Lippincott Willliams & Wilkins, 2001:521–524.

Loring DW, Murr AM, Meador KJ, Lee GP, Gratton CA, Nichols ME, Gallagher BB, King DW, Smith JR. Wada memory testing and hippocampal volume measurements in the evaluation of temporal lobectomy. *Neurology* 1993;43:1789–1793.

Loring DW, Strauss E, Hermann BP, Barr WB, Perrine K, Trenerry MR, Chelune G, Westerveld M, Lee GP, Meador KJ, Bowden SC. Differential neuropsychological test sensitivity to left temporal lobe epilepsy. *Journal of the International Neuropsychological Society* 2008;14:394–400.

Moser DJ, Bauer RM, Gilmore RL, Dede DE, Fennell EB, Algina JJ, et al. Electroencephalographic, volumetric, and neuropsychological indicators of seizure focus lateralization in temporal lobe epilepsy. *Archives of Neurology* 2000;57:707–712.

National Association of Epilepsy Centers. Guidelines for essential services, personnel, and facilities in specialized epilepsy centers in the United States. *Epilepsia* 2001;42(6):804–814.

Ogden-Epker M, Cullum CM. Quantitative and qualitative interpretation of neuropsychological data in the assessment of temporal lobectomy candidates. *The Clinical Neuropsychologist* 2001;15:183–195.

Rausch R. Psychological evaluation. In Engel J Jr. (ed.), *Surgical treatment of the epilepsies*. New York: Raven Press, 1987:181–195.

Stroup E, Langfitt J, Berg M, McDermott M, Pilcher W, Como P. Predicting verbal memory decline following anterior temporal lobectomy. *Neurology*, 2003;60: 1266–1273.

Chapter 10

Ahmad Z, Balsamo LM, Sachs BC, Xu B, Gaillard WD. Auditory comprehension of language in young children: neural networks identified with fMRI. *Neurology* 2003;60:1598–1605.

Alpherts WCJ, Vermeulen J, van Veelen CWM. The Wada test: prediction of focus lateralization by asymmetric and symmetric recall. *Epilepsy Research* 2000;39:239–249.

Andelman F, Kipervasser S, Reider-Groswasser II, Fried I, Neufeld MY. Hippocampal memory function as reflected by the intracarotid sodium methohexital Wada test. *Epilepsy & Behavior* 2006;9:579–586.

Baxendale SA, Thompson PJ, Duncan JS. The role of the Wada test in surgical treatment of temporal lobe epilepsy: an international survey. *Epilepsia* 2008;49:715–720.

Binder JR, Sabsevitz DS, Swanson SJ, Hammeke TA, Raghavan M, Mueller WM. Use of preoperative functional MRI to predict verbal memory decline after temporal lobe epilepsy surgery. *Epilepsia* 2008;49:1377–1394.

Buchtel HA, Passaro EA, Selwa LM, Deveikis J, Gomez-Hassan D. Sodium methohexital (Brevital) as an anesthetic in the Wada test. *Epilepsia* 2002;43:1056–1061.

Chelune GJ. Hippocampal adequacy versus functional reserve: predicting memory functions following temporal lobectomy. *Archives of Clinical Neuropsychology* 1995;10:413–432.

Gaillard WD. Metabolic and functional neuroimaging. In Wyllie E, Gupta A, Lachhwani DK (eds.), *The treatment of epilepsy: Principles and practice* (4th ed.). Philadelphia: Lippincott Williams & Wilkins, 2006:1041–1058.

Gaillard WD, Balsamo L, Xu B, McKinney C, Papero PH, Weinstein S, Conry J, et al. fMRI language task panel improves determination of language dominance. *Neurology* 2004;63:1403–1408.

Go C, Snead OC. Pharmacologically intractable epilepsy in children: diagnosis and preoperative evaluation. *Neurosurgery Focus* 2008;25(3):E2.

Hamberger MJ, Goodman RR, Perrine K, Tamny T. Anatomic dissociation of auditory and visual naming in the lateral temporal cortex. *Neurology* 2001;56:56–61.

Helmstaedter C. Neuropsychological aspects of epilepsy surgery. *Epilepsy & Behavior* 2004;5:S45–S55.

Jokeit H, Daamen M, Zang H, Janszky J, Ebner A. Seizures and forgetting in patients with left temporal lobe epilepsy. *Neurology* 2001;57:125–126.

Kilgore WDS, Glosser G, Casasanto DJ, French JA, Alsop DC, Ditre JA. Functional MRI and the Wada test provide complimentary information for predicting postoperative seizure control. *Seizure* 1999;8:450–455.

Kneebone AC, Chelune GJ, Naugle RI, Dinner DS, Awad IA. Intracarotid amobarbital procedures as a predictor of material-specific memory change after anterior temporal lobectomy. *Epilepsia* 1995;36:857–865.

Lee GP, Park YD, Westerveld M, Hempel A, Loring DW. Effect of Wada methodology in predicting memory impairment in pediatric epilepsy surgery candidates. *Epilepsy & Behavior* 2002;3:439–447.

Lee GP, Park YD, Westerveld M, Hempel A, Blackburn LB, Loring DW. Wada memory performance predict seizure outcome after epilepsy surgery in children. *Epilepsia* 2003;44:936–943.

Loring DW, Meador KJ, Lee GP, Flanigin HF, King DW, Smith JR. Crossed aphasia in a patient with complex partial seizures: evidence from intracarotid sodium Amytal testing, functional cortical mapping, and neuropsychological assessment. *Journal of Clinical & Experimental Neuropsychology* 1990;12:340–354.

Loring DW, Meador KJ, Lee GP, King DW. *Amobarbital effects and lateralized brain function: the Wada test.* New York: Springer-Verlag, 1992.

Loring DW, Meador KJ, Lee GP, Nichols ME, King DW, Gallagher BB, Murro AM, Smith JR. Wada memory performance predicts seizure outcome following anterior temporal lobectomy. *Neurology* 1994;44:2322–2324.

Loring DW, Hermann BP, Meador KJ, Lee GP, Gallagher BB, King DW, Murro AW, Smith JR, Wyler AR. Amnesia after unilateral temporal lobectomy. *Epilepsia* 1994;35:757–763.

Loring DW, Meador KJ, Lee GP, Nichols ME, King DW, Murro AM, Park YD, Smith JR. Wada memory and timing of stimulus presentation. *Epilepsy Research* 1997;26:461–464.

Luders HO, Awad I, Wyllie E, Schaffler L. Functional mapping of language abilities with subdural electrode grids. In Wyler AR, Hermann BP (eds.), *The surgical management of epilepsy*. Boston: Butterworth-Heinemann, 1994: 70–77.

Milner B, Branch C, Rasmussen T. Study of short-term memory after intracarotid injection of sodium Amytal. *Transactions of the American Neurological Association* 1962;87:224–226.

Morris HH, Luders H, Lesser RP, Dinner DS, Hahn J. Transient neuropsychological abnormalities (including Gerstmann's syndrome) during cortical stimulation. *Neurology* 1984;34:877–883.

Morris RG, Polkey CE, Cox T. Independent recovery of memory and language functioning during the intracarotid sodium Amytal test. *Journal of Clinical & Experimental Neuropsychology* 1998;20:433–444.

Ojemann GA. Intraoperative functional mapping at the University of Washington, Seattle. In Engel J Jr. (ed.), *Surgical treatment of the epilepsies*. New York: Raven Press, 1987:635–639.

Ojemann GA. Cortical stimulation and recording in language. In Kertesz A (ed.), *Localization and neuroimaging in neuropsychology*. San Diego, CA: Academic Press, 1994:33–55.

Ojemann GA, Dodrill CB. Verbal memory deficits after left temporal lobectomy for epilepsy. *Journal of Neurosurgery* 1985;62:101–107.

Ojemann GA, Engel J Jr. Acute and chronic intracranial recording and stimulation. In Engel J Jr. (ed.), *Surgical treatment of the epilepsies*. New York: Raven Press, 1987:263–288.

Ojemann GA, Ojemann JG, Lettich E, Berger M. Cortical language localization in left dominant hemisphere: an electrical stimulation mapping investigation in 117 patients. *Journal of Neurosurgery* 1989;71:316–326.

Perrine K, Westerveld M, Sass KJ, Spencer D, Devinsky O, Dogali M, Nelson PK, Luciano DJ. Wada memory disparities predict seizure laterality and postoperative seizure control. *Epilepsia* 1995;36:851–856.

Rausch R, Walsh G. Right-hemisphere language dominance in right-handed epileptic patients. *Archives of Neurology* 1984;41:1077–1080.

Ravdin LD, Perrine K, Haywood CS, Gershengorn J, Nelson PK, Devinsky O. Serial recovery of language during the intracarotid amobarbital procedure. *Brain & Cognition* 1997;33:151–160.

Richardson MP, Strange BA, Duncan JS, Dolan RJ. Memory fMRI in left hippocampal sclerosis: optimizing the approach to predicting postsurgical memory. *Neurology* 2006;66:699–705.

Rosenbaum T, DeToledo J, Smith DB, Kramer RE, Stanulis RG, Kennedy RM. Preoperative assessment of language laterality is necessary in all epilepsy surgery candidates: a case report. *Epilepsia* 1989;30:712.

Schevon CA, Carlson C, Zaroff CM, Weiner HJ, Doyle WK, Miles D, Lajoie J et al. Pediatric language mapping: sensitivity of neurostimulation and Wada testing in epilepsy surgery. *Epilepsia* 2007;48:539–545.

Silbergeld DL. Cortical mapping. In Luders HO, Comair YG (eds.), *Epilepsy surgery* (2nd ed.). Philadelphia: Lippincott Williams & Wilkins, 2001:633–635.

Synder PJ, Novelly RA, Harris LJ. Mixed speech dominance in the intracarotid Amytal procedure: validity and criteria issues. *Journal of Clinical & Experimental Neuropsychology* 1990;12:629–643.

Swanson SJ, Sabsevitz DS, Hammeke TA, Binder JR. Functional magnetic resonance imaging of language in epilepsy. *Neuropsychology Review* 2007;17:491–504.

Szabo CA, Wyllie E. Intracarotid amobarbital testing for language and memory dominance in children. *Epilepsy Research* 1993;15:239–246.

Tracy JI, Shah S. Presurgical functional brain mapping and neurocognitive testing in epilepsy. In Morgan JE, Ricker JH (eds.), *Textbook of Clinical Neuropsychology*. New York: Taylor & Francis, 2008:466–498.

Trenerry MR, Loring DW. The intracarotid amobarbital procedure. In Wyllie E, Gupta A, Lachhwani DK (eds.), *The treatment of epilepsy: Principles and practice* (4th ed.). Philadelphia: Lippincott Williams & Wilkins, 2006:1032–1040.

Wada J, Rasmussen T. Intracarotid injection of sodium Amytal for the lateralization of cerebral speech dominance: experimental and clinical observations. *Journal of Neurosurgery* 1960;17:266–282.

Chapter 11

Arfanakis K, Hermann BP, Rogers BP, Carew JD, Seidenberg M, Meyerand ME. Diffusion tensor MRI in temporal lobe epilepsy. *Magnetic Resonance Imaging* 2002;20:511–519.

Asadi-Pooya AA, Sperling MR. Strategies for surgical treatment of epilepsies in developing countries. *Epilepsia* 2008;49:381–385.

Benbadis SR, Wyllie E, Bingaman WE. Intracranial electroencephalography and localization studies. In Wyllie E, Gupta A, Lachhwani DK (eds.), *The treatment of epilepsy: Principles and practice* (4th ed.). Philadelphia: Lippincott Williams & Wilkins, 2006:1059–1067.

Binnie CD, Polkey CE (eds.). Commission on Neurosurgery of the International League Against Epilepsy (ILAE) 1993–1997: recommended standards. *Epilepsia* 2000;41:1346–1439.

Black PM, Holmes G, Lombroso CT. Corpus callosum section for intractable epilepsy in children. *Pediatric Neurosurgery* 1992;18:298–304.

Brodie MJ. Medical therapy of epilepsy: when to initiate treatment and when to combine? *Journal of Neurosurgery* 2005;252:125–130.

Cendes F, Caramanos Z, Andermann F, Dubeau F, Arnold DL. Proton magnetic resonance spectroscopic imaging and magnetic resonance imaging volumetry in lateralization of temporal lobe epilepsy: a series of 100 patients. *Annals of Neurology* 1997;42:737–746.

Delvin AM, Cross JS, Harkness W, Chong WK, Harding B, Vargha-Khadem F, Neville BG. Clinical outcomes of hemispherectomy for epilepsy in childhood and adolescence. *Brain* 2003;126:556–566.

Elisevich K, Smith BJ. Surgery for parieto-occipital epilepsy. In Luders HO, Comair YG (eds.), *Epilepsy surgery* (2nd ed.). Philadelphia: Lippincott Williams & Wilkins, 2001:705–718.

Engel J Jr. Outcome with respect to epileptic seizures. In Engel J Jr. (ed.), *Surgical treatment of the epilepsies*. New York: Raven Press, 1987:553–571.

Eriksson SH, Rugg-Gunn FJ, Symms MR, Barker GJ, Duncan JS. Diffusion tensor imaging in patients with epilepsy and malformations of cortical development. *Brain* 2001;124:617–626.

Fisher RS & Handforth A. Reassessment: Vagus nerve stimulation for epilepsy: A report of the Therapeutics and Technology Assessment Subcommittee of the American Academy of Neurology. *Neurology* 1999;53:666–669.

Fountas KN, Smith JR, Murro AM, Politsky J, Park YD, Jenkins PD. Implantation of a closed-loop stimulation in the management of medically refractory focal epilepsy: a technical note. *Stereotactic & Functional Neurosurgery* 2005;83:153–158.

Gates JR, dePaola L. Corpus callosum section. In Sliovon S, Dreifuss F, Fish, D, et al. (eds.), *The treatment of epilepsy*. London: Blackwell Scientific, 1996:722–738.

Gleissner U, Helmstaedter C, Schramm J, Elger CE. Memory outcome after selective amygdalohippocampectomy in patients with temporal lobe epilepsy: one-year follow-up. *Epilepsia* 2004;45:960–962.

Grote CL, VanSlyke P, Hoeppner JB. Language outcome following multiple subpial transection for Landau-Kleffner syndrome. *Brain* 1999;122:561–566.

Hamer HM, Morris HH, Mascha EJ, Karafa MT, Bingaman WE, Bej MD, Burgess RC, Dinner DS, et al. Complications of invasive video-EEG monitoring with subdural grid electrodes. *Neurology* 2002;58:97–103.

Hetherington HP, Pan JW, Spencer DD. 1H and 31P spectroscopy and bioenergetics in the lateralization of seizures in temporal lobe epilepsy. *Journal of Magnetic Resonance* Imaging 2002;16:477–483.

Jensen I. Temporal lobe surgery around the world: results, complications, and mortality. *Acta Neurologica Scandinavia* 1975;52:354–373.

Kantarci K, Shin C, Britton JW, So EL, Cascino GD, Jack CR Jr. Comparative diagnostic utility of 1H MRS and DWI in evaluation of temporal lobe epilepsy. *Neurology* 2002;58:1745–1753.

Knake S, Grant PE. Magnetic resonance imaging techniques in the evaluation of epilepsy surgery. In Wyllie E, Gupta A, Lachhwani DK (eds.), *The treatment of epilepsy: Principles and practice* (4th ed.). Philadelphia: Lippincott Williams & Wilkins, 2006:1009–1029.

Kuzniecky, RI. Introduction to structural imaging. In Luders HO, Comair YG (eds.), *Epilepsy surgery* (2nd ed.). Philadelphia: Lippincott Williams & Wilkins, 2001:201–207.

Kwan P, Brodie MJ. Early identification of refractory epilepsy. *New England Journal of Medicine* 2000;342:314–319.

Kwan P, Brodie MJ. Issues of medical intractability for surgical candidacy. In Wyllie E, Gupta A, Lachhwani DK (eds.), *The treatment of epilepsy: Principles and practice* (4th ed.). Philadelphia: Lippincott Williams & Wilkins, 2006:983–992.

Lewis PJ, Siegel A, Siegel AM, Studholme C, Sojkova J, Roberts DW, et al. Does performing image registration and subtraction in ictal brain SPECT help localize neocortical seizures? *Journal of Nuclear Medicine* 2000;41:1619–1626.

Montes JL, Farmer J-P, Andermann F, Poulin C. Hemispherectomy: Medications, technical approaches, and results. In Wyllie E, Gupta A, Lachhwani DK (eds.), *The treatment of epilepsy: Principles and practice* (4th ed.). Philadelphia: Lippincott Williams & Wilkins, 2006:1111–1124.

Morrell F, Whisler WW, Smith MC. Multiple subpial transection in Rasmussen's encephalitis. In Andermann F (ed.), *Chronic encephalitis and epilepsy: Rasmussen's syndrome*. Boston: Butterworth-Heinemann, 1991:219–233.

Moriarty GL, Penovich PE, Gates JR, et al. Parietal lobe surgery for intractable epilepsy. *Epilepsia* 1997;38(Suppl.8):76.

Munari C, Tassi L, Cardinale F, et al. Surgical treatment for frontal lobe epilepsy. In Luders HO, Comair YG (eds.), *Epilepsy surgery* (2nd ed.). Philadelphia: Lippincott Williams & Wilkins, 2001:689–697.

Oliver A, Awad IA. Extratemporal resections. In Engel J Jr. (ed.), *Surgical treatment of the epilepsies* (2nd ed.). New York: Raven Press, 1993:489–500.

Philips J, Sakas DE. Anterior corpus callosotomy for intractable epilepsy: Outcome in a series of twenty patients. *British Journal of Neurosurgery* 1996;10:351–356.

Pilcher WH, Roberts DW, Flanigin HF, et al. Complications of epilepsy surgery. In Engel J Jr. (ed.), *Surgical treatment of the epilepsies* (2nd ed.). New York: Raven Press, 1993:565–581.

Polkey CE. Preoperative tailoring of temporal lobe resections. In Engel J Jr. (ed.), *Surgical treatment of the epilepsies* (2nd ed.). New York: Raven Press, 1993:473–480.

Roberts DW, Siegel AM. Corpus callosotomy. In Luders HO, Comair YG (eds.), *Epilepsy surgery* (2nd ed.). Philadelphia: Lippincott Williams & Wilkins, 2001:747–756.

Smith MC, Byrne R, Kanner AM. Corpus callosotomy and multiple subpial transections. In Wyllie E, Gupta A, Lachhwani DK (eds.), *The treatment of epilepsy: Principles and practice* (4th ed.). Philadelphia: Lippincott Williams & Wilkins, 2006:1159–1168.

Spencer DD, Ojemann GA. Overview of therapeutic procedures. In Engel J Jr. (ed.), *Surgical treatment of the epilepsies* (2nd ed.). New York: Raven Press, 1993:455–471.

Sperling MR. Depth electrodes. In Luders HO, Comair YG (eds.), *Epilepsy surgery* (2nd ed.). Philadelphia: Lippincott Williams & Wilkins, 2001:597–611.

Strickland S, Park Y, Serrano E, Lee G, Lee K, Lee M. Parietal lobe epilepsy surgery in children: predictors of seizure relief. *Epilepsia* 2007;48(Suppl.6): 139–140.

Swartz BE, Delgado-Escueta AV, Walsh GO, Rich JR, Dwan PS, DeSalles AA, Kaufman MH. Surgical outcomes in pure frontal lobe epilepsy and foci that mimic them. *Epilepsy Research* 1998;29:97–108.

van Emde Boas W, Parra J. Long-term noninvasive video-electroencephalographic monitoring in temporal lobe epilepsy. In Luders HO, Comair YG (eds.), *Epilepsy surgery* (2nd ed.). Philadelphia: Lippincott Williams & Wilkins, 2001:413–429.

Wheless JW. Vagus nerve stimulation therapy. In Wyllie E, Gupta A, Lachhwani DK (eds.), *The treatment of epilepsy: Principles and practice* (4th ed.). Philadelphia: Lippincott Williams & Wilkins, 2006:969–980.

Wiebe S, Blume WT, Girvin JP, Eliasziw M. A randomized controlled trial of surgery for temporal-lobe epilepsy. *New England Journal of Medicine* 2001;345:311–318.

Weiser HG. Selective amygdalohippocampectomy: indications and follow-up. *Canadian Journal of Neurological Science* 1991;18:617–627.

Wieser HG, Yasargil MG. Selective amygdalohippocampectomy as a surgical treatment of mediobasal limbic epilepsy. *Surgical Neurology* 1984;17:445–457.

Williamson PD, Boon PA, Thadani VM, Darcey TM, Spencer DD, Spencer SS, Novelly RA, Mattson RH. Parietal lobe epilepsy: diagnostic considerations and results of surgery. *Annuals of Neurology* 1992;31:193–201.

Wylar AR. Subdural strip electrodes in surgery of epilepsy. In Luders HO (ed.), Epilepsy surgery. New York: Raven Press, 1991:395–398.

Wylar AR, Hermann BP, Somes G. Extent of medial temporal resection on outcome from anterior temporal lobectomy: A randomized prospective study. *Neurosurgery* 1995;37:982–990.

Yoon HH, Kwon HL, Mattson RH, Spencer DD, Spencer SS. Long-term seizure outcome in patients initially seizure-free after resective epilepsy surgery. *Neurology* 2003;61:445–450.

APPENDIX I

Ardila A, Lopez MV. Paroxysmal aphasias. *Epilepsia* 1988;29:630–634.

Bancaud J, Brunet-Bourgin F, Chauvel P, Halgren, E. Anatomical origin of déjà-vu and vivid 'memories' in human temporal lobe epilepsy. *Brain* 1994;117:71–90.

Browne TR, Holmes GL. *Handbook of epilepsy* (3rd ed.). Philadelphia: Lippincott, Williams & Wilkins, 2004.

Butler CR, Graham KS, Hodges JR, Kapur N, Wardlaw JM, Zeman AZJ. The syndrome of transient epileptic amnesia. *Annals of Neurology* 2007;61: 587–598.

Chee MW, Kotagal P, Van Ness PC, Gragg L, Murphy D, Luders HO. Lateralizing signs in intractable partial epilepsy: Blinded multiple-observer analysis. *Neurology* 1993;43:2519–2525.

Delgado-Escueta AV, Serratosa JM, Medina MT. Myoclonic seizures and progressive myoclonus epilepsy syndromes. In Wylllie E (ed.), *The treatment of epilepsy: Principles and practice* (2nd ed.). Philadelphia: Lippincott, Williams & Wilkins, 1996:467–483.

Dietl T, Bien C, Urbach H, Elger C, Kurthen M. Episodic depersonalization in focal epilepsy. *Epilepsy & Behavior* 2005;7:311–315.

Farrell K. Tonic and atonic seizures. In Wylllie E (ed.), *The treatment of epilepsy: Principles and practice* (3rd ed.). Philadelphia: Lippincott, Williams & Wilkins, 2001:405–413.

Gastaut H, Broughton R. *Epileptic seizures: Clinical and electrographic features, diagnosis, and treatment.* Springfield, IL: Charles C. Thomas, 1972.

Gloor P. Experiential phenomena of temporal lobe epilepsy. *Brain* 1990;120:183–192.

Isnard J, Guenot M, Ostrowsky K, Sindou M, Mauguiere F. The role of the insular cortex in temporal lobe epilepsy. *Annals of Neurology* 2000;48: 614–623.

Kellinghaus C, Luders HO, Wyllie E. Classification of seizures. In Wyllie E, Gupta A, Lachhwani DK (eds.), *The treatment of epilepsy: Principles and practice* (4th ed.). Philadelphia: Lippincott, Williams, & Wilkins, 2006:217–228.

Kotagal P, Arunkumar GS. Lateral frontal lobe seizures. *Epilepsia* 1998;39 (Suppl 4):S62–S68.

Oxbury JM, Duchowny M. Diagnosis and classification. In Oxbury JM, Polkey CE, Duchowny M (eds.), *Intractable focal epilepsy*. London: WB Saunders, 2000:11–23.

Palmini AL, Gloor P, Jones-Gotman M. Pure amnestic seizures in temporal lobe epilepsy. *Brain* 1992;115(Pt 3):749–769.

Roger J, Dravet C, Bureau M. The Lennox-Gastaut syndrome. *Cleveland Clinic Journal of Medicine* 1989;56(Suppl 2):S172–S180.

Stefan H, Pauli E, Kerling F, Schwarz A, Koebnick C. Autonomic auras: Left hemispheric predominance of epileptic generators of cold shivers and goose bumps? *Epilepsia* 2002;43:41–45.

Stefan H, Snead III, OC. Absence seizures. In Engel J Jr, Pedley TA (eds.), *Epilepsy: A comprehensive textbook*. Philadelphia: Lippincott-Raven, 1997: 579–590.

Tatum WO, Farrell K. Atypical absence, myoclonic, tonic, and atonic seizures. In Wyllie E, Gupta A, Lachhwani DK (eds.), *The treatment of epilepsy: Principles and practice* (4th ed.). Philadelphia: Lippincott, Williams, & Wilkins, 2006:317–331.

Vigevano F, Fusco L, Kazuichi Y. Tonic seizures. In Engel J Jr, Pedley TA (eds.), *Epilepsy: A comprehensive textbook*. Philadelphia: Lippincott-Raven, 1997:617–625.

APPENDIX II

Bancaud J, Talairach J. (1992). Clinical semiology of frontal lobe seizures. In Chauvel P, Delgdo-Escueta AV, Halgren E, Bancaud J (eds.), *Frontal lobe seizures and epilepsies*. New York: Raven Press, 1992:3–58.

Bebek N, Gurses C, Gokyigit A, Baykan B, Ozkara C, Dervent A. Hot water epilepsy: Clinical and electrophysiologic findings based on 21 cases. *Epilepsia* 2001;42:1180–1184.

Berroya AG, McIntyre J, Webster R, Lah S, Sabaz M, Lawson J, Bleasel AF, Bye AM. Speech and language deterioration in benign rolandic epilepsy. *Journal of Child Neurology* 2004;19:53–58.

Bien CG, Widman G, Urbach H, Sassen R, Kuczaty S, Wiestler OD, Schramm J, Elger CE. The natural history of Rasmussen's encephalitis. *Brain* 2002;125:1751–1759.

Browne TR, Holmes GL. *Handbook of epilepsy* (3rd ed.). Philadelphia: Lippincott, Williams & Wilkins, 2004.

Caplan R, Curtiss S, Chugani HT, Vinters HV. Pediatric Rasmussen encephalitis: Social communication, language, PET and pathology before and after hemispherectomy. *Brain and Cognition* 1996;32:45–66.

Commission on Classification and Terminology, International League Against Epilepsy. Proposal for revised classification of epilepsy and epileptic syndromes. *Epilepsia* 1985;26:268–278.

Commission on Classification and Terminology of the International League Against Epilepsy. Proposal for revised classification of epilepsies and epileptic syndromes. *Epilepsia* 1989;30:389–399.

Erba G. Shedding light on photosensitivity, one of epilepsy's most Complex conditions. http://www.epilepsyfoundation.org/epilepsyusa/photosensitivity 20060306.cfm, 2006.

Filho GM, Jackowski AP, Lin K, Guaranha MSB, Guilhoto LM, Silva HH, Caboclo LO, Carrete H, Bressan RA, Yacubian EM. Personality traits related to juvenile myoclonic epilepsy: MRI reveals prefrontal abnormalities through a voxel-based morphometry strudy. *Epilepsy & Behavior* 2009;15: 202–207.

Hayashi T, Ichiyama T, Nishikawa M, Isumi H, Furukawa S. Pocket Monsters, a popular television cartoon, attacks Japanese children. *Annals of Neurology* 1998;44:427–428.

Hirose S, Mitsudome A, Okada M., Kaneko S. Epilepsy Study Group, Japan. Genetics of idiopathic epilepsies. *Epilepsia* 2005;46(Suppl 1):38–43.

Ishida S, Yamashita Y, Matsuishi T, Ohshima M, Ohshima H, Kato H, Maeda H. Photosensitive seizures provoked while viewing "Pocket Monsters," a made-for-television animation program in Japan. *Epilepsia* 1998;39:1340–1344.

Koutroumanidis M, Koepp MJ, Richardson MP, Camfield C, Agathonikou A, Ried S., Papadimitriou A, Plant GT, Duncan JS, Panayiotopoulos CP. The variants of reading epilepsy: A clinical and video EEG study of 17 patients with reading-induced seizures. *Brain* 1998;121:1409–1427.

Northcott E, Connolly AM, McIntyre J, Christie J, Berroya A, Taylor A, Batchelor J, Aaron G, Soe S, Bleasel AF, Lawson JA, Bye AM. Longitudinal assessment of neuropsychological function in children with benign rolandic epilepsy. *Journal of Child Neurology* 2006;21:518–522.

Panayiotopoulos CF. *The epilepsies: Seizures, syndromes and management.* Chipping Norton, Oxfordshire, UK: Bladon Medical Publishing, 2005.

Radhakrishnan K, Silbert PL, Klass DW. Reading epilepsy. An appraisal of 20 patients diagnosed at the Mayo Clinic, Rochester, Minnesota, between 1949 and 1989, and delineation of the epileptic syndrome. *Brain* 1995;118(Pt 1):75–89.

Scheffer IE, Bhatia KP, Lopes-Condes I., Fish DR, Marsden CD, Andermann E, Andermann F, Desbiens R, Keene D, Cendes F, Manson JI, Constantinou JEC, Mclntosh A, Berkovic SF. Autosomal dominant nocturnal frontal lobe epilepsy: A distinctive clinical disorder. *Brain* 1995;118 (Pt 1):61–73.

Sharma A, Cameron D. Reasons to consider a plasma screen television – photosensitive epilepsy. *Epilepsia* 2007;48(10):1528.

Williamson P, Spencer D, Spencer S, Novelly R, Mattson R. Complex partial seizures of extra temporal origin. *Annals of Neurology* 1985;18:497–504.

Williamson PD, Boon PA, Thadani VM, Darcey TM, Spencer DD, Spencer SS, Novelly RA, Mattson RH. Parietal lobe epilepsy: Diagnostic considerations and results of surgery. *Annuals of Neurology* 1992;31:193–201.

Zifkin B, Andermann F. Epilepsy with reflex seizures. In Wyllie E, Gupta A, Lachhwani DK (eds.), *The treatment of epilepsy: Principles and practice* (4th ed.). Philadelphia: Lippincott, Williams, & Wilkins, 2006:463–475.

APPENDIX III

Lee GP, Park YD, Westerveld M, Hempel A, Loring DW. Effect of Wada methodology in predicting memory impairment in pediatric epilepsy surgery candidates. *Epilepsy & Behavior* 2002;3:439–447.

Loring DW, Meador KJ, Lee GP, King DW. *Amobarbital effects and lateralized brain function: The Wada test.* New York: Springer-Verlag, 1992.

Index

Note: Page number followed by "f" and "t" refers to figures and tables, respectively.

AAN. *See* American Academy of Neurology
AAP. *See* American Academy of Pediatrics
Absence seizures, 24, 26–27
  in children, 81
Academic Performance Rating Scales, 123
Acquired epileptic aphasia (Landau-Kleffner syndrome), 54–55, 115
"Acute symptomatic" seizures. *See* Provoked seizures
ADHD. *See* Attention-deficit hyperactivity disorder
AEDs. *See* Antiepileptic drugs
AES. *See* American Epilepsy Society
Akinetic seizure, 255
Alprazolam, 141
  for anxiety disorders, 141
Alzheimer disease, 38, 101
American Academy of Neurology (AAN), 125
American Academy of Pediatrics (AAP), 58
American Epilepsy Society (AES), 79, 125
Amobarbital (Amytal)
  for Wada memory assessment, 193
Amygdalohippocampectomy
  for language disorder, 189
  selective, 240–41
Anterior cingulate gyrus seizure, 266
Anterior temporal lobectomy (ATL), 227–30
  complications of, 230
  outcome of, 228–30, 229t
Antiepileptic drugs (AEDs), 9, 74–91, 76–77t
  adverse effects of, 74, 78–88, 103–4
  for benign myoclonic epilepsy of childhood, 268
  for bipolar disorder, 88, 137–38
  for children with epilepsy, 81–82
  cognitive and behavioral side effects of, 83–88

  for complex febrile seizures, 57
  compliance with, 90–91
  for depression, 136–37
  during epilepsy surgery, 212–13
  for epilepsy with myoclonic absences, 271
  for idiopathic generalized epilepsy syndromes, 51
  for Landau-Kleffner syndrome, 55
  for Lennox-Gastaut syndrome, 53
  for parietal seizures, 82
  pregnancy and teratogenic effects, 80–81
  for psychogenic nonepileptic seizures, 152–53
  for psychotic disorders, 143
  for reading epilepsy, 258
  for startle epilepsy, 264
  for sudden unexplained death in epilepsy, 63
  suicidality and, 88–90
  for supplementary motor area seizures, 265
  switching, 79
  for West syndrome (infantile spasms), 101, 270
  for women with epilepsy, 79–80
Anxiety disorders
  generalized anxiety disorder, 140–41
  obsessive-compulsive disorder, 140
  pharmacologic treatment for, 141
  prevalence of, 138
  symptoms of, 139–40, 139t
ATL. *See* Anterior temporal lobectomy
Atonic (astatic) seizures, 255–56
Attention, 108–9
  focused, 108
Attention-deficit hyperactivity disorder (ADHD), 109
Atypical absence seizures, 252–53
Auras, 19–20, 20t. *See also* Complex partial (focal) seizures

339

Automatisms, 21–22. *See also* Complex partial (focal) seizures
Autosomal dominant nocturnal frontal lobe epilepsy, 259–60

Barbiturates, 78
BDAE. *See* Boston Diagnostic Aphasia Examination
Bear-Fedio Personality Inventory, 145
BECTS. *See* Benign epilepsy of childhood with centrotemporal spikes
Behavior Rating Inventory of Executive Function (BRIEF), 118
Behavior Rating Inventory of Executive Function Adult (BRIEF-A), 118
Benign epilepsy of childhood with centrotemporal spikes (BECTS), 82, 257
Benign myoclonic epilepsy of childhood, 268
Benign neonatal convulsions, 268
  familial, 268
  nonfamilial, 268
Benign Rolandic epilepsy. *See* Benign epilepsy of childhood with centrotemporal spikes
Benzodiazepine, 141
  for anxiety disorders, 141
Bilateral mesial temporal lobe (MTL) dysfunction, 242–46
Bipolar disorder
  antiepileptic drugs for, 88
  pharmacologic treatment for, 137–38
  prevalence of, 137
  symptoms of, 137
Boston Diagnostic Aphasia Examination (BDAE)
  for postoperative language impairment risk assessment, 170
Boston Naming Test, 114, 115, 169
Brain infections, 37–38
Brain tumor, 37
BRIEF. *See* Behavior Rating Inventory of Executive Function
Brother-Sister Questionnaire, 123

California Verbal Learning Test, 159
Carbamazepine (Tegretol), 74, 78, 141
  for anxiety disorders, 141
  for benign epilepsy of childhood with centrotemporal spikes, 82
  for bipolar disorder, 88
  for children with epilepsy, 81, 83
  for complex partial seizures, 75

effects during pregnancy, 80, 81
for generalized tonic-clonic seizures, 75
for parietal seizures, 82
for women with epilepsy, 79, 80
Central apnea syndrome, 62
Cerebrovascular disease, 34–35
Childhood absence epilepsy (pyknolepsy), 52
Childhood epilepsy with occipital paroxysms, 258
Childhood learning disabilities, 119
Chronic progressive epilepsia partialis continua of childhood, 260
Classification, of epilepsy syndromes, 257–71
Clonazepam, 141
  for anxiety disorders, 141
Clonic seizures, 254
Cognitive testing
  for psychogenic nonepileptic seizures, 159–60
Committee to Revise the Guidelines for Services, Personnel and Facilities at Specialized Epilepsy Centers, 11
Complex febrile seizures, 57–58
Complex partial (focal) seizures, 15, 17–18t, 19–24. *See also* Partial (focal or localized) seizures
  auras, 19–20, 20t
  automatisms, 21–22
Computed tomography imaging, for epilepsy, 71
Concentration, 108. *See also* Attention
Corpus callosotomy
  complications of, 238
  outcome of, 237–38
Cortical malformations
  MRI scanning for, 217
Creutzfeldt-Jacob disease, 254
Cryptogenic generalized epilepsy syndromes, 53. *See also* Generalized epilepsy syndromes
  with myoclonic absences, 271
  with myoclonic-astatic seizures, 271
  West syndrome (infantile spasms), 101, 270–71
Cryptogenic localization-related epilepsy syndromes, 50. *See also* Localization-related epilepsy syndromes

Déjà-vu, 47, 250
Delis-Kaplan Executive Function System (D-KEFS), 117

Depression
    pharmacologic treatment for, 136–37
    prevalence of, 135
    symptoms of, 135–36
Depth electrodes, 223–24
Diffusion tensor imaging (DTI)
    for epilepsy during inpatient video-EEG monitoring, 218
D-KEFS. *See* Delis-Kaplan Executive Function System
Dorsolateral prefrontal lobe seizures, 265
Down syndrome, 35
Driving and epilepsy
    seizure-free period requirements, 125–26
    expectations to, 126–27
DTI. *See* Diffusion tensor imaging

Eating epilepsy, 263
ECSWS. *See* Epilepsy with continuous spike-waves during slow wave sleep
EEG. *See* Electroencephalogram
EFA. *See* Epilepsy Foundation of America
ELDQOL. *See* Epilepsy and Learning Disabilities Quality of Life scale
Electrocortical stimulation mapping, 11, 199–205
Electroencephalography (EEG), 65–71. *See also* Invasive video-electroencephalography monitoring, for epilepsy; Noninvasive video-electroencephalography monitoring, for epilepsy
    for atypical absence seizures, 253
    for autosomal dominant nocturnal frontal lobe epilepsy, 259
    for benign childhood epilepsy with centrotemporal spikes, 257
    for childhood epilepsy with occipital paroxysms, 258
    for epilepsy with myoclonic absences, 271
    for generalized tonic-clonic seizures, 269
    for hot water epilepsy, 259
    interictal scalp recordings, 66–67, 66f
    for juvenile myoclonic epilepsy, 269
    for myoclonic seizures, 253
    patterns in generalized epilepsies, 69–71
        generalized paroxysmal fast activity, 70–71
        multiple spike-and-wave pattern, 69
        slow spike-and-wave pattern, 69–70
        3-Hz generalized spike-and-wave pattern, 69, 70f

patterns in partial epilepsies, 67–69
    intermittent rhythmic δ activity, 68–69
    periodic lateralized epileptiform discharges, 68, 68f
for psychological and psychiatric, 134
for reading epilepsy, 258
for simple partial seizures
    with affective signs, 251
    with illusions, 252
    with motor signs, 247, 248
for startle epilepsy, 264
for West syndrome (infantile spasms), 270
Electromyography (EMG)
    for atonic (astatic) seizures, 255–56
    for tonic seizures, 254, 255
EMG. *See* Electromyography
EMU. *See* Epilepsy monitoring unit
Epidemiology, of epilepsy, 5–6
Epilepsia partialis continua, 248
Epilepsy, defined, 3, 13
Epilepsy and Learning Disabilities Quality of Life scale (ELDQOL), 123
Epilepsy disorders
    classification of, 13–38
    epidemiology of, 31–33
    etiology of, 33–38, 34t
    incidence of, 32t
    prevalence of, 31–32
    seizures, duration of, 31
Epilepsy Foundation of America (EFA), 5, 6, 7, 125, 261, 262
Epilepsy monitoring unit (EMU), 214
Epilepsy-specific psychological disorders, 145–49. *See also* Psychological disorders
    affective-somatoform (dysphoric) disorders of epilepsy, 147
    alternative affective-somatoform syndromes, 147
    alternative psychosis, 146
    anticonvulsant-induced psychiatric disorders, 148
    anxiety/phobias, 148
    interictal dysphoric disorder, 147
    interictal psychosis of epilepsy, 146
    personality disorders, 147–48
    postictal dysphoric disorder, 147
    prodromal dysphoric disorder, 147
    psychoses of epilepsy, 146
Epilepsy Surgery Inventory-55 (ESI-55), 121–22

Epilepsy syndromes, 39–63
  epidemiology of, 60
  etiology of, 60–61
  ILAE classification of, 39–60, 41–42t
  nonepileptic seizures, 61–62
  sudden unexplained death in epilepsy, 62–63
Epilepsy with continuous spike-waves during slow wave sleep (ECSWS), 55–56, 101, 115
Ethosuximide (Zarontin), 74, 75
  for children with epilepsy, 81
Etiology, of epileptic seizures, 7, 8t
Etomide
  for Wada memory assessment, 193
Executive cognitive functions, 117–18

Family Enrichment Scale, 123
F-A-S tests, 114
FDA. See U.S. Food and Drug Administration
Febrile convulsions, 56–57
Felbamate (Felbatol), 76, 78
  for Lennox-Gastaut syndrome, 75
FLAIR. See Fluid attenuation inversion recovery
Fluid attenuation inversion recovery (FLAIR)
  imaging for epilepsy, 72, 216
fMRI. See Functional magnetic resonance imaging
Frontal lobectomy
  complications of, 231–32
  outcome of, 231
Frontal lobe epilepsy(ies), 47–48, 264–66
  dorsolateral prefrontal lobe seizures, 265
  frontal opercular seizures, 266
  medial frontal lobe seizures, 266
  orbitofrontal seizures, 265–66
  precentral frontal lobe seizures, 264
  premotor frontal lobe seizures, 264–65
  supplementary motor area seizures, 265
Frontal opercular seizures, 266
Frontal Systems Behavior Scale (FrSBe), 118
FrSBe. See Frontal Systems Behavior Scale
Functional magnetic resonance imaging (fMRI), 196–99

Gabapentin (Neurontin), 76, 77, 78
  for bipolar disorder, 88
  for children with epilepsy, 81, 83
  cognitive and behavioral side effects of, 85
  effects during pregnancy, 80
  for parietal seizures, 82

GAD. See Generalized anxiety disorder
Generalized anxiety disorder (GAD), 140
Generalized seizures (convulsive or nonconvulsive), 4, 24–30, 25–26t, 50, 252–56
  absence, 24, 26–27
  atonic (astatic), 255–56
  atypical absence, 252–53
  clonic, 254
  cryptogenic, 53
  EEG patterns in, 69–71
    generalized paroxysmal fast activity, 70–71
    multiple spike-and-wave pattern, 69
    slow spike-and-wave pattern, 69–70
    3-Hz generalized spike-and-wave pattern, 69, 70f
  idiopathic, 50–53
  myoclonic, 253–54
  symptomatic, 53–54
  tonic, 254–55
  tonic-clonic, 27–30
Glasgow Coma Scale, 274
Grid electrodes, 222–23, 223f

Handedness (motor dominance), 187–88
Head (traumatic brain) injury, 35–36, 36f
Hemispherectomy, 235–36
  complications of, 236
  outcome of, 236
Hemispherotomy, 236
^{1}H-magnetic resonance spectroscopy (^{1}H-MRS)
  for epilepsy during inpatient video-EEG monitoring, 218
^{1}H-MRS. See ^{1}H-magnetic resonance spectroscopy
Hormonal contraceptives
  effects during pregnancy, 79
Hot water epilepsy, 259

Ictal aphasia, 250
Ictal electroencephalography, for generalized tonic-clonic seizures, 29. See also Electroencephalography
Ictal single-photon emission computed tomography (SPECT)
  for epilepsy during inpatient video-EEG monitoring, 219

Idiopathic generalized epilepsy syndromes, 50–53. *See also* Generalized seizures (convulsive or nonconvulsive)
 benign myoclonic epilepsy of childhood, 268
 benign neonatal familial convulsions, 268
 benign neonatal nonfamilial convulsions, 268
 childhood absence epilepsy (pyknolepsy), 52
 juvenile myoclonic epilepsy, 269–70
 juvenile-onset absence epilepsy, 52–53
 tonic-clonic epilepsy, 270
Idiopathic localization-related epilepsy(ies), 43. *See also* Localization-related epilepsy syndromes
 autosomal dominant nocturnal frontal lobe epilepsy, 259–60
 benign epilepsy of childhood with centrotemporal spikes, 257
 childhood epilepsy with occipital paroxysms, 258
 hot water epilepsy, 259
 musicogenic epilepsy, 262
 reading epilepsy, 258–59
ILAE. *See* International League Against Epilepsy
Impact of Pediatric Illness Scale (IPES), 123
Implantation of electrical brain stimulators, 241–42, 241f
Incidence
 of epilepsy, 5–6, 5t, 6t
 of seizures, 5–6, 5t
Inferior parietal lobule seizures, 267. *See also* Parietal lobe epilepsy(ies)
Intelligence, 112–13
Intercept Epilepsy Control System, 242
International League Against Epilepsy (ILAE), 13, 71, 74
 classification of epilepsy syndromes, 39–60, 41–42t
 of generalized seizures (convulsive or nonconvulsive), 24–30, 25–26t
 of partial (focal or localized) seizures, 14, 15–24, 16–18t
Intracarotid amobarbital (Wada) procedure, 10–11, 114, 273–89
 behavioral assessment, 278
 language and memory functions, measuring
  aural comprehension, 274–75
  fluency of speech, 274
  memory items, scoring of, 277–78

memory stimuli, 277
 paraphasias, 276
 reading, 276
 repetition, 275–76
 visual naming, 275
 for memory assessment, 190–96
  in children, 194
  memory loss, prediction of, 191–92
  methohexital (Brevital) for, 193–94
  seizure control, prediction of, 192–93
  seizure onset, lateralization of, 192
  sodium amobarbital (Amytal) for, 193
 for speech and language disorders, 114
Invasive video-electroencephalography monitoring, for epilepsy, 221–24. *See also* Electroencephalography; Noninvasive video-electroencephalography monitoring, for epilepsy
 with depth electrodes, 223–24
 with grid electrodes, 222–23, 223f
 with strip electrodes, 222–23, 223f
IPES. *See* Impact of Pediatric Illness Scale
Isolated status epilepticus, 58–60, 59t

Jacksonian march, 247, 264
Jacksonian seizures, 247
Jamais-vu, 47, 251
JME. *See* Juvenile myoclonic epilepsy
Juvenile myoclonic epilepsy (JME), 82, 269–70
Juvenile-onset absence epilepsy, 52–53

Ketogenic diet, for epilepsy, 73, 91–93
 adverse effects of, 92
 effectiveness of, 92–93
 mechanism of action, 91–92
Kojewnikow syndrome, 43, 260

Lacosamide (Vimpat)
 cognitive and behavioral side effects of, 87–88
Lamotrigine (Lamictal), 76, 77
 for bipolar disorder, 88
 for children with epilepsy, 81, 83
 cognitive and behavioral side effects of, 85
 effects during pregnancy, 79, 80
 for juvenile myoclonic epilepsy, 82
Landau-Kleffner syndrome. *See* Acquired epileptic aphasia

Lateral (neocortical) temporal lobe epilepsy, 46–47. *See also* Temporal lobe epilepsy(ies)
  signs and symptoms of, 47t
Lennox-Gastaut syndrome, 53, 101, 113, 254, 271
Lesionectomy, 234–35
Levetiracetam (Keppra), 76, 77
  for children with epilepsy, 83
  cognitive and behavioral side effects of, 85–86
  effects during pregnancy, 79, 81
  for juvenile myoclonic epilepsy, 82
  for parietal seizures, 82
Localization-related epilepsy syndromes
  cryptogenic, 50
  idiopathic, 43
  symptomatic, 43–50

MAE. *See* Multilingual Aphasia Examination
MAE Controlled Oral Word Association, 114
MAE Token Test, 114
Magnetic resonance imaging (MRI), 71, 216–18
  for detecting cortical malformations, 217
  for detecting mesial temporal lobe sclerosis, 216, 217f
    through temporal lobes, 217–18
  for detecting neoplasms, 217
Magnetic source imaging (MSI), 221
Magnetization-prepared rapid gradient echo (MPRAGE), 72, 216
Magnetoencephalography (MEG)
  for epilepsy, during inpatient video-EEG monitoring, 221
Medial frontal lobe seizures, 266. *See also* Frontal lobe epilepsy(ies)
MEG. *See* Magnetoencephalography
Memory, 110–12
Mental retardation, 113
Mesial temporal lobe epilepsy, 45–46. *See also* Temporal lobe epilepsy(ies)
  signs and symptoms of, 47t
  typical progression of, 46t
Mesial temporal lobe sclerosis (MTS)
  MRI scanning for, 216, 217f
    through temporal lobes, 217–18
Methohexial (Brevital)
  for memory assessment, 193–94
Minnesota Multiphasic Personality Inventory (MMPI-2), 159
MMPI-2. *See* Minnesota Multiphasic Personality Inventory

Mood disorders, 135–38
  bipolar, 137
  depression, 135–37
MPRAGE. *See* Magnetization-prepared rapid gradient echo
MRI. *See* Magnetic resonance imaging
MSI. *See* Magnetic source imaging
MST. *See* Multiple subpial transaction
MTS. *See* Mesial temporal lobe sclerosis
Multilingual Aphasia Examination (MAE), 170
  for postoperative language impairment risk assessment, 170
  visual confrontation naming, 114, 115
Multiple subpial transaction (MST), 238–40
  complications of, 239–40
  outcome of, 239
Musicogenic epilepsy, 263
Myoclonic-astatic seizures, 271
Myoclonic reading epilepsy, 258. *See also* Reading epilepsy
Myoclonic seizures, 253–54

National Association of Epilepsy Centers, 11
NEO Personality Inventory (NEO-PI-R), 159
NEO-PI-R. *See* NEO Personality Inventory
Neoplasms
  MRI scanning for, 217
Neurodegenerative central nervous system diseases, 38
Neuropsychological assessment, of epilepsy
  electrocortical stimulation mapping, 199–205
  functional magnetic resonance imaging, 196–99
  intracarotid amobarbital (Wada) procedure, 183–96, 205–10
  for language assessment, 185–90
    atypical language representation, 186–88
    clinical implications, 188–89
    clinical interpretation, 189–90
    mixed language representation, 186–88
    in patients with non-normal language representation, 190
    recovery after amobartial injection, 186
  method, 184–85
Neuropsychological assessment, in epilepsy surgery, 95–131, 165–81, 166f
  antiepileptic drugs, adverse effects of, 103–4
  baseline assessment, 172–73
  case examples, 127–31, 173–81, 205–210, 242–246
  cognitive deficits in, 105–19, 106t

confounding factors, in test interpretation, 103–5
domains in adults with epilepsy, 96–97
factors contributing to cognitive decline in, 98–103
  age of onset, 100
  duration of seizures, 100
  ethilogy of seizure disorder, 99
  location and extent of lesion, 99
  neuropsychological impairment factors in children, 101–3
  type of seizure, 100
preoperative
  language impairment, risk of, 170–71
  lateralization and localization, 167–69
  memory loss, prediction of, 169–70
  seizure control, prediction of, 172
quality-of-life assessment. in See Quality-of-life (QoL)
selection of tests, 95–98
Neuropsychologist
  role in epilepsy, 7–9
  role in epileptic surgery, 9–12
Nonepileptic seizures, 61–62
Noninvasive video-electroencephalography monitoring, for epilepsy, 214–16. See also Electroencephalography; Invasive video-electroencephalography monitoring, for epilepsy
  during impatient monitoring, 215–16
  during video-electroencephalography-monitored seizures, 214–15

Obsessive-compulsive disorder (OCD), 140
Occipital lobectomy, 233–34
  complications of, 234
  outcome of, 234
Occipital lobe epilepsy, 49–50
OCD. See Obsessive-compulsive disorder
Olanzapine, 143
  for psychotic disorders, 143
Orbitofrontal seizures, 265–66
Oxcarbazepine (Trileptal), 76, 77
  for anxiety disorders, 141
  for benign epilepsy of childhood with centrotemporal spikes, 82
  for children with epilepsy, 81, 83
  cognitive and behavioral side effects of, 86
  for partial seizures, 82

PAI. See Personality Assessment Inventory
Panic disorder, 139–40. See also Anxiety disorder

Paracentral parietal lobe seizures, 267. See also Parietal lobe epilepsy(ies)
Parenting Stress Index, 123
Parietal lobe epilepsy(ies), 48–49, 266–68
  inferior, 267
  paracentral, 267
  postcentral gyrus seizures, 266–67
  superior, 267
Parieto-occipital lobectomy
  complications of, 233
  outcome of, 232–33
Partial complex seizure with secondary generalization, 4. See also Partial seizure
Partial continuous epilepsy. See Epilepsia partialis continua
Partial (focal or localized) seizures, 4, 14, 15–24, 16–18t
  antiepileptic drugs for, 82
  complex, 15, 17–18t, 19–24
  EEG patterns in, 67–69
    intermittent rhythmic δ activity, 68–69
    periodic lateralized epileptiform discharges, 68, 68f
  into generalized tonic-clonic convulsions, 15, 18t
  simple, 15, 16–17t, 18–19
Periodic lateralized epileptiform discharges (PLEDs), 68, 68f
Personality Assessment Inventory (PAI), 159
Personality disorders, 143–44
  prevalence of, 143–44
  symptoms of, 144–45
Personality testing
  for psychogenic nonepileptic seizures, 159
PET. See Position emission tomography
Pharmacologic therapies, for epilepsy, 74–91
Phenobarbital, 74
  for children with epilepsy, 83
  effects during pregnancy, 81
  for women with epilepsy, 79
Phenytoin (Dilantin), 74, 75, 78
  for children with epilepsy, 81, 83
  effects during pregnancy, 80, 81
  for women with epilepsy, 79
Photosensitive seizures, 261–63
Physiologic nonepileptic seizures, 61. See also Nonepileptic seizures
Piers-Harris Children's Self-Concept Scale, 123
PNES. See Psychogenic nonepileptic seizures
PLEDs. See Periodic lateralized epileptiform discharges

Polytherapy
  in psychological and psychiatric, 134
POMS. *See* Profile of Mood States
Portland Digit Recognition Test, 159, 160
Position emission tomography (PET)
  for epilepsy, during inpatient video-EEG monitoring, 218–19
  for generalized tonic-clonic seizures, 28
Postcentral gyrus seizures, 266–67. *See also* Parietal lobe epilepsy(ies)
Precentral frontal lobe seizures, 264. *See also* Frontal lobe epilepsy(ies)
Pregabalin (Lyrica), 76
  cognitive and behavioral side effects of, 87
Premotor frontal lobe seizures, 264–65. *See also* Frontal lobe epilepsy(ies)
Preoperative testing, 10
Primary generalized epilepsy, 4
Primidone (Mysoline), 74
  for children with epilepsy, 83
  for women with epilepsy, 79
Profile of Mood States (POMS), 121
Propofol
  for Wada memory assessment, 193
Provoked seizures, 4
Pseudoseizures, 61. *See also* Nonepileptic seizures
Psychiatric Disorders, 133–49
  risk factors, 134–35
Psychogenic nonepileptic seizures (PNES), 61–62, 151–61. *See also* Nonepileptic seizures
  causes of, 152t
  diagnosis of, 151–53
  etiology of, 154–55
  prevalence of, 153
  psychological and neuropsychological assessment of, 158–60
    cognitive testing, 159–60
    personality testing, 159
  symptoms of, 155–58, 158t
  treatment for, 160–61
Psychological disorders, 133–49
  epilepsy-specific. *See* Epilepsy-specific psychological disorders
  risk factors, 134–35
  temporal lobe epilepsy, interictal behavior syndrome of, 145–49
Psychotic disorders, 141–43
  pharmacologic treatment for, 143
  prevalence of, 142
  symptoms of, 142–43
Pyknolepsy. *See* Childhood absence epilepsy

QoL. *See* Quality-of-life
QOLCE. *See* Quality of Life in Childhood Epilepsy Questionnaire
QOLIE. *See* Quality of Life in Epilepsy
QOLIE-10, 120
QOLIE-31, 120
QOLIE-89, 120, 121
QOLIE-AD-48. *See* Quality of Life in Epilepsy Inventory for Adolescents
Quality-of-life (QoL)
  adolescent, 124
  in adults, 120–22
  assessment in epilepsy, 119–27
  driving issues, 125–27
  pediatric, 122–24
Quality of Life in Childhood Epilepsy Questionnaire (QOLCE), 122–23
Quality of Life in Epilepsy (QOLIE), 120
Quality of Life in Epilepsy Inventory for Adolescents (QOLIE-AD-48), 120–21, 124

RAND 36–item Health Survey (SF-36), 120
Rasmussen's syndrome, 260–61
Reading epilepsy, 258–59
Recurrent seizures, cognitive effects of, 107–8
Reflex epilepsy(ies), 43–44, 261–64
  eating epilepsy, 263
  musicogenic epilepsy, 263
  photosensitive seizures, 261–63
  startle epilepsy, 263–64
Responsive Neurostimulation System (RNS), 241, 242–46
Rey Auditory Verbal Learning Test, 168
Rey Complex Figure Test, 168
Risperidone, 143
  for psychotic disorders, 143
RNS. *See* Responsive Neurostimulation System

Seizure, defined, 3–4
Seizure semiology, 4
Selective amygdalohippocampectomy, 240–41. *See also* Amygdalohippocampectomy
Selective serotonin reuptake inhibitor (SSRI), 137
  for anxiety disorders, 141
  for depression, 137
Severe myoclonic seizures, 101
Simple febrile seizures, 57

Simple partial (or focal) seizures, 15, 16–17t, 18–19, 247–52. *See also* Partial (focal or localized) seizures
  with affective signs, 251–52
  with autonomic symptoms or signs, 249–50
  with cognitive signs, 250–51
  with hallucinations, 252
  with illusions, 252
  with motor signs, 247–48
  with psychic signs, 250
  with somatosensory or special sensory symptoms, 248–49
SISCOM. *See* Subtraction Ictal SPECT Co-registered with MRI
SMA. *See* Supplementary motor area seizures
SPECT. *See* Ictal single-photon emission computed tomography
Speech and language disorders, 113–15
SSRI. *See* Selective serotonin reuptake inhibitor
Startle epilepsy, 263–64
Status epilepticus, 15, 55, 58-60, 59t, 100, 106t, 248, 300
Steroids
  for Rasmussen's syndrome, 260
Stevens-Johnson syndrome, 78, 83
Strip electrodes, 222–23, 223f
Structural neuroimaging in epilepsy, 71–72
Subdural hematomas, 36–37
Subtraction Ictal SPECT Co-registered with MRI (SISCOM)
  for epilepsy during inpatient video-EEG monitoring, 219–20, 220f
Sudden unexplained death in epilepsy (SUDEP), 62–63
SUDEP. *See* Sudden unexplained death in epilepsy
Suicidality
  and antiepileptic drugs, 88–90
    U.S. Food and Drug Administration alert for, 89–90
Superior parietal lobule seizures, 267. *See also* Parietal lobe epilepsy(ies)
Supplementary motor area (SMA) seizures, 265
Surgery for epilepsy, 3, 211–46
  anterior temporal lobectomy, 227–30, 229t
  corpus callosotomy, 237–38
  criteria for evaluation
    AEDs, duration of, 212–13, 212t
    failed drugs, 211–12
    seizure frequency, 213
  diagnostic evaluation for, 213–24
    functional neuroimaging, 218–21
    invasive video-electroencephalography monitoring, 221–24
    noninvasive video-electroencephalography monitoring, 214–16
    structural neuroimaging, 216–18
  frontal lobectomy, 231–32
  hemispherectomy, 235–36
  implantation of electrical brain stimulators, 241–42, 241f
  lesionectomy, 234–35
  multiple subpial transaction, 238–40
  neuropsychological assessment in. *See* neuropsychological assessment, in epilepsy surgery
  neuropsychologist role in, 9–12
  occipital lobectomy, 233–34
  parieto-occipital lobectomy, 232–33
  selective amygdalohippocampectomy, 240–41
  vagal nerve stimulation, 225–27, 226f
Symptomatic generalized epilepsy syndromes, 53–54. *See also* Generalized epilepsy syndromes
Symptomatic localization-related epilepsy syndromes, 43–50. *See also* Localization-related epilepsy syndromes
  frontal lobe epilepsies, 47–48, 264–66
  occipital lobe epilepsy, 49–50
  parietal lobe epilepsies, 48–49, 266–68
  Rasmussen's syndrome, 260–61
  reflex epilepsies, 43–44, 261–64
  temporal lobe epilepsies, 44–47

TCI. *See* Transient Cognitive Impairment
Temporal lobe epilepsy(ies) (TLE), 44–47
  interictal behavior syndrome of, 145–49
  lateral (neocortical), 46–47
  mesial, 45–46, 46t
Therapeutics and Technology Assessment Subcommittee of the American Academy of Neurology, 227
Tiagabine (Gabitril), 76
  for children with epilepsy, 81
  cognitive and behavioral side effects of, 86
TLE. *See* Temporal lobe epilepsy(ies)
Todd's paralysis, 248
Tonic-clonic seizures, 27–30, 270
  complications of, 29–30
  postictal phase, 28–29

Tonic-clonic seizures (*Continued*)
  precipitating factors, 27–28
  premonitory symptoms, 27
  tonic phase, 28
  tonic-clonic phase, 28
Tonic seizures, 254–55
Topiramate (Topamax), 76, 77
  for bipolar disorder, 88
  for children with epilepsy, 83
  cognitive and behavioral side effects of, 86–87
  effects during pregnancy, 80
  for parietal seizures, 82
Tower of London, 117
Trails B, 117
Transient Cognitive Impairment (TCI), 104

Unprovoked seizures, 4
U.S. Food and Drug Administration (FDA), 75, 104
  alert for suicidality, 89–90

Vagal nerve stimulation (VNS), 225–27
  adverse effects of, 227
  current status of, 227
  efficacy of, 227
  mechanism of action, 226–27
  parameters, 226
  placement of, 225, 226f
Valproate (Depakote), 74, 78
  for bipolar disorder, 88
  for children with epilepsy, 83
  for complex partial seizures, 75
  effects during pregnancy, 80, 81
  for generalized tonic-clonic seizures, 75
  for women with epilepsy, 80
Valproic acid, 74
  for anxiety disorders, 141
  for children with epilepsy, 81, 83
  for juvenile myoclonic epilepsy, 82

Vigabatrin (Sabril), 76, 78
  for children with epilepsy, 81
  effects during pregnancy, 80
Vigilance, 108. *See also* Attention
Visual-perceptual and spatial functions, 116
VNS. *See* Vagal nerve stimulation

WAB. *See* Western Aphasia Battery
WAIS. *See* Wechsler Adult Intelligence Scale
WAIS-III, 112
WAIS-R Performance IQ, 168
Washington Psychosocial Inventory (WPSI), 122
WCST. *See* Wisconsin Card Sorting Test
WCST-Categories, 168, 169
Wechsler Adult Intelligence Scale (WAIS), 103
Wechsler Intelligence Scale for Children (WISC), 117
Western Aphasia Battery (WAB), 170
  for postoperative language impairment risk assessment, 170
West syndrome (infantile spasms), 101, 113, 270–71
Wide Range Achievement Test-III Reading, 169
WISC. *See* Wechsler Intelligence Scale for Children
Wisconsin Card Sorting Test (WCST), 117, 118
WMS-II, 168
WMS-III, 169
WMS-R, 168
Word Memory Test, 159, 160
World Health Organization
  prevalence of epilepsy, 6
WPSI. *See* Washington Psychosocial Inventory

Zonisamide (Zonegran), 76
  cognitive and behavioral side effects of, 87

www.ingramcontent.com/pod-product-compliance
Ingram Content Group UK Ltd.
Pitfield, Milton Keynes, MK11 3LW, UK
UKHW021316180426
11947UKWH00015B/1267